MEMS/NEMS Sensors

MEMS/NEMS Sensors: Fabrication and Application

Special Issue Editor

Goutam Koley

MDPI • Basel • Beijing • Wuhan • Barcelona • Belgrade

MDPI

Special Issue Editor
Goutam Koley
Clemson University
USA

Editorial Office
MDPI
St. Alban-Anlage 66
4052 Basel, Switzerland

This is a reprint of articles from the Special Issue published online in the open access journal *Micromachines* (ISSN 2072-666X) from 2018 to 2019 (available at: https://www.mdpi.com/journal/micromachines/special_issues/MEMS_NEMS_Sensors_Fabrication_Application)

For citation purposes, cite each article independently as indicated on the article page online and as indicated below:

LastName, A.A.; LastName, B.B.; LastName, C.C. Article Title. *Journal Name* **Year**, *Article Number*, Page Range.

ISBN 978-3-03921-634-5 (Pbk)
ISBN 978-3-03921-635-2 (PDF)

Contents

About the Special Issue Editor

Goutam Koley received his B. Tech. degree from the Indian Institute of Technology Kharagpur, M.S. from the University of Massachusetts Lowell, and Ph.D. from Cornell University. He is currently a Samuel R. Rhodes Endowed Professor in the department of Electrical and Computer Engineering and Director of the Nanoscale Electronics and Sensors Laboratory at Clemson University. Dr. Koley's research focuses on the development of sensors and sensing systems, utilizing micro and nanoscale materials and devices, for chemical, biological, environmental and wearable sensing applications. He has authored over 75 peer-reviewed journal articles, 150 conference presentations and seminars, 18 invited talks, 4 book chapters, and has 7 issued patents. Dr. Koley is a reviewer for more than 30 international journals, serves in the Editorial board of *Sensors* and *Micromachines* (MDPI), and is a reviewer for many US and international funding agencies. He is a co-founder and shareholder of several start-up companies, a senior member of the IEEE and a member of the APS.

micromachines

MDPI

Editorial

Editorial for the Special Issue on MEMS/NEMS Sensors: Fabrication and Application

Goutam Koley

Department of Electrical and Computer Engineering, Clemson University, Clemson, SC 29634, USA;
gkoley@clemson.edu

Received: 20 August 2019; Accepted: 20 August 2019; Published: 22 August 2019

check for
updates

MEMS sensors are currently undergoing a phase of exciting technological development, not only enabling advancements in traditional applications such as accelerometers and gyroscopes, but also in emerging applications such as microfluidics, thermoelectromechanical, and harsh environment sensors. While traditional MEMS sensors have found wide applications in motion sensing, navigation, and robotics, emerging MEMS sensors are likely to open up applications in the rapidly expanding fields of wearables, internet-of-things, point-of-care detection, and harsh environment monitoring. Novel applications, enabled by advancements in system miniaturization, design innovation and cutting edge fabrication techniques promise an exciting era for MEMS-based sensors and systems development. However, to fully realize their potential several challenges still need to be overcome. Among these challenges, long-term sensor reliability and performance parameter modeling for expedient and robust designs are significant. Additionally, there are issues of cost and power consumption, especially for mass applications requiring small size and weight.

There are 17 papers published in this Special Issue focusing on a wide range of MEMS sensor applications and fabrication methodologies. Almost a third of the papers, [1–5], and [6], present various accelerometer and gyroscope designs and their performance evaluation. Three of the papers, [7,8], and [9], explore novel fabrication methodologies for MEMS devices. The remaining papers cover various novel MEMS sensors and actuators focusing on inertial micro-switch [10], micro hot plates [11], near IR spectrometry [12], magnetic microactuator [13], resonant microfluidic chip [14], high temperature pressure sensors [15], thermoelectric power sensors [16], and a review of the photonic crystal nanobeams for sensing [17].

In particular, Yang et al. [1] proposed a z-axis magnetoresistive accelerometer with electrostatic force feedback in a three-layer design, taking advantage of the change in a magnetic field caused by input acceleration, which is measured by a pair of magnetoresistive sensors at the top layer. They achieved a good sensitivity of 8.85 mV/g for a plate gap of 1 mm. Qin et al. investigated the effect of anisotropy in single-crystal silicon vibrating ring gyroscope, and found out that the frequency split is much more for the [100] direction compared to [111] direction for n = 2 mode, concluding that fabrication in the latter direction is preferable [2]. Liu et al. presented an ASIC-based design process for a monolithic CMOS MEMS accelerometer [3]. They also presented a low-noise and low zero-g offset design of MEMS accelerometer using a low noise chopper circuit and telescopic architecture, which significantly reduced noise and zero-g offset, but increased the power requirement [4]. Jia et al. addressed an important problem of frequency mismatch in MEMS gyroscopes, and presented an approach for reducing it by designing a dual-mass gyroscope that utilizes a quadrature modulation signal [5]. A maximum frequency mismatch of less than 0.3 Hz was demonstrated using their design. Fang et al. proposed a novel adaptive control algorithm incorporating a back-stepping technique to compensate for model uncertainties, disturbances, and unknown parameters in micro-gyroscopes, which are very pertinent issues in their performance optimization [6].

On the fabrication techniques for the MEMS sensors, Smiljanić et al. reported on the deep wet etching of Si substrate in various crystallographic directions and performed theoretical modeling of the etch profiles, which agreed well with the experimental results [7]. Wu et al. presented an innovative fabrication method for a catalytic gas sensor based on a Pt coil addressing the non-uniformity of pellistor material at the inside surface of the coil [8]. Using a droplet-based coating methodology they demonstrated uniformly coated and reliable pellistor sensors. Kim et al. presented a femtosecond laser-based micro-welding technique for bonding glass and fabricate reliable microfluidic channels. They compared the microfluidic channels fabricated using this method with those fabricated using a glue-based technique, highlighting their relative ease of fabrication and reliability [9].

On the new device applications side, Peng et al. presented an inertial microswitch with a very low threshold of 5 g and high threshold accuracy, leveraging squeeze film damping [10]. Liu et al. presented novel designs of micro hot-plates with significantly improved temperature non-uniformity [11]. Huang et al. reported on a novel MEMS-based infrared spectrometer operating in the range of 800–1800 nm with a wavelength resolution of 10 nm, which compared favorably with similar commercial systems [12]. Feng et al. designed, simulated and fabricated a linear magnetic microactuator with bistable behavior with less than 1 ms response time [13]. An LC resonant circuit-based sensor for detecting metallic debris in hydraulic fuel is proposed by Yu et al., where they were able to successfully demonstrate selective detection of iron and copper particles with diameters down to tens of microns [14]. Gajula et al. designed a GaN circular membrane-based pressure sensor capable of operating at high temperatures. The pressure sensors exhibited high sensitivity at temperatures in excess of 200 °C, which is a significant improvement over their Si counterparts [15]. Zhang et al. presented a MEMS-based thermoelectric power sensor for measuring microwave power using a floating slug design to minimize microwave power loss. The sensor was implemented with GaAs MMIC technology and exhibited very good sensitivity up to 25 GHz [16]. Finally, Qiao et al. presented a comprehensive review of photonic crystal nanobeam-based sensors providing a ready reference for researchers interested in this area. They specifically focused on the sensing of refractive index changes, nanoparticle sensing, optomechanical sensing, and temperature sensing [17].

I would like to take this opportunity to thank all the authors for submitting their papers to this Special Issue. I would also like to thank all the reviewers for dedicating their time and helping to improve the quality of the submitted papers.

Conflicts of Interest: The author declares no conflict of interest.

References

1. Yang, B.; Wang, B.; Yan, H.; Gao, X. Design of a Micromachined Z-axis Tunneling Magnetoresistive Accelerometer with Electrostatic Force Feedback. *Micromachines* **2019**, *10*, 158. [CrossRef] [PubMed]
2. Qin, Z.; Gao, Y.; Jia, J.; Ding, X.; Huang, L.; Li, H. The Effect of the Anisotropy of Single Crystal Silicon on the Frequency Split of Vibrating Ring Gyroscopes. *Micromachines* **2019**, *10*, 126. [CrossRef]
3. Liu, Y.; Wen, K. Implementation of a CMOS/MEMS Accelerometer with ASIC Processes. *Mcromachines* **2019**, *10*, 50. [CrossRef] [PubMed]
4. Liu, Y.; Wen, K. Monolithic Low Noise and Low Zero-g Offset CMOS/MEMS Accelerometer Readout Scheme. *Micromachines* **2018**, *9*, 637. [CrossRef]
5. Jia, J.; Ding, X.; Gao, Y.; Li, H. Automatic Frequency Tuning Technology for Dual-Mass MEMS Gyroscope Based on a Quadrature Modulation Signal. *Micromachines* **2018**, *9*, 511. [CrossRef]
6. Fang, Y.; Fei, J.; Yang, Y. Adaptive Backstepping Design of a Microgyroscope. *Micromachines* **2018**, *9*, 338. [CrossRef] [PubMed]
7. Smiljanić, M.; Lazić, Ž.; Radjenović, B.; Radmilović-Radjenović, M.; Jović, V. Evolution of Si Crystallographic Planes-Etching of Square and Circle Patterns in 25 wt% TMAH. *Micromachines* **2019**, *10*, 102. [CrossRef] [PubMed]
8. Wu, L.; Zhang, T.; Wang, H.; Tang, C.; Zhang, L. A Novel Fabricating Process of Catalytic Gas Sensor Based on Droplet Generating Technology. *Micromachines* **2019**, *10*, 71. [CrossRef] [PubMed]

9. Kim, S.; Kim, J.; Joung, Y.; Choi, J.; Koo, C. Bonding Strength of a Glass Microfluidic Device Fabricated by Femtosecond Laser Micromachining and Direct Welding. *Micromachines* **2018**, *9*, 639. [CrossRef] [PubMed]

10. Peng, Y.; Wu, G.; Pan, C.; Lv, C.; Luo, T. A 5 g Inertial Micro-Switch with Enhanced Threshold Accuracy Using Squeeze-Film Damping. *Micromachines* **2018**, *9*, 539. [CrossRef] [PubMed]

11. Liu, Q.; Ding, G.; Wang, Y.; Yao, J. Thermal Performance of Micro Hotplates with Novel Shapes Based on Single-Layer SiO_2 Suspended Film. *Micromachines* **2018**, *9*, 514. [CrossRef] [PubMed]

12. Huang, J.; Wen, Q.; Nie, Q.; Chang, F.; Zhou, Y.; Wen, Z. Miniaturized NIR Spectrometer Based on Novel MOEMS Scanning Tilted Grating. *Micromachines* **2018**, *9*, 478. [CrossRef] [PubMed]

13. Feng, H.; Miao, X.; Yang, Z. Design, Simulation and Experimental Study of the Linear Magnetic Microactuator. *Micromachines* **2018**, *9*, 454. [CrossRef] [PubMed]

14. Yu, Z.; Zeng, L.; Zhang, H.; Yang, G.; Wang, W.; Zhang, W. Frequency Characteristic of Resonant Micro Fluidic Chip for Oil Detection Based on Resistance Parameter. *Micromachines* **2018**, *9*, 344. [CrossRef] [PubMed]

15. Gajula, D.; Jahangir, I.; Koley, G. High Temperature AlGaN/GaN Membrane Based Pressure Sensors. *Micromachines* **2018**, *9*, 207. [CrossRef] [PubMed]

16. Zhang, Z.; Ma, Y. DC-25 GHz and Low-Loss MEMS Thermoelectric Power Sensors with Floating Thermal Slug and Reliable Back Cavity Based on GaAs MMIC Technology. *Micromachines* **2018**, *9*, 154. [CrossRef] [PubMed]

17. Qiao, Q.; Xia, J.; Lee, C.; Zhou, G. Applications of Photonic Crystal Nanobeam Cavities for Sensing. *Micromachines* **2018**, *9*, 541. [CrossRef] [PubMed]

micromachines

MDPI

Review

Applications of Photonic Crystal Nanobeam Cavities for Sensing

Qifeng Qiao [1,2,3], Ji Xia [1], Chengkuo Lee [2,3] and Guangya Zhou [1,3,*]

1   Department of Mechanical Engineering, National University of Singapore, Singapore 117579, Singapore;
     e0204977@u.nus.edu (Q.Q.); e0267876@u.nus.edu (J.X.)
2   Department of Electrical and Computer Engineering, National University of Singapore, Singapore 117583,
     Singapore; elelc@nus.edu.sg
3   Center for Intelligent Sensors and MEMS (CISM), National University of Singapore,
     Singapore 117608, Singapore
*   Correspondence: mpezgy@nus.edu.sg; Tel.: +65-6516-1235

Received: 4 September 2018; Accepted: 19 October 2018; Published: 23 October 2018

check for
updates

Abstract: In recent years, there has been growing interest in optical sensors based on microcavities due to their advantages of size reduction and enhanced sensing capability. In this paper, we aim to give a comprehensive review of the field of photonic crystal nanobeam cavity-based sensors. The sensing principles and development of applications, such as refractive index sensing, nanoparticle sensing, optomechanical sensing, and temperature sensing, are summarized and highlighted. From the studies reported, it is demonstrated that photonic crystal nanobeam cavities, which provide excellent light confinement capability, ultra-small size, flexible on-chip design, and easy integration, offer promising platforms for a range of sensing applications.

Keywords: photonic crystal cavity; photonic crystal nanobeam cavity; optical sensor; refractive index sensor; nanoparticle sensor; optomechanical sensor; temperature sensor

1. Introduction

Currently, optical sensors are among the most widely used types of sensing platforms for various applications in every aspect of life, including industry, society, and the military. Optical sensors have advantages over other types of sensors including small size, usability in harsh environments, remote sensing, immunity to interference, etc. With the recent advance of studies on optical microcavities [1], optical sensors can also be realized through on-chip microcavities. Through resonant recirculation, light can be confined into a small volume by optical microcavities, among which photonic crystal cavity is a promising candidate for sensing applications due to its small mode volume and strong light field confinement [2]. They have received much attention in recent years due to the flexible structure, easy on-chip integration, outstanding light confinement capability, and compact size [3].

In this review, we focus on the sensing applications of photonic crystal nanobeam cavities [4–27], which are attractive for their ultra-small physical footprint and extremely low effective mass. This review is organized as follows. Section 2 gives a brief overview of photonic crystals and other types of optical cavities. Section 3 outlines the sensing principle and highlights some key developments in refractive index sensing based on nanobeam cavities. Section 4 presents applications on nanoparticle sensing and analyzes the techniques for nanoparticle capture. Section 5 outlines the mechanisms for optomechanical sensing and at the same time introduce their applications. Section 6 highlights the principles and applications of temperature sensing using photonic crystal nanobeam cavities. Finally, we sum up the whole review.

2. Optical Cavity

Optical microcavities are capable of confining light in small mode volumes. Based on them, sensors are possible to achieve unprecedentedly high sensitivity and low detection limits. The resonance characteristics of a microcavity, such as resonance wavelength and line width, can be significantly affected by slight physical and chemical variations in the optical mode region.

In a photonic cavity, only light at its resonance wavelengths can be strongly coupled into it. For a whispering gallery mode (WGM) cavity or a Fabry-Pérot (F-P) cavity, the resonance wavelength λ_r is defined by:

$$\lambda_r = Ln_{eff}/m \tag{1}$$

where n_{eff} is the effective refractive index of the cavity, L is the round-trip optical path length and m is an integer. A resonance wavelength shift can be induced if there is a change of effective refractive index in the optical mode region. As light propagates through the cavity, a dramatic dip or peak in the intensity of the transmitted light occurs, which can be monitored in the transmission spectrum. For general sensing applications, measurements on the shift of resonance wavelength ($\Delta\lambda$) are most frequently implemented (Figure 1). The transmission spectrum of a side-coupled optical resonator is shown in Figure 1a. In Figure 1b, the transmission spectrum of an optical resonator that is directly in the optical path (i.e., input-cavity-output configuration) is presented. In addition to resonance wavelength, the quality factor (Q) is used to compare losses in optical microcavities, which is defined as:

$$Q = \lambda_r/\delta\lambda \tag{2}$$

where $\delta\lambda$ is the resonance linewidth. The Q factor indicates the photon lifetime within the cavity. Q factors can range from 10^3 to 10^{10} [28] for different cavity designs. For high-Q microcavities, continuous and repetitive sampling of analytes at the resonator surface contributes to the high sensitivity of optical microcavities by significantly increasing the effective optical path length for light matter interaction.

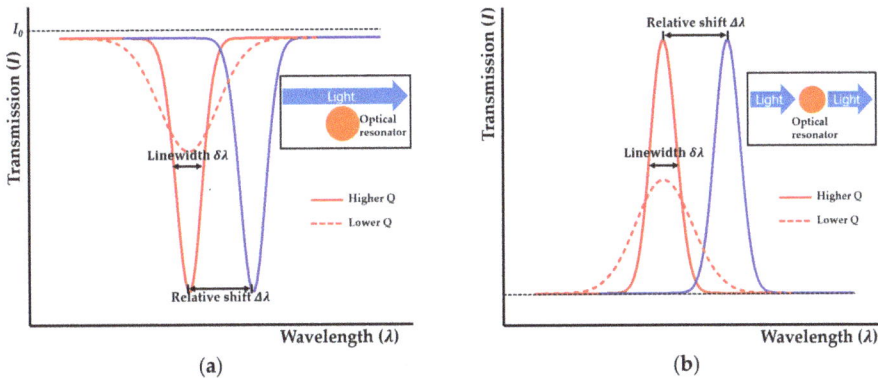

Figure 1. Transmission spectrum of an optical microcavity: (**a**) a characteristic dip at resonance wavelength for side-coupled cavity; (**b**) a characteristic peak at resonance wavelength for the input-cavity-output configuration. The resonance wavelength shifts from the red line to blue line due to the physical or chemical variations in optical mode region. In addition, the decrease of Q factor causes the expansion of spectral resonance line width ($\delta\lambda$).

Following extensive studies on optical cavity designs and technological advance in device fabrication, a variety of micro- and nano-photonic cavity structures have been developed. Below, some of the typical ones are introduced.

Photonic crystal (PhC) microcavities possess regions of varying materials with different refractive indices arranged in a periodic structure and exhibit abundant optical properties of slowing down and confining the light. The periodic spatial arrangements of contrasting dielectric media create a photonic bandgap (PBG) [29,30]. Through periodic modulation of the dielectric constant in one, two, or three orthogonal directions in a structure, PhC can be obtained. 1D PhCs can be formed by placing alternating dielectric stacks periodically [31] or by etching a row of holes in a perfect waveguide (in Figure 2a). 2D PhCs can be realized either by growing high aspect ratio dielectric rods or by etching holes in a higher dielectric material periodically in two dimensions [32]. The latter is most commonly adopted due to its easy fabrication process. In 3D PhCs, a complete PBG can be obtained and the refractive index is modulated in all three directions [33,34]. The fabrications for this type of structures are challenging. The micro-/nano-photonic cavities constructed using these three types of PhCs are schematically shown in Figure 2.

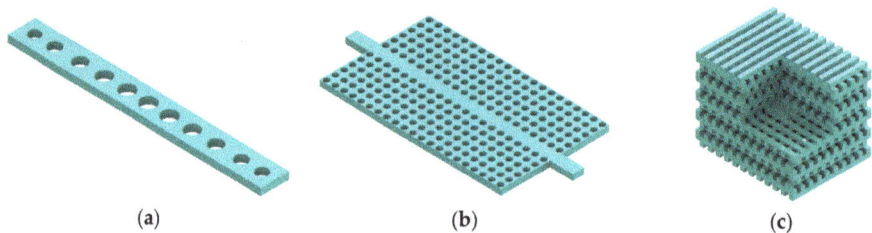

(a)   (b)   (c)

Figure 2. Schematic diagrams of cavities using (**a**) 1D, (**b**) 2D, and (**c**) 3D photonic crystals (PhCs).

In 1997, Foresi et al. [35] demonstrated the proof of concept of a nanobeam cavity through integrating PBG structures directly into a silicon waveguide on a silicon-on-insulator (SOI) wafer. This nanobeam cavity had a modal volume (V) of 0.055 μm^3 and a Q factor of 265 at resonance wavelength 1.56 μm. The high Q/V ratio and large bandgap of this proposed PBG waveguide microcavity made it advantageous over traditional stacked mirror cavities. In general, the Q factor and modal volume are used to characterize optical cavities, and high Q/V is desirable for many applications such as filter, laser, and high Purcell factor [36]. Later, much work has been carried out experimentally and numerically in order to increase the Q/V ratio through subtle tuning of the hole geometry around the cavity defect [37–43]. Importantly, Lalanne et al. proposed the Bloch mode engineering concepts [42] and revealed two physical mechanisms of the fine-tuning of holes geometry of PhC cavities [43]. The first mechanism could be realized through engineering the mirrors and thus reducing the out-of-plane far field radiation. The other mechanism involved recycling, which could be understood as an interference between leaky modes and fundamental modes. Moreover, the authors modified a classical F-P cavity model with consideration of energy recycling through leaky waves to physically interpret the second mechanism. As shown through the analytical model, the recycling mechanism complied with a phase-matching condition. Through subtle tapering of the hole size, the modal mismatch effects between cavity defect space and PhC mirrors could be reduced, which could increase the Q/V ratio by several orders of magnitude. These studies provide significant physical insights on the optimization of nanobeam cavity and are fundamental for further development of nanobeam cavity design.

Based on the Bloch mode engineering concepts [42,44], in 2006 Velha et al. [45] achieved a nanobeam cavity with Q factor of 8900. Furthermore, in 2007 Velha et al. [46] tried to adjust the length of the cavity defect as a function of the number of mirror holes. Consequently, they obtained a modal volume of 0.6 $(\lambda/n)^3$ and a Q factor of 58,000. In addition, many studies have been carried out on the design concepts and methodologies for the optimization of nanobeam cavities [47–54]. In general, three elements are modulated in the design process, the PhC mirror, cavity length, and taper (in Figure 3). Notomi et al. [48] proposed mode-gap-based cavities numerically in 2008, and later Kuramochi et al. [51] experimentally demonstrated the ladder nanobeam cavity and stack nanobeam cavity on an SOI wafer with a Q factor of higher than 10^5 and a small modal volume. Later, a deterministic method was proposed for a nanobeam cavity with high Q/V [52,53]. The authors adopted photonic band calculations rather than a trial-based method in the design process, which could save on computation costs and improve the design efficiency. In this way, the final cavity resonance with a small deviation from the predetermined wavelength could be achieved.

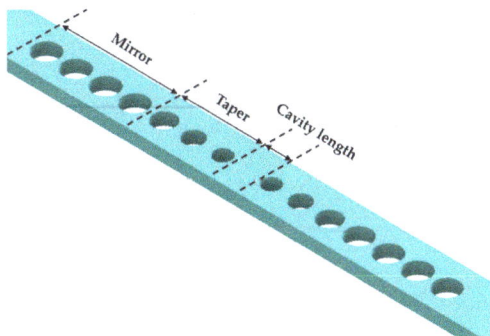

Figure 3. Three elements involved in the optimization of PhC nanobeam cavities.

In addition, there are some other types of optical microcavities using different confinement methods. As shown in Figure 4a, two planar mirrors are placed parallelly to form an F-P cavity. In this way, resonant photons can be bounded between mirrors so that light can be confined to the optical cavity. Due to the unique "air gap" offered by the F-P cavity, it has some advantages over other optical cavities, such as tunable cavity length and easy interaction with analytes. In WGM microcavities (Figure 4b), the circular structure of dielectric material can confine optical field strongly through total internal reflection [55]. WGM cavities with an ultra-high optical Q (exceeding 10^8) [28] provide a remarkable advantage over other microcavities in terms of extremely low loss and long photon lifetime.

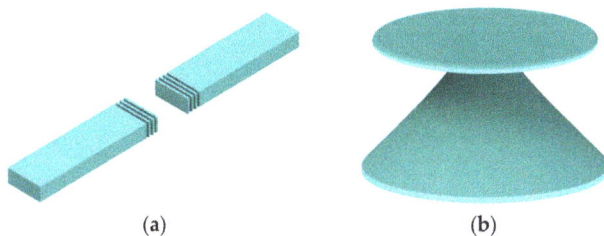

(a) (b)

Figure 4. (a) Schematic diagram of Fabry-Pérot (F-P) cavity with DBRs; (b) schematic diagram of whispering gallery mode (WGM) cavity.

3. Refractive Index Sensing

Refractive index (RI) sensing is one of the most prominent commercialized sensing technologies, and there is a vast amount of literature on RI sensors. Various examples of RI sensors using photonic structures such as ring resonators [56,57], long-period fiber gratings [58,59], surface plasmon resonators [60], and PhCs have been proposed in recent years. In this section, we focus on RI sensors using PhC nanobeam cavities.

3.1. Sensing Principle

For sensing applications, the resonance wavelength shift of a PhC cavity induced by RI changes in the optical mode region can be measured to evaluate the RI variations. Section 3 gives an overview of RI sensors for detection of the homogenous change of background RI in the optical mode region, which is usually used for the determination of RI changes in liquid or gas samples.

With the use of perturbation theory, the frequency shift $\Delta\omega$ caused by a small perturbation of dielectric function $\Delta\varepsilon$ can be obtained as follows [32]:

$$\Delta\omega = -\frac{\omega}{2}\frac{\int d^3\mathbf{r}\Delta\varepsilon(\mathbf{r})|E(\mathbf{r})|^2}{\int d^3\mathbf{r}\varepsilon(r)|E(\mathbf{r})|^2} + O(\Delta\varepsilon^2), \tag{3}$$

where $E(\mathbf{r})$ is the mode profile of the perfectly linear and unperturbed dielectric function $\varepsilon(\mathbf{r})$. Writing $\Delta\varepsilon \approx \varepsilon \cdot 2\Delta n/n$ and considering the homogenous change of RI, which means that $\Delta n/n$ is all the same in the perturbed region and zero in the unperturbed region. An intuitive interpretation of Equation (3) is then given as [32]:

$$\frac{\Delta\omega}{\omega} \approx -\frac{\Delta n}{n}\left(fraction\ of \int \varepsilon|E|^2\ in\ the\ perturbed\ regions\right). \tag{4}$$

It is indicated in Equation (4) that the frequency change is proportional to the fraction of electric field energy confined in the perturbed region.

Furthermore, for detection of a nanoparticle or a single molecule in the vicinity of the photonic cavity, the cavity resonance wavelength shift can be given as follows [15]:

$$\frac{\delta\lambda}{\lambda} = \frac{3(\varepsilon_p - \varepsilon_s)}{\varepsilon_p + 2\varepsilon_s}\frac{|\mathbf{E}_{mol}|^2}{2\int \varepsilon|\mathbf{E}|^2 d\mathbf{r}}V_{mol}, \tag{5}$$

where $\delta\lambda$ is the shift of resonance wavelength, ε_s is the permittivity of the background environment, ε_p is the permittivity of the nanoparticle, $\mathbf{E}_{mol}$ is the electric field at the nanoparticle location, $\int \varepsilon|\mathbf{E}|^2 d\mathbf{r}$ is the overall optical mode energy inside cavity, and V_{mol} is the volume of the nanoparticle. From Equations (4) and (5), it is indicated clearly that a cavity with a small mode volume and a strong optical field concentrated at the perturbed (or sensing) region is preferred to yield a large resonance shift for high sensitivity.

To quantitatively compare and evaluate the capability of the RI sensor, the concepts of sensitivity and detection limit are introduced [61]. The sensitivity is defined as the resonance shift with unit variation of sample RI, and the detection limit is defined as the minimal variation of RI that can be measured precisely. Various RI sensors based on PhC nanobeam cavities have been put forward to achieve outstanding refractive sensing capability in the past decade. The developments of these designs will be reviewed in detail.

3.2. Sensing Applications

Initial work on PhC RI sensors adopted primarily 2D PhC cavities [62,63]. Chow et al. [62] experimentally demonstrated the measurement of the shift in resonant wavelength of a 2D PhC microcavity induced by the change of ambient RI. Commercially available optical fluids having different refractive indices were used for experimental characterization. The proposed sensor with a Q factor of 400 demonstrated a sensitivity of 200 nm/RIU and a detection limit of 0.002 RIU. A growing number of studies on RI sensors have focused on 1D PhC cavities in the last decade due to the fact that PhC nanobeam cavities have demonstrated ultra-high Q factors experimentally [48]. What is more, PhC nanobeam cavities show clear advantages in miniaturization and on-chip integration of optical sensors due to their small effective masses and physical footprints.

3.2.1. Efforts on Sensitivity

As introduced in Section 2, most of the initial work on nanobeam cavity design has been motivated by the development of optical communications and data processing. Thus, many studies have focused on the improvement of Q/V ratio, which is crucial for functional devices such as optical filter and optical switch. Meanwhile, some efforts have been carried out to use these nanobeam cavities as sensors. For the use of sensing, the design objective is not limited to high Q/V ratio, and many studies have been put forward to optimize the cavity design for the sensing performance.

Wang et al. [5] demonstrated the experiment of RI sensing with the use of a single nanobeam cavity in 2010. As shown in Figure 5, the PhC nanobeam waveguides were designed to be two parallelly suspended nanobeams with a small slot between them. Moreover, these two nanobeams were patterned with 1D holes for strong light confinement in the slot region. Significantly, the light field is confined in the low RI region. Due to a large overlap with the potential analyte, the fabricated sample could obtain a sensitivity of 700 nm/RIU and a Q factor of 500 at 1386.5 nm. The parameters of the PhC structure were determined using a photonic bands software package [64]. Furthermore, the proposed sensor was evaluated theoretically through the use of a 3D finite-difference time-domain (FDTD) method [65] and analyzed experimentally after fabrication, which were useful approaches for the characterization of the proposed sensor. For the experimental demonstration, the proposed sensor was fabricated on an InGaAsP membrane of 220 nm thickness with embedded InAs quantum dots (QD). This could easily yield photoluminescence (PL) after excitation with a continuous wave (CW) diode laser [66]. After being dispersed through a monochromator, the PL signal was detected with a liquid-nitrogen-cooled InGaAs camera.

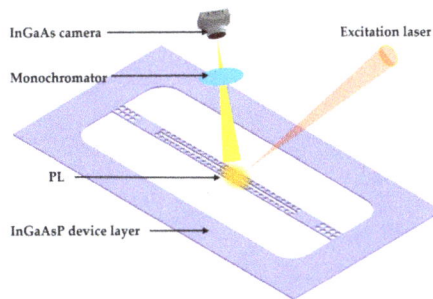

Figure 5. Refractive index (RI) sensor based on PhC slot nanobeam slow light waveguide.

With the development of nanobeam cavity design methods, there have been some ultra-high Q nanobeam cavities emerging. Typically, high-Q cavities are preferred for RI sensing due to the advantageous detection limit. However, due to the lack of light-matter interaction, an ultra-high Q factor nanobeam cavity [67] provided a sensitivity of 83 nm/RIU. This also pointed out the importance of sufficient overlap between optical mode field and analytes (typically having a low index) for sensitivity.

In order to achieve high sensitivity, Yao and Shi [6] designed a 1D PhC stack mode-gap cavity with width modulation, which localized 35% of the electric field in the low index region. Consequently, the measurement of sensitivity was reported as 269 nm/RIU. The proposed sensor had a wide sensing range. After being immersed in a water-ethanol mixture in their experiment, the proposed sensor still maintained a Q factor of about 27,000 across a 50 nm wide spectral band. Furthermore, their structure design was appropriate for the implementations of sensing in a flowing sample. When a PhC nanobeam cavity used for RI sensing, the structures of resonators (high index material) usually isolate the void space (low index region) into pieces, which may potentially block the flow channel of the sample [68]. It can be seen in Figure 6a that the structure of this PhC nanobeam provides enough channels for the flow of the samples.

To further increase sensitivity, some efforts have introduced discontinuity into photonic structures. As the introduced discontinuity results in a high-index-contrast interface in the optical field, the electric field will be much stronger in the low-index region as a large discontinuity to satisfy the continuity of the electric flux density, as stated by the Maxwell Equations [69]. In 2013, Xu et al. [70] made use of the discontinuity based on the modulated width stack cavity design. As shown in Figure 6b, a slot was introduced between periodic arrays of stacks. This design combines the characteristics of a slot waveguide (light field concentrated in the slot region) and a 1D PhC cavity (light field enhanced and confined in the cavity region). The middle slot created a high-index-contrast interface in the optical mode. In this way, the majority of the light field was confined in the slot (low index region) and strongly interacted with analytes. A high sensitivity of 410 nm/RIU could be obtained in the experiments with NaCl solutions of different concentrations. In experiments, the slot tapers and ridge tapers were used to efficiently guide the light into and out of the sensing region. Through the use of numerical simulations, the authors also discovered that the sensitivity increased and the Q factor decreased exponentially with the expansion of the slot width. After a tradeoff between Q factor and sensitivity was made, the Q factor remained at around 10^4 in their experiments with the cavity in the NaCl solution. Moreover, in 2015 Yang et al. [71] reported a slot PhC nanobeam cavity, as shown in Figure 6c. With the parabolically tapered air holes and an air slot introduced in the middle of the nanobeam, the nanobeam cavity could confine optical field robustly in the low RI region between air holes. Therefore, an ultra-high Q factor of 2.67×10^7 and sensitivity of 750.89 nm/RIU could be obtained simultaneously according to their simulation results. What is more, this proposed geometry significantly provided ultra-small mode volume around 0.01 $(\lambda/n_{air})^3$, which made it a potential platform for energy-efficient single particle detection.

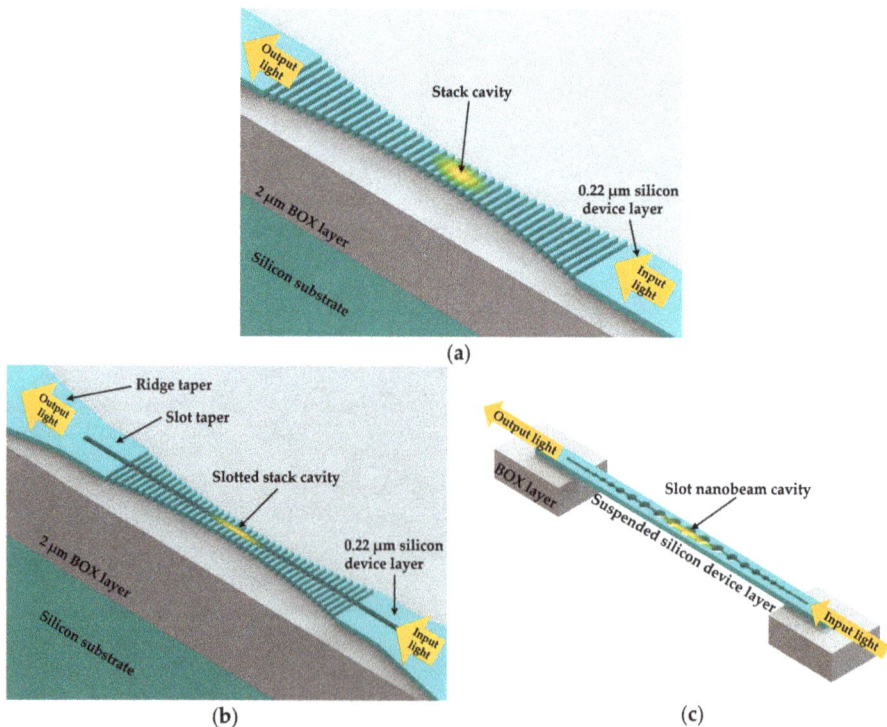

Figure 6. RI sensors aimed at increasing light matter interaction: (**a**) width-modulated stack nanobeam cavity; (**b**) slotted width-modulated stack nanobeam cavity; (**c**) slotted tapered-hole nanobeam cavity.

3.2.2. The Pursuit of Both High Q Factor and High Sensitivity

As mentioned above, a discontinuity in photonic structures could be adopted to enhance the sensitivity for a dielectric mode (optical field confined in the dielectric waveguide medium) PhC nanobeam. However, this leads to an intrinsic tradeoff between the quality factor and sensitivity for the design of a dielectric mode PhC nanobeam RI sensor [70]. The figure of merit (*FOM*) was proposed by Leif et al. [72] for the evaluation on sensing performance of RI sensors. It can be understood as:

$$FOM = S \cdot Q/\lambda_{res}, \tag{6}$$

where λ_{res} is the resonant wavelength of the cavity, Q is the quality factor of the cavity, and S is the sensitivity of the cavity. However, for the dielectric mode nanobeam cavity, *FOM* is limited by the tradeoff between the sensitivity (S) and quality factor (Q). The high sensitivity of RI sensors requires a large overlap between the optical mode field and analytes, which means that the optical mode should be confined in the analyte region (usually a low-index region), while the high Q factor of RI sensors requires strong confinement of optical mode in the waveguide medium (usually the high-index region). The high-Q cavities are desirable for a low detection limit of RI sensing, because a high Q factor leads to narrow resonance linewidth and hence small resonance shift resolution. Therefore, some studies have been carried out aiming at simultaneously achieving ultra-high quality factor and sensitivity [7,73,74].

A photonic nanobeam structure with simultaneously ultra-high Q and S for RI sensing was reported in 2013 by Yang et al. [7]. There were small gaps between these parallel PhC nanobeam cavities. After guided directly into these gaps through coupler tapers, the light was confined in the low-index region. With the use of a deterministic high-Q nanobeam cavity design method [53], the resonance could be predicted with the numerical simulation on band-edge frequency of a single unit cell with low computational cost. As the simulation results demonstrated, the nanoslotted parallel multibeam cavity achieved a Q factor of 10^7 and a sensitivity of 800 nm/RIU at telecom wavelength range in liquid with negligible absorption. As a follow-up study, Yang et al. [74] experimentally demonstrated the sensing performance of the proposed sensor in 2014. The device was fabricated on an SOI wafer without release of the buried oxide (BOX) layer, as shown in Figure 7. In the measurement of quadra-beam PhC cavity in ethanol/water solution, the sensor obtained a high-Q factor of 7015 and a high sensitivity of 451 nm/RIU experimentally and achieved an *FOM* of 2060. Furthermore, an ultra-low concentration of 10 ag/mL streptavidin could be detected with this proposed sensor in experiments.

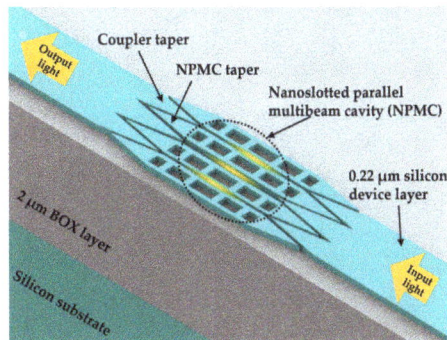

Figure 7. Nanoslotted parallel quadra-beam cavity.

Due to the intrinsic tradeoff between Q factor and sensitivity of dielectric-mode nanobeam cavity, some researchers extended studies to air mode cavities besides the common focus on dielectric mode cavities. As one of the drawbacks of the multibeam cavities is the relatively large footprint, single nanobeam cavities are expected to have better sensing performance. Quan and Loncar [53] proposed a deterministic design method for a single air mode nanobeam cavity. The air mode cavity could localize optical field in the region of low RI, and thus the air mode cavity with both high Q factor and high sensitivity became an appropriate choice for RI sensing. With use of the same design principle for the ultra-high Q dielectric-mode nanobeam cavity, the mode at the air band edge could be pulled down into the bandgap to create an ultra-high Q air mode nanobeam cavity. As introduced in this study, the dielectric mode nanobeam cavity was achieved through decreasing the size of holes from center to end, while the air mode nanobeam cavity was obtained by increasing the size of holes from center to end. Liang and Quan [8] experimentally demonstrated the air mode PhC nanobeam cavity, which had the advantages of both ultra-high Q (2.5×10^5) and ultra-small mode volume (0.01 (λ/n_{air})^3) at telecom wavelength, in 2015. As shown in Figure 8a, the lengths of these rectangular slots were tapered from the end to the middle, with constant width and periodicity. The proposed sensor was fabricated on an SOI wafer with input and output waveguide. The ultra-small mode volume of this proposed sensor enabled its application to single nanoparticle detection. What is more, the air mode nanobeam cavity can also be realized by tapering the nanobeam width rather than the hole size [53]. Figure 8b presents a RI sensor based on this kind of nanobeam cavity, which was reported by Yang et al. in 2015 [9]. The air mode nanobeam cavity greatly increased the interaction between the optical field and the analytes; thus, the single nanobeam could realize a high sensitivity of 537.8 nm/RIU and a high Q factor of 5.16×10^6, as indicated in their simulation results. Based on the similar sensing principle, Huang et al. [75] also reported a tapered width nanobeam cavity with elliptical holes in 2016. Moreover, in 2015 Fegadolli et al. [75] utilized the air-mode nanobeam cavity integrated with a NiCr microheater to demonstrate an RI sensor with a local heating function. The proposed sensor presented a sensitivity of 98 nm/RIU and a heating temperature range of 98 °C, which could be used for sensing applications that required local temperature control [76].

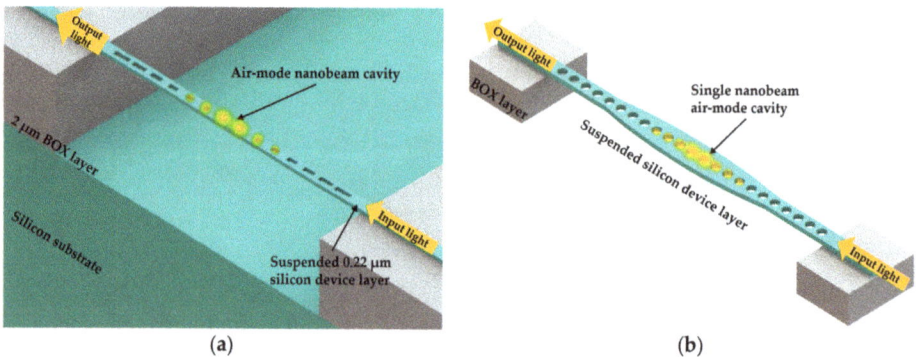

Figure 8. Air mode nanobeam cavities that localize light in low index region: (**a**) nanobeam cavity with modulated air holes; (**b**) nanobeam cavity with modulated width.

In addition to the work described above, some studies have indicated that the optical field in the low-index sensing region could also be enhanced with coupled optical resonators, such as a side-coupled nanobeam cavity, microring resonator, etc. [77–80]. Furthermore, research has been carried out to enhance the single nanobeam cavity design in order to achieve the ultra-small mode volume suitable for nanoscale RI sensing [81–84]. Along with much work on telecom wavelength range, Liu et al. [85] experimentally demonstrated a Si_3N_4 PhC nanobeam cavity for highly sensitive RI sensing at visible spectrum resonance wavelength. Visible light sensors have the unique advantage of avoiding high light absorption of water in the telecom near-infrared region. In addition, Xu et al. [86] theoretically presented a nanobeam cavity sensor design with a resonant wavelength of 4132 nm in the mid-infrared region for RI sensing. The proposed sensor achieved sensitivity as high as 2280 nm/RIU theoretically.

3.2.3. Multiplex RI Sensing

Besides much research work on the detection of single analyte, there are studies focused on multiplex RI sensing, including nanobeam cavity sensor array [4] and single nanobeam cavity [11,87].

Mandal and Erickson [4] demonstrated a sensor array based on PhC nanobeam cavities in 2008. A nanobeam cavity with Q factor of 8900 was achieved by Velha et al. [88] in 2006 using Bloch-mode engineering concepts. Based on the same design principles, the authors adopted FDTD simulations to optimize the cavity and achieved a Q factor of 2000, which was satisfactory for concept illustration in this study. In their design, the nanobeam cavities side-coupled to the input/output waveguide were designed to be at different resonances and separated in different fluid channels. Thus, each analyte could be detected simultaneously with the resonance shift of its corresponding nanobeam cavity by monitoring the output of the waveguide. As shown in Figure 9a, the proposed device was fabricated in a 250 nm thick silicon device layer on an SOI wafer. To demonstrate the nanobeam cavity sensor array, the authors used soft lithography with PDMS to form fluid channels that could separate each nanobeam cavity along the waveguide. The authors injected fluids into the channels and measured the resonance shifts of the corresponding nanobeam cavities in the multi-peak transmission spectrum of the device. Moreover, the effects of functionalized surfaces on low mass detection were also investigated numerically. In addition, some researchers have also demonstrated the improved sensor arrays based on nanobeam cavities side-coupled in series with a waveguide as well as parallel nanobeam cavities directly coupled to the input and output waveguides, as shown in Figure 9c [27,89,90]. As for the multiplex RI sensing with the use of multiple parallel nanobeam cavities, it is possible for the sensing signal of each nanobeam cavity to interfere with others due to the existence of multiple resonances of each nanobeam cavity. Therefore, several studies have been carried out on filtering out the resonances of unwanted orders of a single nanobeam cavity without sacrificing the sensing performance [91–93]. The schematic of a typical approach is shown in Figure 9b. With the use of a PhC nanobeam bandgap filter, the free spectral range of each sensing channel could be increased in the wavelength-multiplexed sensing scheme so that crosstalk among multiple channels could be avoided [10], as shown in Figure 9c.

What is more, the single nanobeam cavity can be used to realize complex RI sensing (detection of both real and imaginary parts of RI) [11,87]. In the field of RI sensing based on PhC nanobeam cavity, the majority of studies have focused on the real part of RI, while Zhang et al. [11] in 2016 demonstrated the possibility of multi-element mixture detection based on the combination of both real and imaginary parts of RI detection. In their demonstration of the detection of a $D_2O/H_2O/EtOH$ mixture, the authors achieved a sensitivity of 58 nm/RIU for the real part and 139 nm/RIU for the imaginary part, with a satisfactory detection limit. The authors suggested that the changes in real and imaginary parts of RI resulted in linear changes of the resonance wavelength and mode linewidth, respectively. Thus, the proposed sensor was capable of detecting ternary mixture concentrations with two unknown parameters. After the characterization of the sensor responses of resonance wavelength shift and linewidth variation to the known concentrations in specific calibration binary mixtures, the detection of unknown concentrations of a ternary mixture could be realized with the measurements of the total resonance wavelength shift and linewidth change. The dielectric nanobeam cavity with both high transmission and high Q factor [94] was fabricated with a silicon device layer of 260 nm thick above the 2 μm thick BOX layer on a SOI wafer, and light was coupled into the cavity through a grating coupler and tapered waveguide.

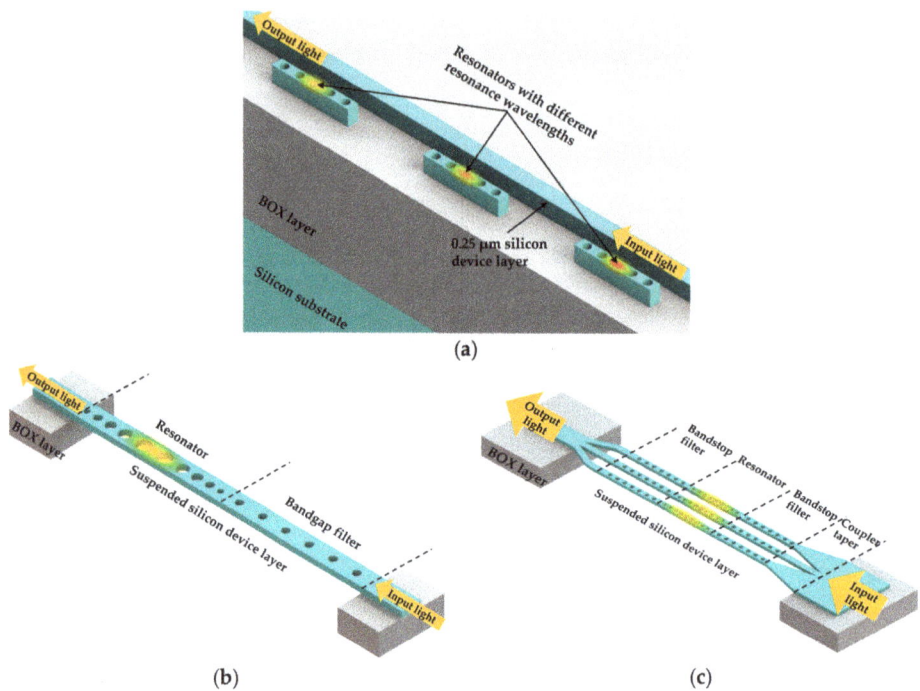

Figure 9. Proposed sensors for multiplex sensing: (**a**) waveguide side-coupled nanobeam cavity array; (**b**) schematic of nanobeam cavity integrated with bandgap filter; (**c**) parallel multiplex sensing array integrated with band stop filters.

3.3. Discussion

Many efforts have been carried out to optimize the nanobeam cavity design for RI sensing. For RI sensing applications, there are design objectives other than the high Q/V ratio. In order to simultaneously achieve outstanding sensitivity and detection limit, studies have worked to increase the light-matter interaction without sacrificing the Q factor in the nanobeam cavity design. Many meaningful approaches have been presented, such as the introduction of discontinuity into PhC structures, coupling of light into mirror gaps, and design of air mode cavity.

In summary, with the advancements of nanobeam cavity design and fabrication in recent years, the PhC nanobeam cavities have become an excellent platform for RI sensing due to the high Q factor, small mode volume, and large overlap between light field and analytes. Moreover, the advantage of the small footprint makes it suitable for multiple channel sensing using a wavelength multiplexing scheme and facilitates further on-chip integration.

4. Nanoparticle Sensing

In the previous section about RI sensors, the studies introduced mainly focus on the homogenous change of RI induced by the sample. Despite some designs already possessing nanoparticle detection ability, we feel that nanoparticle sensing using nanobeam cavities deserves special attention here. For the sensing of nanoparticles such as biomolecules, specific capture of nanoparticles on the sensor/cavity surface is required to induce a significant change in cavity characteristics. As shown in Figure 10, it is hard to induce a detectable resonance shift in a nanobeam cavity without surface binding capability, especially for single nanoparticle sensing.

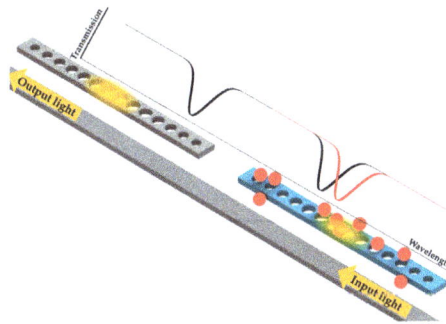

Figure 10. Schematic of nanoparticle sensing.

In recent years, there has been great progress in nanoparticle sensing using micro- and nanoscale optical approaches. To realize next-generation clinical diagnostic tests, much work has been carried out on approaches to sensitive detection of biomolecules, such as DNA, proteins, and other nanoparticles [95]. The detection limit of these nanoparticle sensors is expected to be ultra-low for single molecule capability. Moreover, it is crucial to achieve this detection in an aqueous environment with a selective detection capability among multiple species of nanoparticles. Among all the detection methods reported, the most prominent include those based on resonant cavities [96,97], surface plasmon resonance [98,99], interferometry [100,101], and photonic crystals [102]. Some nanoparticle sensing mechanisms such as mode shift, mode broadening, and mode splitting have been demonstrated using WGM cavities [103–105]. PhC nanobeam cavities, which can provide an ultra-small mode volume and high Q factor [1], are well suited for light field interaction with ultra-small nanoparticles and biomolecules, and hence show promise in this field.

The sensitivity of the nanoparticle sensor is given as the ratio of the resonance shift induced to the surface density of the captured nanoparticles [61]. This clearly indicates demand for ultra-small mode volume cavity to interact with the target nanoparticles. Moreover, the ability of the sensor surface to capture nanoparticles is important for the sensing performance. This nanoparticle capturing ability can be realized through functionalized coatings on sensing resonators or other techniques such as trapping using optical forces. To realize optimal and selective binding of nanoparticles on the sensor surface, specified functionalized coatings are widely used in many studies. For example, the antibody-antigen locking mechanism is adopted for the binding of target biomolecules to the surface. On the other hand, different techniques such as optical trapping also exist to capture the target nanoparticles. With the use of it, there is no need for coating on the resonator surface; instead we use optical trapping force to achieve the same task. The following will briefly discuss the nanoparticle sensors using these methods.

4.1. Functionalized Coating Surface

The antibody-antigen locking mechanism has been widely adopted for the functionalization of the nanobeam cavity surface. In 2009 Mandal et al. [13] presented a biomolecular sensor with a 1D PhC nanobeam cavity array. The authors demonstrated the capability of this proposed sensor for wavelength-multiplexed sensing with monoclonal antibodies to interleukin 4, 6, 8 on three adjacent nanobeam cavities. As shown in Figure 11, these nanobeam cavities were given different types of functionalized surfaces to capture the corresponding biomolecules. Wavelength-scale mode volume could be obtained due to the optimized PhC structures [88]. The detection limit on the order of 63 ag total bound mass could be achieved according to the estimation using a polyelectrolyte "layer-by-layer" growth model. The proposed device was fabricated in a 250 nm thick silicon device layer on a SOI wafer. Moreover, the authors adopted a polyelectrolyte multilayer deposition method to determine the proposed sensor response to the bound mass. In this way, the maximum distance for the biomolecules to be captured and detected on the surface was determined, and the appropriate surface conjugation method could be chosen.

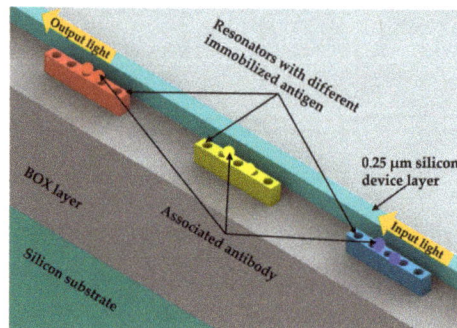

Figure 11. Three waveguide-coupled nanobeam cavities with different immobilized antigens on their surfaces.

Liang et al. [14] also took advantage of the antibody-antigen locking mechanism. In 2013 the authors indicated that the resonance shift would increase with a decrease in optical mode volume. Thus, a PhC nanobeam cavity with high Q on the order of 10^4 and small mode volume of wavelength scale was used for further investigations. The target carcinoembryonic antigen (CEA) was a biomarker for tumors in the colon cancer treatment process. Before selectively detecting CEA, the authors first modified the nanobeam surface with captured anti-CEA. The resonance shift could be clearly observed starting from a CEA concentration of 0.1 pg/mL. However, a CEA concentration above 10 µg/mL could not induce any measurable resonance shift due to the surface saturation of physical absorption. It was remarkable that the scalable deep UV lithography on SOI wafer was adopted for fabrication. Compared with commonly used E-beam lithography process, the deep ultraviolet (UV) lithography had the advantages of high-volume production, low cost, and a fast process. It was indicated that these sample PhC chips provided a mean quality factor of 9000, which proved that scalable deep UV lithography was a reliable approach to achieve sensitive and low-cost biosensing tools.

Quan et al. [15] used a single PhC nanobeam with ultra-small mode volume to investigate the detection on a single streptavidin molecule, which was only 5 nm in diameter [106]. In their experiments, the authors demonstrated that the proposed sensor was capable of detecting polystyrene particles with radii as small as 12.5 nm in DI water. To further demonstrate single molecule sensing, the authors modified the sensor surface with biotin for effective capture. They demonstrated detection of 2 pM concentration of streptavidin molecules in phosphate buffered saline (PBS) solution. Moreover, it was indicated that single molecule sensing could be improved with enhanced cavity design and laser wavelength stability in their experiments. The proposed device was fabricated on a SOI wafer with a 220 nm thick silicon device layer and a 3 µm thick BOX layer. As shown in Figure 12a, a polymer fiber-waveguide coupler was used to effectively couple the light from fiber to silicon waveguide. In this study, it was meaningful that the authors combined the perturbation theory [107] with the field distribution achieved through FDTD simulation for the theoretical prediction of the resonance shift resulted by a streptavidin molecule.

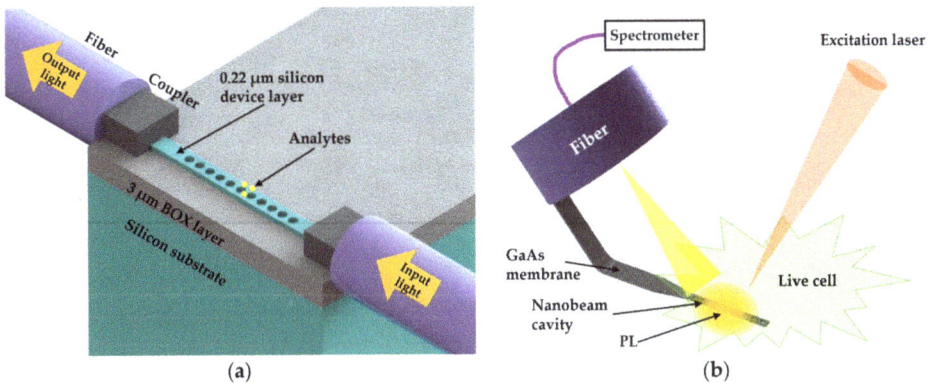

Figure 12. (**a**) Nanobeam cavity sensor for single streptavidin molecule; (**b**) nanobeam cavity sensor inside a live cell.

In 2013, Shambat et al. [16] employed a PhC nanobeam cavity as a living cellular nanoprobe. The authors developed the design of nanobeam cavity based on [50]. To demonstrate the protein sensing ability, they modified the mechanical design of the probe to make it rigid enough for experiments in beakers. After chemically functionalizing the surface to achieve streptavidin-biotin binding, they demonstrated the probe capability of protein sensing. The capability of streptavidin sensing with nanoprobe could enable the possibility of further studies on label-free sensing in live cells. As shown in Figure 12b, the nanobeam cavity was fabricated on the tip of a thin GaAs membrane that was bonded to the edge of a fiber. In their experiment, a laser was used to pump the cell sample through an objective lens. As the GaAs membrane contained layers of QDs at high density, photoluminescence (PL) would be emitted, which formed a resonant mode in the cavity. PL readout was available through the fiber bonded with the nanobeam cavity. In experiments, the authors demonstrated the sound optical performance of nanobeam resonator inside the living cell. Interestingly, it was found that a cell with inserted nanobeam could survive over one week with normal cellular activities.

Nanoparticle detection has also been reported in a gaseous environment. A gas sensor for chemical sensing based on a waveguide-coupled nanobeam cavity with chemical functionalization on the surface was presented by Chen et al. in 2014 [108]. With a fluoroalcohol polysiloxanes polymer coated on the device surface, reversible and robust binding with a target MeS molecule could be achieved. However, the excessive optical absorption of the coating caused a great reduction in Q factor at the same time. As a result, a detection limit of 1.5 ppb was demonstrated in ambient environment.

4.2. Unfunctionalized Coating Surface

As introduced above, outstanding results of nanoparticle detection including enhanced selectivity have been achieved through resonator surface binding. This typically requires an additional process to capture the nanoparticles, including antibodies, chemical, and other types of functionalized surfaces. However, the shortcoming of this approach has also been recognized.

With the unwanted optical absorption of the coating, the performance of the microcavity degrades. Moreover, the optical sensors cannot be reused due to the chemical reaction on the surface, which also increases the sensing cost. To overcome these drawbacks, research has been carried out to investigate the nanoparticle manipulation with optical trapping forces based on the PhC nanobeam cavities without functionalized coatings [17–19]. With an appropriate design, a PhC nanobeam cavity can confine a strong light field in a small mode volume. The intense evanescent field with a large gradient yields promising optical forces to capture nanoparticles on the resonator surface. As shown in Figure 13, the waveguide-coupled nanobeam cavity [18] is capable of capturing nanoparticles in an aqueous flowing sample.

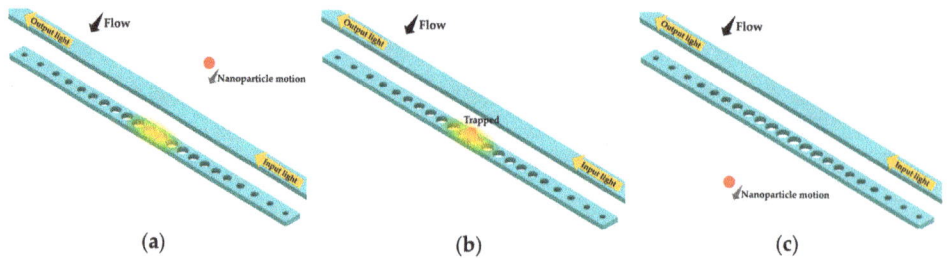

Figure 13. Schematics of the trap and release for nanoparticle in time sequence: (**a**) nanoparticle in flowing sample with laser power on; (**b**) nanoparticle trapped in the area where the optical field is the strongest; (**c**) nanoparticle released after laser power off.

A PhC nanobeam cavity with a central nanoslot for nanoparticle trapping was presented numerically by Lin et al. in 2009 [17]. Compared with traditional waveguide trapping devices, the optical trapping force of the proposed cavity was greatly enhanced. In 2010 Mandal et al. [18] demonstrated the optical trapping of dielectric nanoparticles as small as 62 nm and 48 nm based on a nanobeam cavity side-coupled with a waveguide. Previously, the capability of biomolecular detection based on this resonator design was demonstrated by the same group using the functionalized surface approach [13]. In their study [18], they presented the capability of trapping, storing, and rejecting nanoparticles in a microfluid channel using the optical trapping force, which enabled the possibility of an integrated biomolecule detection platform for simultaneous probing, sensing, and manipulation. However, it was also indicated that the optical trapping for a single molecule had not yet been realized due to the low Q factor of the nanobeam resonator resulting from the optical absorption of water.

Research efforts have therefore been made to reduce the optical absorption of water. Chen et al. [19] proposed a PhC nanobeam cavity tweezer for the manipulation of nanoparticles in 2012. The device was designed with silicon nitride, working at a resonance wavelength of 1064 nm. In this way, the optical absorption of water was greatly reduced compared with a typical design based on silicon material working at around 1550 nm wavelength. Additionally, silicon nitride has less RI contrast between the nanobeam cavity and background medium, which could extend the tail of the evanescent field. This proposed device enabled the manipulation of nanoparticles as small as Wilson disease proteins, QDs, and 22 nm polymer particles. What is more, the authors investigated the temperature effect during operation due to the optical absorption in the region of the strongest electric field. This would generate a temperature gradient and further reduce the migration of nanoparticles towards the warmest region [109]. The increase of temperature in the sample solution caused by the silicon nitride device was found to be lower than 0.3 K with the use of finite element method simulation under experimental conditions.

Using optical trapping force, Lin et al. [110] in 2013 also demonstrated the sensing of polystyrene nanoparticles and green fluorescent protein (GFP) based on a reusable silicon waveguide-coupled nanobeam cavity. For the protein sensing, the functionalized polystyrene particles instead of functionalized resonator surface were used as carriers. These carriers were coated with antibodies to aggregate target protein molecules into clusters. Due to their large dimensions, the clusters could easily be trapped on the cavity surface through optical trapping force. These nanoparticles coated with antibodies were added to a GFP solution of different concentrations. As the probability of a nanoparticle combining with a GFP molecule was higher in the high-concentration GFP samples, there would be a large fraction of clusters and a small fraction of single particles after the carriers were added in. Since the single particle and cluster would induce different step sizes of resonance shifts, the GFP concentration could be statistically analyzed through the measurement of the percentage of single particles. The proposed device was fabricated on a SOI wafer with a silicon device layer of thickness 220 nm. Moreover, it was interesting that the authors experimentally compared the sensing performance of a waveguide-coupled micro-donut resonator and a nanobeam cavity for polystyrene nanoparticles. Even though the micro-donut resonator had a higher Q factor of 9000 compared with the nanobeam cavity Q factor of only 2000, the nanobeam cavity could generate a larger resonance shift and detect smaller nanoparticles having a diameter of 110 nm due to its smaller mode volume.

In addition, research has been carried out on the detection of gold nanoparticles due to their use as biosensing labels. The sensitive detection of gold nanoparticles enables the detection of potential labeled analytes, such as DNA, proteins, and antigens. Schmidt et al. [12] demonstrated the detection of gold nanoparticles as small as 10 nm in diameter with a density of 1.25 particles per 0.04 μm^2. A nanobeam cavity with a Q factor of about 180 was designed based on [35,69]. The strongly confined light field in the nanobeam cavity enhanced the effective cross section of gold nanoparticles. The proposed device was fabricated on a SOI wafer. In their experimental demonstration, a small amount of 10 nm gold particles were added to a water-based solution and subsequently deposited on the top surface of the cavity after the solution evaporated. It was indicated that there was a transmission loss due to the absorption of gold nanoparticles.

Liang and Quan [8] demonstrated the detection of 1.8 nm diameter single gold nanoparticles in 2015. The proposed device was fabricated on a SOI wafer with 220 nm thick silicon device layer. The authors used two methods, namely piezospray and electrospray, in their experimental demonstration for single particle detection. After they are evaporated and deposited on the PhC nanobeam cavity with piezospray, the gold particles would be trapped in the center of the nanobeam cavity where the optical field was the strongest due to the optical trapping effect. In this way, a single-step resonance shift could be measured. Moreover, the authors adopted the electrospray method to demonstrate their detection capability in a flowing sample. The electrospray-generated aerosol nanoparticles had high kinetic energies that could not be easily trapped by the optical field. Thus, the detection was based on further statistical analysis of distributed resonance shifts instead of a single-step resonance shift.

Research is also ongoing for developing new sensing mechanisms aiming at single molecule detection [111,112]. For example, in 2012 Lin et al. [112] theoretically proposed a photothermal sensor that was capable of detection of a single molecule. There were two parallel nanobeam cavities in their design, with one acting as a pump and the other as a probe. The light pumped in the nanobeam cavity was tuned to the characteristic absorption line of target molecules. Thus, the temperature of the suspended nanobeam cavities would increase due to the heat generated from the absorption process. Furthermore, the resonance shift induced by the thermo-optic effect could be measured in the probe nanobeam cavity to evaluate the molecule concentration. The detection limit of gas concentration was numerically calculated to be 1.7 ppb.

4.3. Discussion

In general, for both the RI sensing and nanoparticle sensing applications introduced above, the transmission spectrum of the nanobeam cavity is measured to monitor the resonance shift. Most of the studies are in pursuit of nanobeam cavities with high Q factor and high sensitivity in principle. However, the sensing performance of these highly sensitive nanobeam cavity sensors may be influenced by several factors in practice [15]. Several factors, such as temperature change, chip oxidation, and solvent deposition, may cause considerable resonance fluctuations due to the highly sensitive performance. Thus, some studies have been carried out on the on-chip stabilization of cavity resonance. In [6], the authors characterized the ambient temperature effect on the fluctuations of resonance shift during measurements. Moreover, some researchers mentioned the use of on-chip thermal-stabilized reference nanobeam cavity [26,27]. Furthermore, some researchers paid attention to the sample environment effects on the degradation of sensor behavior [16,113]. In [16], the authors made use of an alumina/zirconia coating to prevent the photoinduced oxidation of device. Due to the dense layer of hydroxyl groups on the surface of silica, the silica resonators are easily to be degraded through the attraction of water in the air. To avoid degradation, the authors made use of a SiO_XN_Y layer to fabricate the device [113]. In this way, the resonance and Q factor of the cavity could be stabilized. In [114], an on-chip NEMS actuator was used to stabilize the optical mode wavelength.

Normally, the resonance shift is monitored in real time for the measurement of nanoparticle sensing. An overall slope of resonance shifts versus time can be used to analyze target nanoparticles in liquid sample. However, both the analyte and solvent contribute to the slope. Since the nanoparticle binding on the cavity surface could result in resonance jump during the monitoring process, many researchers observed the fluctuations on the slope to distinguish the resonance shift caused by target nanoparticle binding on the surface (analyte) and homogenous RI change in solution (solvent) [115,116]. In this way, different types and sizes of nanoparticles could be distinguished based on the particular step size of resonance shifts in the fluctuations. In [115], the authors analyzed the resonance shift fluctuations and showed a histogram of resonance shift versus nanoparticle size. In the histogram, polystyrene nanobeads with a radius of 12.5 nm, 25 nm, or 50 nm could be inferred from resonance shift maxima. In [8,14,15], the authors also used this strategy to analyze the resonance shifts.

In summary, the ability of PhC nanobeam cavities to capture nanoparticles on cavity surfaces is essential for the further improvement of nanoparticle sensing capability. To detect flowing nanoparticles in aqueous samples, large optical gradient forces are demanded for successful optical trapping of nanoparticles. Hence, further optimization of cavity design might be needed. Moreover, there is a need for further studies on the functionalized coating of the sensor to reduce optical absorption, simplify the functionalization process, and improve sustainability.

5. Optomechanical Sensors

In addition to the chemical/nanoparticle sensors described in the above sections, PhC nanocavities can also be utilized for physical sensors, which rely on the interaction between optical field and nanomechanical motion [117]. The explorations of such optomechanical interactions at the nanoscale potentially enable the highly sensitive optical detection of displacement, force, and mass [118].

There are several typical optomechanical systems in these studies. Various types of optical resonator are used, including F-P cavities, WGM resonators, and photonic crystal cavities. Due to their properties of simple structure and high finesse, F-P cavities are widely used at the micro scale. As an on-chip optomechanical system, two mirrors are integrated on a chip including a fixed mirror and a movable mirror. A distributed Bragg reflectors (DBRs)-based F-P cavity is one of the most typical structures [119,120]. As shown in Figure 14a, one DBR is mechanically movable and the other is fixed. However, miniaturization of such F-P cavities at the micro/nano scale generally leads to limited Q factor and sensitivity in sensing applications. WGM cavities represent a more viable solution for on-chip miniaturization. There are generally two ways to introduce optomechanical interactions in WGM cavities. One is to modify its own structure, such as double-layered structures [121], and the other is to design a combined structure by introducing an extra mechanical resonator such as a cantilever or a bridge beam coupled to the WGM cavity [122,123], as shown in Figure 14b.

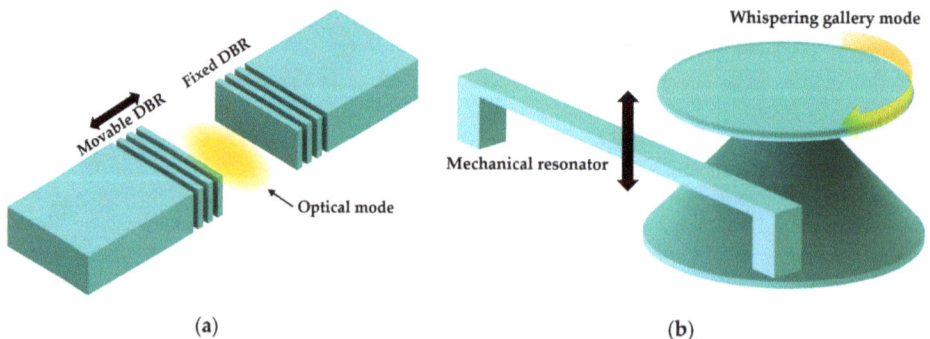

(a)   (b)

Figure 14. Schematics of typical optomechanical systems based on F-P cavity and WGM cavity: (**a**) an F-P cavity coupled with a mechanical resonator; (**b**) a WGM cavity coupled with a mechanical resonator.

PhC cavity devices exhibit high optical quality factors and ultra-small mode volumes, which promise strong optomechanical effects. These optomechanical effects have been demonstrated using both 2D [124,125] and 1D [20,126] PhC cavities. Compared with 2D PhC cavities, 1D PhC nanobeam cavities offer a potentially smaller physical footprint and more flexibility in design, which make them attractive for optomechanical sensing.

In optomechanical systems, optical field and mechanical motion are coupled. Thus, the mechanical motion can be easily detected through the measurement of the light transmitted through the optical cavity. For a microcavity of optical resonance frequency (ω_C) and cavity length (x), the optomechanical coupling coefficient can be defined as $g_{om} = d\omega_C/dx$, which represents the relationship between the resonance frequency shift and mechanical deformation of the cavity. Many sensing applications such as acceleration, magnetic field, etc. are based on displacement sensing because these physical variations can be converted into the displacement of a sensing component. Below, we briefly highlight the potential configurations that could be used for sensing nanoscale displacements using PhC nanobeam cavities.

There are various approaches to form optomechanical sensors based on PhC nanobeam cavities, which typically fall into three categories, single nanobeam cavity, coupled nanobeam cavities, and nanobeam cavities coupled with mechanical resonators. The single nanobeam optomechanical sensors can be realized with a slice introduced in the middle or a split along the beam length direction in the center, as shown in Figure 15a. This separates the nanobeam cavity into two parts, one fixed and the other movable, and the motion of the movable part can be induced by various physical measurands. This approach creates a deformable PhC nanobeam cavity, of which resonance wavelength and Q factor can be affected by the mechanical motion of the movable part. The coupled nanobeam optomechanical sensor can be achieved through arranging two PhC nanobeam cavities in parallel formation, allowing them to be optically coupled as shown in Figure 15b. One nanobeam cavity can be fixed, while the other can be movable. The mechanical movement changes the coupling strength, thereby generating resonance shifts of the symmetric and anti-symmetric supermodes of the coupled cavities. In the third configuration, the PhC nanobeam cavity is coupled with a mechanical structure or resonator, as shown in Figure 15c. Such a mechanical resonator can be extrinsic, as shown on the left side, or intrinsic, as shown on the right side. The motion of the mechanical resonator thus modulates the PhC nanobeam cavity in terms of inducing resonance shift or Q factor change. With a proper design, the on-chip nanobeam cavities are able to achieve sensitive mechanical motion detection for torsion, rotation, and translation. Similar sensing mechanisms can be constructed based on other types of optical resonators [127–131]. To keep the paper concise, we only focus on various sensors using single nanobeam cavity and coupled nanobeam cavities below.

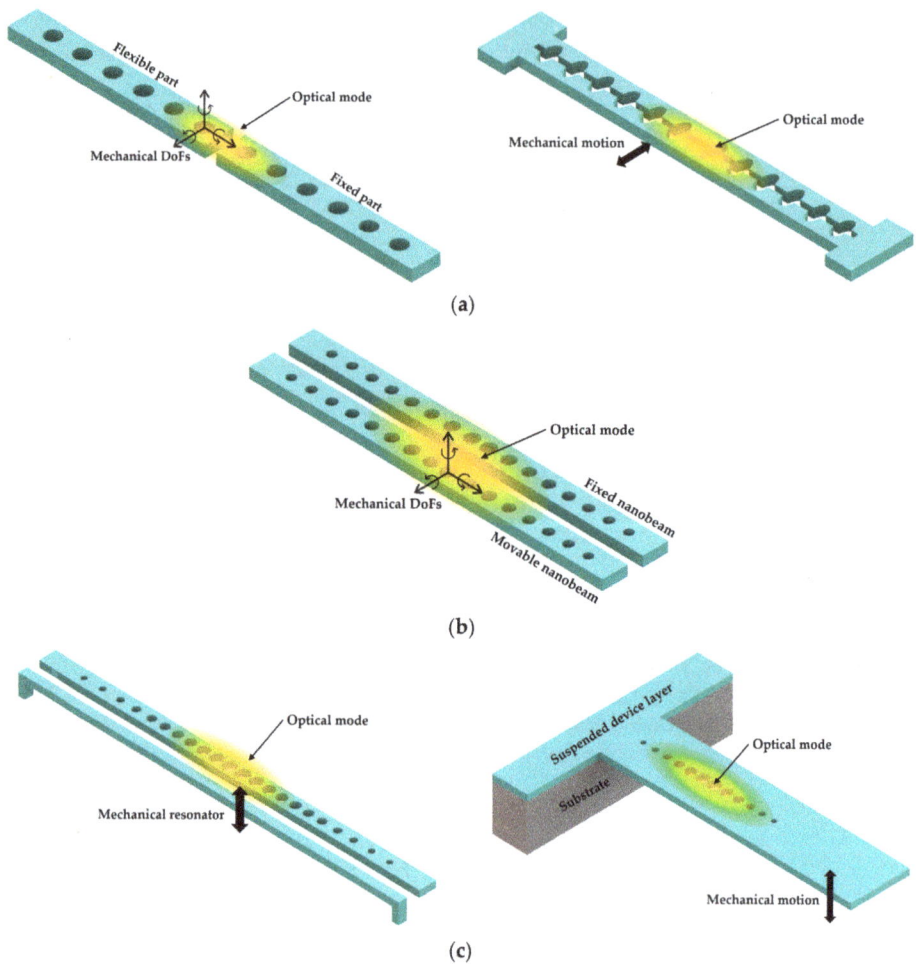

Figure 15. Schematics of potential optomechanical systems based on PhC nanobeam cavities: (**a**) single nanobeam cavity; (**b**) coupled nanobeam cavities; (**c**) nanobeam cavity coupled with a mechanical structure.

5.1. Single Nanobeam Cavity

Wu et al. [22] demonstrated an optomechanical torque sensor based on a PhC split-beam cavity in 2014. As shown in Figure 16a, there were two patterned nanobeams serving as optical mirrors to confine high-Q mode between them. The nanobeams were also suspended as cantilever mechanical resonators to support independent mechanical motion. Through engineering the holes from elliptical to circle shape [132], the authors achieved the confinement of light field in the central gap between these two nanobeams. With experiments in low vacuum and ambient conditions, the sensitivities of the optomechanical torque detection were determined to be 1.3×10^{-21} Nm·Hz$^{-1/2}$ and 1.2×10^{-20} Nm·Hz$^{-1/2}$, respectively. The demonstrated sensitivity was comparable with the magnetic tweezer torque sensor reported in [133]. The proposed device was fabricated on an SOI wafer with 220 nm thick silicon device layer and 3 µm thick BOX layer. A dimpled optical fiber taper was used to couple the light into the cavity. With the unique property of split beam cavities, the authors investigated not only the dispersive coupling resulting from the cavity gap modified by mechanical motions but also the dissipative optomechanical coupling [134]. The latter was strongly dependent on the photon decay rate inside the cavity gap. These interferences between dispersive and dissipative coupling were observed through measuring the transmission fluctuations. In the transmission spectrum, the optical response with dispersive coupling was resonance shift, and the optical response with dissipative coupling was a change of line width.

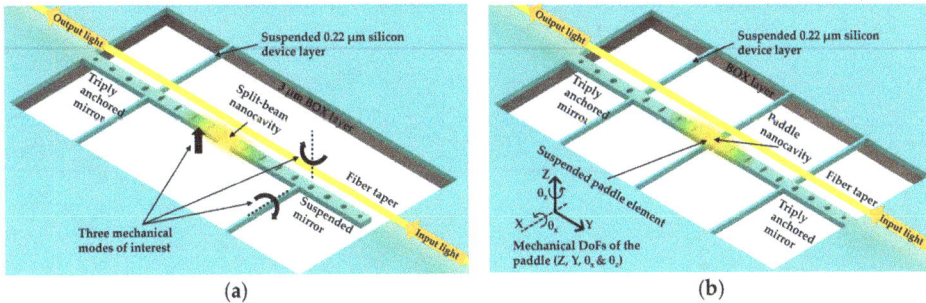

Figure 16. Optomechanical sensors based on split nanobeam cavity: (a) nanocavity torque sensor; (b) optomechanical paddle cavities.

Later, Kaviani et al. [23] presented strong nonlinear optomechanical coupling by modifying the abovementioned split-beam nanocavities. As illustrated in Figure 16b, with a combination of the design principles of the membrane-in-the-middle (MiM) cavities [129] and nanobeam optomechanical cavities [20], a "paddle" component was suspended inside the mirror gap between two suspended PhC nanobeam optical mirrors. In this way, the mechanical resonance of the paddle element could modify the dynamics of the optical mode confined inside the mirror gap. Due to the unique properties of optical confinement, wide free spectral range of mechanical resonance, and low mass of the whole design, the proposed optomechanical nanobeam cavity theoretically enabled the observation of thermally driven motion at a temperature around 50 mK in the device.

Leijssen and Verhagen [24] reported a sliced PhC nanobeam optomechanical system in 2015. As shown in Figure 17, the sliced PhC nanobeam cavity was split down the middle, which formed two doubly clamped beams joined at their supports. Due to the slice being introduced as a subwavelength dielectric discontinuity, a high concentration of light energy in the subwavelength gap could be realized, which enabled a high rate of optomechanical coupling with low mass mechanical components. After analyzing the experimental results of optical radiation pressure and motion transduction, the authors reported that the photon-phonon coupling rate was as high as 11.5 MHz. The proposed device was fabricated on a SOI wafer with a silicon device layer of 200 nm thick. The free-space readout method was used for experimental detection, which provided a coupling rate comparable with the standard fiber taper coupling approach. The laser beam was focused on the cavity through an aspheric lens. To reject the directly reflecting light and detect the light coupled to the cavity, a polarizing beamsplitter was used. These investigations on large optomechanical interaction can potentially realize the detection of thermally driven displacement with noise even below the standard quantum limit [135].

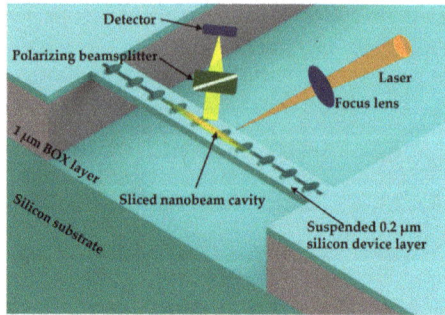

Figure 17. Sliced nanobeam optomechanical system with a free-space experimental setup.

5.2. Coupled Nanobeam Cavities

Eichenfield et al. [20] demonstrated the potential use of coupled PhC nanobeam cavities for optomechanical sensing. These two coupled PhC nanobeam cavities were named a "zipper cavity" [136] by the authors due to the fact that they work like a mechanical fastener, allowing sensing and actuation via the confined optical cavity field. With the use of the optical fiber taper coupler for excitation and probing, the Si_3N_4 zipper cavity was measured to have a high Q factor in the range of 10^4 to 10^5 in experiments. Compared with high-finesse glass microtoroid structures or F-P cavities [137], the optomechanical coupling accomplished with the zipper cavity was increased greatly. Besides the sensitive detection of mechanical displacement achieved, the mechanical motion could also be driven through the zipper cavity's internal optical field.

Using the zipper cavity as displacement readout, in 2012 Krause et al. [21] demonstrated an optomechanical accelerometer with excellent performance. The zipper cavity device realized a bandwidth over 20 kHz, an acceleration detection resolution of 10 µg·$Hz^{-1/2}$, and a dynamic range over 40 dB, which were comparable to commercial sensors. The acceleration resolution could be quantified as noise-equivalent acceleration, which required a maximization of mass and mechanical Q factor product. However, there is a natural tradeoff between band width and resolution for most of the commercial accelerometers. The intrinsically low mechanical Q factor required a large test mass for better resolution, but a large test mass would limit the device resonance frequency and thus reduce the bandwidth. On the other hand, the zipper cavity device was capable of achieving both high resolution and wide bandwidth because the nanogram test mass with nanotether suspension brought both low mass and high mechanical Q factor of 10^6. Moreover, the large optical radiation pressure force in optomechanical zipper devices enabled the control of sensor bandwidth with the optical spring effect [137]. As shown in Figure 18a, the proposed device was fabricated in a SiN layer of 400 nm thickness above a 500 µm silicon layer. The fiber taper was used to couple lightinto cavity.

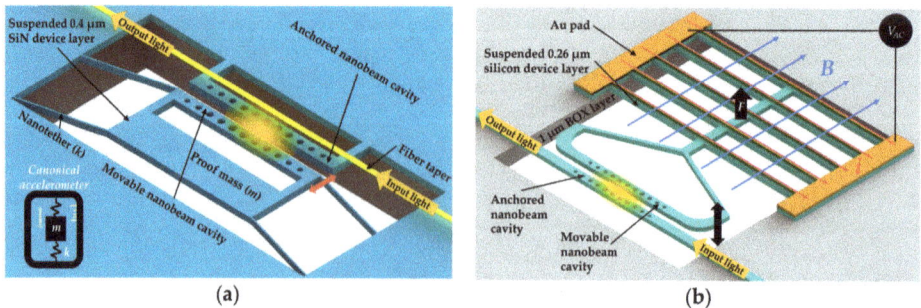

Figure 18. Optomechanical sensors based on coupled nanobeam cavities: (**a**) high-solution optomechanical accelerometer; (**b**) magnetic field sensor.

In addition, a magnetic field sensor based on similar coupled nanobeam cavities with wide operation bandwidth of 160 Hz and small footprint was demonstrated by Du et al. [25] in 2017. The authors experimentally demonstrated the sensitivity of 22.9 mV/T and resolution of 48.1 $\mu T \cdot Hz^{-1/2}$. As illustrated in Figure 18b, one of the coupled nanobeam cavities was fixed for guiding the input and output light, and the other was suspended and connected with a silicon bridge structure deposited in a thin gold layer. After being applied with an AC voltage at the structure's mechanical resonance frequency, the gold wires as current carriers yielded a mechanical oscillation of the bridge structure in the magnetic field parallel to the device surface, as shown. This could further induce the resonance shift of a selected supermode of the coupled cavities, and subsequently affect the light intensity output.

5.3. Discussion

Given the above studies, there are various sensing mechanisms based on the PhC nanobeam cavities. In general, the RI sensors introduced in previous sections are appropriate to use for biochemical sensing. These well-designed nanobeam sensors are highly sensitive to RI changes even in an aqueous environment, which makes them useful for water-based clinical samples. On the other hand, the optomechanical sensors mentioned in this section are preferred as physical sensors. Through proper design of on-chip device structure, the target physical quantity can be converted into on-chip mechanical motion. Then, the optical measurements of the optomechanical cavity can be used to detect the mechanical motion and thus quantify the target physical signal. With the use of optomechanical sensors, various physical signals, such as acceleration, magnetic field, torsion, temperature, etc., are possible to detect with ultra-high sensitivity and a low detection limit. To further develop the optomechanical sensors, it is crucial to optimize the design of the optomechanical systems to yield large optomechanical coupling coefficients.

6. Temperature Sensors

Temperature sensors play a significant role in various application areas, such as automobiles [138], environmental control in buildings [139], medicine [140], and manufacturing [141]. For the past century, the resistance thermometer has been a prevailing choice for accurate measurement [142]. Though temperature uncertainties below 10 mK can be realized using resistance thermometers, they require frequent, time-consuming, and expensive calibrations due to the sensitivity to mechanical shocks. On the other hand, due to the immunity to mechanical shock and electromagnetic interference, there has been growing interest in optical temperature sensors as a substitute to resistance thermometers in recent years. Moreover, temperature sensors based on photonic structures have the advantages of a small footprint, flexible on-chip integration, complementary metal-oxide-semiconductor (CMOS) compatibility, and fast response.

Photonic temperature sensors utilize a combination of thermo-optic effect and thermal expansion. Hence, the RI and structure of PhC cavity are altered by temperature variations, which further induce resonance shifts of the cavity. Thus, the temperature variations can be evaluated from the measurement of the resonance shift. Both the thermo-optic effect $\Delta\lambda_T$ and the thermal expansion effect $\Delta\lambda_L$ contribute to the overall resonance wavelength shift $\Delta\lambda$, which is given as [143,144]:

$$\Delta\lambda = \Delta\lambda_T + \Delta\lambda_L = \alpha_W \frac{n_{eff}}{n_g}\lambda\Delta T + \frac{\sigma_T}{n_g}\lambda\Delta T, \text{ where } \sigma_T = \frac{\partial n_{eff}}{\partial T} \tag{7}$$

where α_w is the thermal expansion coefficient, n_g and n_{eff} are the group index and effective index of the photonic resonator, respectively, ΔT is the temperature variation, λ is the resonance wavelength, and σ_T is the rate of effective index change with temperature.

Many photonic temperature sensors have been reported to have outstanding performance, including waveguide Bragg gratings [145], micro-ring resonators [146,147], and PhC nanobeam cavities [26,27,90,148]. Due to the small footprint and easy on-chip integration, a PhC nanobeam cavity is an outstanding candidate for temperature sensing. In 2015, Klimov et al. [149] demonstrated silicon PhC nanobeam cavity thermometers. The waveguide-coupled nanobeam cavities were cladded separately with a silicon dioxide layer and a PMMA layer, which demonstrated their corresponding sensitivity of 83 pm/°C and 68 pm/°C, respectively. The proposed device was fabricated on an SOI wafer with a 220 nm thick silicon device layer and 3 μm thick BOX layer. The one with a PMMA layer had lower sensitivity due to the negative thermo-optic coefficient of PMMA. The sensitivity of the PhC temperature sensor is mainly limited by the thermo-optic coefficient of the material used. Silicon, the most common material for photonic devices, has a thermo-optic coefficient of only 1.8×10^{-4} [150].

Taking advantage of the Vernier effect, in 2016 Kim and Yu [147] demonstrated a temperature sensor based on cascaded ring resonators with a high sensitivity of 293.9 pm/°C, but the large envelope peak fitting error made the detection limit 0.18 °C. Zhang et al. [26] utilized the parallel nanobeam cavities [151,152] as shown in Figure 19. They demonstrated a temperature sensor with a sensitivity of 162.9 pm/°C. The detection limit was calculated to be 0.08 °C using the definition provided in [61]. To better utilize the thermo-optic property of materials, the authors designed two different types of nanobeam cavity [6,153]. One of them was cladded with SU-8 and the other had no cladding. Through a proper design, the SU-8 cladded nanobeam cavity could localize 70% of light field in the SU-8 region. This contributed to the increase of the resonance blue shift due to the negative thermo-optic coefficient of SU-8. Meanwhile, the other nanobeam cavity was designed to confine light in the silicon core, which would yield a red shift. The schematic of the device is shown in Figure 19. The sensitivities of these two nanobeam cavities were reported to be −99 pm/°C and 63.9 pm/°C, respectively. Thus, the overall sensitivity of the proposed device was increased to 162.9 pm/°C. What is more, based on the different responses to surrounding RI change and the temperature variation of the two cascaded nanobeam cavities, the simultaneous sensing of temperature and RI was demonstrated by Liu and Shi in 2017 [27].

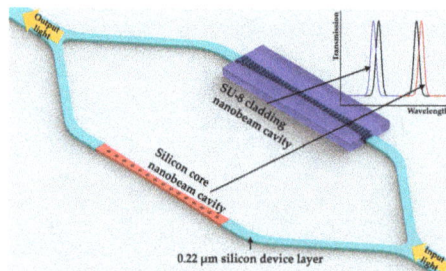

Figure 19. Sensitive temperature sensor based on nanobeam cavities (silicon core of positive thermo-optic coefficient in red and SU8 cladding of negative TO coefficient in blue).

The ultra-small footprint of a nanobeam cavity makes it an ideal candidate as a temperature reference sensor for on-chip integration with other types of sensors. Due to the fluctuations of ambient temperature, signals for sensors might deviate from their ideal calibrated performance. Thus, additional approaches such as the use of a thermo-electric cooler and cooling fluid [154] are often required to eliminate the deviation induced by the temperature difference, which obviously increases the cost and design complexity. On the other hand, with an on-chip integrated temperature sensor, the temperature fluctuations can be compensated for effectively during operations.

7. Conclusions

To be implemented in practice, these PhC nanobeam sensors pose technical challenges in manufacturing. The ability to pattern features smaller than 100 nm is required for the realization of these PhC nanobeam sensors for near-infrared wavelengths. As both of the planar PhC structures and the commercial semiconductor devices rely on planar pattern transfer, it makes the commercially available fabrication processes ideally suited for the planar PhC devices. Moreover, as the mass production of 7 nm semiconductor devices has started in 2018, the commercial processes can satisfy the dimension requirement of PhC structures. Leveraging the CMOS-compatible technology, the silicon photonics chip can be fabricated cost-effectively. However, efforts still need to be made to realize the high-yield manufacture of these photonics sensors. The stability of the PhC resonator is significantly influenced by fabrication errors. Fabrication errors in the position and size of the PhC structures may result in fluctuations of resonance wavelength and *Q* factor [155]. Thus, repeatable and stable silicon photonics processes are expected for sensor manufacturing. Moreover, as optical testing for reliability is crucial for the manufacturability, further work from foundries will be needed to develop photonics testing capability [156]. In addition, challenges remain in the packaging for these photonics sensors, such as fiber coupling, laser source, and efficient thermal management [157]. As silicon photonics has been one of the outstanding technical solutions for many applications, such as optical sensing, optical communications, and on-chip optical interconnections, there is a need for further studies and industry efforts to realize high-yield chip manufacturing.

In this paper, a comprehensive review of the PhC nanobeam cavity-based sensors is presented. In recent years, a significant amount of work on PhC nanobeam cavity sensors has been carried out, which demonstrates the promising role of PhC nanobeam cavities among various other types of optical cavities for sensing applications. Here, only the uses of PhC nanobeam cavities in RI sensing, nanoparticle sensing, optomechanical sensing, and temperature sensing are highlighted. We summarized their sensing principles, typical designs, and key developments in the recent past. Other sensing applications of the PhC nanobeam cavities also exist, and their working principles and design methods are similar to the approaches reviewed here. In the future, with technological advancements in cavity design and fabrication process, sensors based on PhC nanobeam cavities with further enhanced sensing performance can be expected. Additionally, due to their advantages (small footprint, ultra-sensitive optical readout, and ease of on-chip integration), new types of sensor-based PhC nanobeam cavities might be identified and explored. In conclusion, PhC nanobeam cavities offer a suitable platform for sensing applications and have significant potential in a range of practical areas.

Author Contributions: Conceptualization, Q.Q. and G.Z.; investigation, Q.Q.; resources, Q.Q. and J.X.; writing—original draft preparation, Q.Q. and J.X.; writing—review and editing, Q.Q., C.L. and G.Z.; supervision, G.Z.; funding acquisition, C.L.

Funding: This research was funded by National Research Foundation (NRF) grant number NRF-CRP15-2015-02 at National University of Singapore. And The APC was funded by NRF-CRP15-2015-02.

Acknowledgments: The authors acknowledge the financial support from the research grant NRF-CRP15-2015-02 "Piezoelectric Photonics Using CMOS-Compatible AlN Technology for Enabling the Next-Generation Photonics ICs and Nanosensors" (WBS: R-263-000-C24-281) at National University of Singapore.

Conflicts of Interest: The authors declare no conflict of interest.

References

1. Vahala, K.J. Optical microcavities. *Nature* **2003**, *424*, 839–846. [CrossRef] [PubMed]
2. Painter, O.; Lee, R.; Scherer, A.; Yariv, A.; O'brien, J.; Dapkus, P.; Kim, I. Two-dimensional photonic band-gap defect mode laser. *Science* **1999**, *284*, 1819–1821. [CrossRef] [PubMed]
3. Goyal, A.K.; Dutta, H.S.; Pal, S. Recent advances and progress in photonic crystal-based gas sensors. *J. Phys. D Appl. Phys.* **2017**, *50*, 203001. [CrossRef]
4. Mandal, S.; Erickson, D. Nanoscale optofluidic sensor arrays. *Opt. Express* **2008**, *16*, 1623–1631. [CrossRef] [PubMed]
5. Wang, B.; Dündar, M.A.; Nötzel, R.; Karouta, F.; He, S.; van der Heijden, R.W. Photonic crystal slot nanobeam slow light waveguides for refractive index sensing. *Appl. Phys. Lett.* **2010**, *97*, 151105. [CrossRef]
6. Yao, K.; Shi, Y. High-Q width modulated photonic crystal stack mode-gap cavity and its application to refractive index sensing. *Opt. Express* **2012**, *20*, 27039–27044. [CrossRef] [PubMed]

7. Yang, D.; Tian, H.; Ji, Y.; Quan, Q. Design of simultaneous high-Q and high-sensitivity photonic crystal refractive index sensors. *J. Opt. Soc. Am. B* **2013**, *30*, 2027–2031. [CrossRef]

8. Liang, F.; Quan, Q. Detecting single gold nanoparticles (1.8 nm) with ultrahigh-Q air-mode photonic crystal nanobeam cavities. *ACS Photonics* **2015**, *2*, 1692–1697. [CrossRef]

9. Yang, D.; Tian, H.; Ji, Y. High-Q and high-sensitivity width-modulated photonic crystal single nanobeam air-mode cavity for refractive index sensing. *Appl. Opt.* **2015**, *54*, 1–5. [CrossRef] [PubMed]

10. Yang, D.; Wang, C.; Ji, Y. Silicon on-chip 1D photonic crystal nanobeam bandstop filters for the parallel multiplexing of ultra-compact integrated sensor array. *Opt. Express* **2016**, *24*, 16267–16279. [CrossRef] [PubMed]

11. Zhang, X.; Zhou, G.; Shi, P.; Du, H.; Lin, T.; Teng, J.; Chau, F.S. On-chip integrated optofluidic complex refractive index sensing using silicon photonic crystal nanobeam cavities. *Opt. Lett.* **2016**, *41*, 1197–1200. [CrossRef] [PubMed]

12. Schmidt, B.; Almeida, V.; Manolatou, C.; Preble, S.; Lipson, M. Nanocavity in a silicon waveguide for ultrasensitive nanoparticle detection. *Appl. Phys. Lett.* **2004**, *85*, 4854–4856. [CrossRef]

13. Mandal, S.; Goddard, J.M.; Erickson, D. A multiplexed optofluidic biomolecular sensor for low mass detection. *Lab Chip* **2009**, *9*, 2924–2932. [CrossRef] [PubMed]

14. Liang, F.; Clarke, N.; Patel, P.; Loncar, M.; Quan, Q. Scalable photonic crystal chips for high sensitivity protein detection. *Opt. Express* **2013**, *21*, 32306–32312. [CrossRef] [PubMed]

15. Quan, Q.; Floyd, D.L.; Burgess, I.B.; Deotare, P.B.; Frank, I.W.; Tang, S.K.; Ilic, R.; Loncar, M. Single particle detection in CMOS compatible photonic crystal nanobeam cavities. *Opt. Express* **2013**, *21*, 32225–32233. [CrossRef] [PubMed]

16. Shambat, G.; Kothapalli, S.R.; Provine, J.; Sarmiento, T.; Harris, J.; Gambhir, S.S.; Vuckovic, J. Single-cell photonic nanocavity probes. *Nano Lett.* **2013**, *13*, 4999–5005. [CrossRef] [PubMed]

17. Lin, S.; Hu, J.; Kimerling, L.; Crozier, K. Design of nanoslotted photonic crystal waveguide cavities for single nanoparticle trapping and detection. *Opt. Lett.* **2009**, *34*, 3451–3453. [CrossRef] [PubMed]

18. Mandal, S.; Serey, X.; Erickson, D. Nanomanipulation using silicon photonic crystal resonators. *Nano Lett.* **2010**, *10*, 99–104. [CrossRef] [PubMed]

19. Chen, Y.F.; Serey, X.; Sarkar, R.; Chen, P.; Erickson, D. Controlled photonic manipulation of proteins and other nanomaterials. *Nano Lett.* **2012**, *12*, 1633–1637. [CrossRef] [PubMed]

20. Eichenfield, M.; Camacho, R.; Chan, J.; Vahala, K.J.; Painter, O. A picogram-and nanometre-scale photonic-crystal optomechanical cavity. *Nature* **2009**, *459*, 550–555. [CrossRef] [PubMed]

21. Krause, A.G.; Winger, M.; Blasius, T.D.; Lin, Q.; Painter, O. A high-resolution microchip optomechanical accelerometer. *Nat. Photonics* **2012**, *6*, 768–772. [CrossRef]

22. Wu, M.; Hryciw, A.C.; Healey, C.; Lake, D.P.; Jayakumar, H.; Freeman, M.R.; Davis, J.P.; Barclay, P.E. Dissipative and dispersive optomechanics in a nanocavity torque sensor. *Phys. Rev. X* **2014**, *4*, 021052. [CrossRef]

23. Kaviani, H.; Healey, C.; Wu, M.; Ghobadi, R.; Hryciw, A.; Barclay, P.E. Nonlinear optomechanical paddle nanocavities. *Optica* **2015**, *2*, 271–274. [CrossRef]

24. Leijssen, R.; Verhagen, E. Strong optomechanical interactions in a sliced photonic crystal nanobeam. *Sci. Rep.* **2015**, *5*, 15974. [CrossRef] [PubMed]

25. Du, H.; Zhou, G.; Zhao, Y.; Chen, G.; Chau, F.S. Magnetic field sensor based on coupled photonic crystal nanobeam cavities. *Appl. Phys. Lett.* **2017**, *110*, 061110. [CrossRef]

26. Zhang, Y.; Liu, P.; Zhang, S.; Liu, W.; Chen, J.; Shi, Y. High sensitivity temperature sensor based on cascaded silicon photonic crystal nanobeam cavities. *Opt. Express* **2016**, *24*, 23037–23043. [CrossRef] [PubMed]

27. Liu, P.; Shi, Y. Simultaneous measurement of refractive index and temperature using cascaded side-coupled photonic crystal nanobeam cavities. *Opt. Express* **2017**, *25*, 28398–28406. [CrossRef]

28. Armani, D.; Kippenberg, T.; Spillane, S.; Vahala, K. Ultra-high-Q toroid microcavity on a chip. *Nature* **2003**, *421*, 925–928. [CrossRef] [PubMed]

29. Yablonovitch, E.; Gmitter, T.; Leung, K. Photonic band structure: The face-centered-cubic case employing nonspherical atoms. *Phys. Rev. Lett.* **1991**, *67*, 2295. [CrossRef] [PubMed]

30. Yablonovitch, E.; Gmitter, T.; Meade, R.; Rappe, A.; Brommer, K.; Joannopoulos, J. Donor and acceptor modes in photonic band structure. *Phys. Rev. Lett.* **1991**, *67*, 3380–3383. [CrossRef] [PubMed]

31. Kuchyanov, A.; Chubakov, P.; Plekhanov, A. Highly sensitive and fast response gas sensor based on a light reflection at the glass-photonic crystal interface. *Opt. Commun.* **2015**, *351*, 109–114. [CrossRef]
32. Joannopoulos, J.D.; Johnson, S.G.; Winn, J.N.; Meade, R.D. *Photonic Crystals: Molding the Flow of Light*; Princeton University Press: Princeton, NJ, USA, 2011.
33. Ogawa, S.; Imada, M.; Yoshimoto, S.; Okano, M.; Noda, S. Control of light emission by 3D photonic crystals. *Science* **2004**, *305*, 227–229. [CrossRef] [PubMed]
34. Takahashi, S.; Suzuki, K.; Okano, M.; Imada, M.; Nakamori, T.; Ota, Y.; Ishizaki, K.; Noda, S. Direct creation of three-dimensional photonic crystals by a top-down approach. *Nat. Mater.* **2009**, *8*, 721–725. [CrossRef] [PubMed]
35. Foresi, J.; Villeneuve, P.R.; Ferrera, J.; Thoen, E.; Steinmeyer, G.; Fan, S.; Joannopoulos, J.; Kimerling, L.; Smith, H.I.; Ippen, E. Photonic-bandgap microcavities in optical waveguides. *Nature* **1997**, *390*, 143–145. [CrossRef]
36. Purcell, E.M. Spontaneous emission probabilities at radio frequencies. In *Confined Electrons and Photons*; Springer: Berlin, Germany, 1995; p. 839.
37. Johnson, S.G.; Fan, S.; Mekis, A.; Joannopoulos, J. Multipole-cancellation mechanism for high-Q cavities in the absence of a complete photonic band gap. *Appl. Phys. Lett.* **2001**, *78*, 3388–3390. [CrossRef]
38. Palamaru, M.; Lalanne, P. Photonic crystal waveguides: Out-of-plane losses and adiabatic modal conversion. *Appl. Phys. Lett.* **2001**, *78*, 1466–1468. [CrossRef]
39. Vučković, J.; Lončar, M.; Mabuchi, H.; Scherer, A. Design of photonic crystal microcavities for cavity QED. *Phys. Rev. E* **2001**, *65*, 016608. [CrossRef] [PubMed]
40. Srinivasan, K.; Painter, O. Momentum space design of high-Q photonic crystal optical cavities. *Opt. Express* **2002**, *10*, 670–684. [CrossRef] [PubMed]
41. Akahane, Y.; Asano, T.; Song, B.-S.; Noda, S. High-Q photonic nanocavity in a two-dimensional photonic crystal. *Nature* **2003**, *425*, 944–947. [CrossRef] [PubMed]
42. Lalanne, P.; Hugonin, J.P. Bloch-wave engineering for high-Q, small-V microcavities. *IEEE J. Quantum Electron.* **2003**, *39*, 1430–1438. [CrossRef]
43. Lalanne, P.; Mias, S.; Hugonin, J.-P. Two physical mechanisms for boosting the quality factor to cavity volume ratio of photonic crystal microcavities. *Opt. Express* **2004**, *12*, 458–467. [CrossRef] [PubMed]
44. Sauvan, C.; Lecamp, G.; Lalanne, P.; Hugonin, J.-P. Modal-reflectivity enhancement by geometry tuning in photonic crystal microcavities. *Opt. Express* **2005**, *13*, 245–255. [CrossRef] [PubMed]
45. Velha, P.; Rodier, J.-C.; Lalanne, P.; Hugonin, J.-P.; Peyrade, D.; Picard, E.; Charvolin, T.; Hadji, E. Ultra-high-reflectivity photonic-bandgap mirrors in a ridge SOI waveguide. *New J. Phys.* **2006**, *8*, 204. [CrossRef]
46. Velha, P.; Picard, E.; Charvolin, T.; Hadji, E.; Rodier, J.-C.; Lalanne, P.; Peyrade, D. Ultra-high Q/V Fabry-Perot microcavity on SOI substrate. *Opt. Express* **2007**, *15*, 16090–16096. [CrossRef] [PubMed]
47. McCutcheon, M.W.; Loncar, M. Design of a silicon nitride photonic crystal nanocavity with a Quality factor of one million for coupling to a diamond nanocrystal. *Opt. Express* **2008**, *16*, 19136–19145. [CrossRef] [PubMed]
48. Notomi, M.; Kuramochi, E.; Taniyama, H. Ultrahigh-Q nanocavity with 1D photonic gap. *Opt. Express* **2008**, *16*, 11095–11102. [CrossRef] [PubMed]
49. Zain, A.R.; Johnson, N.P.; Sorel, M.; De La Rue, R.M. Ultra high quality factor one dimensional photonic crystal/photonic wire micro-cavities in silicon-on-insulator (SOI). *Opt. Express* **2008**, *16*, 12084–12089. [CrossRef] [PubMed]
50. Deotare, P.B.; McCutcheon, M.W.; Frank, I.W.; Khan, M.; Lončar, M. High quality factor photonic crystal nanobeam cavities. *Appl. Phys. Lett.* **2009**, *94*, 121106. [CrossRef]
51. Kuramochi, E.; Taniyama, H.; Tanabe, T.; Kawasaki, K.; Roh, Y.G.; Notomi, M. Ultrahigh-Q one-dimensional photonic crystal nanocavities with modulated mode-gap barriers on SiO$_2$ claddings and on air claddings. *Opt. Express* **2010**, *18*, 15859–15869. [CrossRef] [PubMed]
52. Quan, Q.; Deotare, P.B.; Loncar, M. Photonic crystal nanobeam cavity strongly coupled to the feeding waveguide. *Appl. Phys. Lett.* **2010**, *96*, 203102. [CrossRef]
53. Quan, Q.; Loncar, M. Deterministic design of wavelength scale, ultra-high Q photonic crystal nanobeam cavities. *Opt. Express* **2011**, *19*, 18529–18542. [CrossRef] [PubMed]
54. Wei, J.; Sun, F.; Dong, B.; Ma, Y.; Chang, Y.; Tian, H.; Lee, C. Deterministic aperiodic photonic crystal nanobeam supporting adjustable multiple mode-matched resonances. *Opt. Lett.* **2018**, in press.

55. Matsko, A.B.; Ilchenko, V.S. Optical resonators with whispering gallery modes I: Basics. *IEEE J. Sel. Top. Quantum Electron.* **2006**, *12*, 3–14. [CrossRef]
56. Arnold, S.; Khoshsima, M.; Teraoka, I.; Holler, S.; Vollmer, F. Shift of whispering-gallery modes in microspheres by protein adsorption. *Opt. Lett.* **2003**, *28*, 272–274. [CrossRef] [PubMed]
57. Hanumegowda, N.M.; Stica, C.J.; Patel, B.C.; White, I.; Fan, X. Refractometric sensors based on microsphere resonators. *Appl. Phys. Lett.* **2005**, *87*, 201107. [CrossRef]
58. Mortensen, N.A.; Xiao, S.; Pedersen, J. Liquid-infiltrated photonic crystals: Enhanced light-matter interactions for lab-on-a-chip applications. *Microfluid. Nanofluid.* **2008**, *4*, 117–127. [CrossRef]
59. Rindorf, L.; Jensen, J.B.; Dufva, M.; Pedersen, L.H.; Høiby, P.E.; Bang, O. Photonic crystal fiber long-period gratings for biochemical sensing. *Opt. Express* **2006**, *14*, 8224–8231. [CrossRef] [PubMed]
60. Abdulhalim, I.; Zourob, M.; Lakhtakia, A. Surface plasmon resonance for biosensing: A mini-review. *Electromagnetics* **2008**, *28*, 214–242. [CrossRef]
61. White, I.M.; Fan, X. On the performance quantification of resonant refractive index sensors. *Opt. Express* **2008**, *16*, 1020–1028. [CrossRef] [PubMed]
62. Chow, E.; Grot, A.; Mirkarimi, L.; Sigalas, M.; Girolami, G. Ultracompact biochemical sensor built with two-dimensional photonic crystal microcavity. *Opt. Lett.* **2004**, *29*, 1093–1095. [CrossRef] [PubMed]
63. Lee, M.; Fauchet, P.M. Two-dimensional silicon photonic crystal based biosensing platform for protein detection. *Opt. Express* **2007**, *15*, 4530–4535. [CrossRef] [PubMed]
64. Johnson, S.G.; Joannopoulos, J.D. Block-iterative frequency-domain methods for Maxwell's equations in a planewave basis. *Opt. Express* **2001**, *8*, 173–190. [CrossRef] [PubMed]
65. Oskooi, A.F.; Roundy, D.; Ibanescu, M.; Bermel, P.; Joannopoulos, J.D.; Johnson, S.G. MEEP: A flexible free-software package for electromagnetic simulations by the FDTD method. *Comput. Phys. Commun.* **2010**, *181*, 687–702. [CrossRef]
66. Chen, X.; Uttamchandani, D.; Trager-Cowan, C.; O'donnell, K. Luminescence from porous silicon. *Semicond. Sci. Technol.* **1993**, *8*, 92. [CrossRef]
67. Quan, Q.; Vollmer, F.; Burgess, I.B.; Deotare, P.B.; Frank, I.; Tang, S.; Illic, R.; Loncar, M. Ultrasensitive on-chip photonic crystal nanobeam sensor using optical bistability. In Proceedings of the Quantum Electronics and Laser Science Conference, Baltimore, MD, USA, 22–27 May 2011.
68. Xu, T.; Zhu, N.; Xu, M.Y.-C.; Wosinski, L.; Aitchison, J.S.; Ruda, H. Pillar-array based optical sensor. *Opt. Express* **2010**, *18*, 5420–5425. [CrossRef] [PubMed]
69. Almeida, V.R.; Xu, Q.; Barrios, C.A.; Lipson, M. Guiding and confining light in void nanostructure. *Opt. Lett.* **2004**, *29*, 1209–1211. [CrossRef] [PubMed]
70. Xu, P.; Yao, K.; Zheng, J.; Guan, X.; Shi, Y. Slotted photonic crystal nanobeam cavity with parabolic modulated width stack for refractive index sensing. *Opt. Express* **2013**, *21*, 26908–26913. [CrossRef] [PubMed]
71. Yang, D.; Zhang, P.; Tian, H.; Ji, Y.; Quan, Q. Ultrahigh- Q and Low-Mode-Volume Parabolic Radius-Modulated Single Photonic Crystal Slot Nanobeam Cavity for High-Sensitivity Refractive Index Sensing. *IEEE Photonics J.* **2015**, *7*, 1–8.
72. Sherry, L.J.; Chang, S.-H.; Schatz, G.C.; Van Duyne, R.P.; Wiley, B.J.; Xia, Y. Localized surface plasmon resonance spectroscopy of single silver nanocubes. *Nano Lett.* **2005**, *5*, 2034–2038. [CrossRef] [PubMed]
73. Rahman, M.G.A.; Velha, P.; Richard, M.; Johnson, N.P. Silicon-on-insulator (soi) nanobeam optical cavities for refractive index based sensing. In Proceedings of the Optical Sensing and Detection II, Brussels, Belgium, 16–19 April 2012.
74. Yang, D.; Kita, S.; Liang, F.; Wang, C.; Tian, H.; Ji, Y.; Lončar, M.; Quan, Q. High sensitivity and high Q-factor nanoslotted parallel quadrabeam photonic crystal cavity for real-time and label-free sensing. *Appl. Phys. Lett.* **2014**, *105*, 063118. [CrossRef]
75. Huang, L.; Zhou, J.; Sun, F.; Fu, Z.; Tian, H. Optimization of One Dimensional Photonic Crystal Elliptical-Hole Low-Index Mode Nanobeam Cavities for On-Chip Sensing. *J. Light. Technol.* **2016**, *34*, 3496–3502. [CrossRef]
76. Bartlett, J.M.; Stirling, D. A short history of the polymerase chain reaction. In *PCR Protocols*; Springer: Berlin, Germany, 2003; pp. 3–6.
77. Yaseen, M.T.; Yang, Y.C.; Shih, M.H.; Chang, Y.C. Optimization of high-Q coupled nanobeam cavity for label-free sensing. *Sensors* **2015**, *15*, 25868–25881. [CrossRef] [PubMed]
78. Zhou, J.; Tian, H.; Huang, L.; Fu, Z.; Sun, F.; Ji, Y. Parabolic tapered coupled two photonic crystal nanobeam slot cavities for high-FOM biosensing. *IEEE Photonics Technol. Lett.* **2017**, *29*, 1281–1284. [CrossRef]

79. Meng, Z.-M.; Li, Z.-Y. Control of Fano resonances in photonic crystal nanobeams side-coupled with nanobeam cavities and their applications to refractive index sensing. *J. Phys. D Appl. Phys.* **2018**, *51*, 095106. [CrossRef]

80. Peng, F.; Wang, Z.; Guan, L.; Yuan, G.; Peng, Z. High-Sensitivity Refractive Index Sensing Based on Fano resonances in a Photonic Crystal Cavity-Coupled Microring Resonator. *IEEE Photonics J.* **2018**, 1. [CrossRef]

81. Lin, T.; Zhang, X.; Zhou, G.; Siong, C.F.; Deng, J. Design of an ultra-compact slotted photonic crystal nanobeam cavity for biosensing. *J. Opt. Soc. Am. B* **2015**, *32*, 1788–1791. [CrossRef]

82. Sun, F.; Zhou, J.; Huang, L.; Fu, Z.; Tian, H. High quality factor and high sensitivity photonic crystal rectangular holes slot nanobeam cavity with parabolic modulated lattice constant for refractive index sensing. *Opt. Commun.* **2017**, *399*, 56–61. [CrossRef]

83. Saha, P.; Sen, M. A slotted photonic crystal nanobeam cavity for simultaneous attainment of ultra-high Q-factor and sensitivity. *IEEE Sens. J.* **2018**, *18*, 3602–3609. [CrossRef]

84. Yu, P.; Qiu, H.; Yu, H.; Wu, F.; Wang, Z.; Jiang, X.; Yang, J. High-Q and high-order side-coupled air-mode nanobeam photonic crystal cavities in silicon. *IEEE Photonics Technol. Lett.* **2016**, *28*, 2121–2124. [CrossRef]

85. Liu, W.; Yan, J.; Shi, Y. High sensitivity visible light refractive index sensor based on high order mode Si$_3$N$_4$ photonic crystal nanobeam cavity. *Opt. Express* **2017**, *25*, 31739–31745. [CrossRef] [PubMed]

86. Xu, P.; Yu, Z.; Shen, X.; Dai, S. High quality factor and high sensitivity chalcogenide 1D photonic crystal microbridge cavity for mid-infrared sensing. *Opt. Commun.* **2017**, *382*, 361–365. [CrossRef]

87. Deng, C.-S.; Li, M.-J.; Peng, J.; Liu, W.-L.; Zhong, J.-X. Simultaneously high-Q and high-sensitivity slotted photonic crystal nanofiber cavity for complex refractive index sensing. *J. Opt. Soc. Am. B* **2017**, *34*, 1624–1631. [CrossRef]

88. Velha, P.; Rodier, J.-C.; Lalanne, P.; Hugonin, J.-P.; Peyrade, D.; Picard, E.; Charvolin, T.; Hadji, E. Ultracompact silicon-on-insulator ridge-waveguide mirrors with high reflectance. *Appl. Phys. Lett.* **2006**, *89*, 171121. [CrossRef]

89. Yang, D.; Wang, B.; Chen, X.; Wang, C.; Ji, Y. Ultracompact on-chip multiplexed sensor array based on dense integration of flexible 1-D photonic crystal nanobeam cavity with large free spectral range and high Q-factor. *IEEE Photonics J.* **2017**, *9*, 1–12. [CrossRef]

90. Wang, J.; Fu, Z.; Sun, F.; Wang, Z.; Wang, C.; Tian, H. Multiplexing dual-parameter sensor using photonic crystal multimode nanobeam cavities. *Opt. Commun.* **2018**, *427*, 382–389. [CrossRef]

91. Yang, D.; Wang, C.; Ji, Y. Silicon on-chip 1D photonic crystal nanobeam bandgap filter integrated with nanobeam cavity for accurate refractive index sensing. *IEEE Photonics J.* **2016**, *8*, 1–8. [CrossRef]

92. Yang, Y.; Yang, D.; Ji, Y. Ultra-compact photonic crystal integrated sensor formed by series-connected nanobeam bandstop filter and nanobeam cavity. In Proceedings of the High Power Lasers, High Energy Lasers, and Silicon-based Photonic Integration, Beijing, China, 19 October 2016.

93. Sun, F.; Fu, Z.; Wang, C.; Ding, Z.; Wang, C.; Tian, H. Ultra-compact air-mode photonic crystal nanobeam cavity integrated with bandstop filter for refractive index sensing. *Appl. Opt.* **2017**, *56*, 4363–4368. [CrossRef] [PubMed]

94. Tian, F.; Zhou, G.; Du, Y.; Chau, F.S.; Deng, J.; Tang, X.; Akkipeddi, R. Energy-efficient utilization of bipolar optical forces in nano-optomechanical cavities. *Opt. Express* **2013**, *21*, 18398–18407. [CrossRef] [PubMed]

95. Vollmer, F.; Yang, L. Review Label-free detection with high-Q microcavities: A review of biosensing mechanisms for integrated devices. *Nanophotonics* **2012**, *1*, 267–291. [CrossRef] [PubMed]

96. Armani, A.M.; Kulkarni, R.P.; Fraser, S.E.; Flagan, R.C.; Vahala, K.J. Label-free, single-molecule detection with optical microcavities. *Science* **2007**, *317*, 783–787. [CrossRef] [PubMed]

97. Vollmer, F.; Arnold, S.; Keng, D. Single virus detection from the reactive shift of a whispering-gallery mode. *Proc. Natl. Acad. Sci. USA* **2008**, *105*, 20701–20704. [CrossRef] [PubMed]

98. Wark, A.W.; Lee, H.J.; Corn, R.M. Long-range surface plasmon resonance imaging for bioaffinity sensors. *Anal. Chem.* **2005**, *77*, 3904–3907. [CrossRef] [PubMed]

99. Homola, J. Surface plasmon resonance sensors for detection of chemical and biological species. *Chem. Rev.* **2008**, *108*, 462–493. [CrossRef] [PubMed]

100. Ymeti, A.; Greve, J.; Lambeck, P.V.; Wink, T.; van Hövell, S.W.; Beumer, T.A.; Wijn, R.R.; Heideman, R.G.; Subramaniam, V.; Kanger, J.S. Fast, ultrasensitive virus detection using a young interferometer sensor. *Nano Lett.* **2007**, *7*, 394–397. [CrossRef] [PubMed]

101. Brandenburg, A. Differential refractometry by an integrated-optical Young interferometer. *Sens. Actuators B Chem.* **1997**, *39*, 266–271. [CrossRef]

102. Scullion, M.; Di Falco, A.; Krauss, T. Slotted photonic crystal cavities with integrated microfluidics for biosensing applications. *Biosens. Bioelectron.* **2011**, *27*, 101–105. [CrossRef] [PubMed]

103. Zhu, J.; Ozdemir, S.K.; Xiao, Y.-F.; Li, L.; He, L.; Chen, D.-R.; Yang, L. On-chip single nanoparticle detection and sizing by mode splitting in an ultrahigh-Q microresonator. *Nat. Photonics* **2010**, *4*, 46–49. [CrossRef]

104. Shao, L.; Jiang, X.F.; Yu, X.C.; Li, B.B.; Clements, W.R.; Vollmer, F.; Wang, W.; Xiao, Y.F.; Gong, Q. Detection of single nanoparticles and lentiviruses using microcavity resonance broadening. *Adv. Mater.* **2013**, *25*, 5616–5620. [CrossRef] [PubMed]

105. Shen, B.-Q.; Yu, X.-C.; Zhi, Y.; Wang, L.; Kim, D.; Gong, Q.; Xiao, Y.-F. Detection of single nanoparticles using the dissipative interaction in a high-Q microcavity. *Phys. Rev. Appl.* **2016**, *5*, 024011. [CrossRef]

106. Vörös, J. The density and refractive index of adsorbing protein layers. *Biophys. J.* **2004**, *87*, 553–561. [CrossRef] [PubMed]

107. Vollmer, F.; Braun, D.; Libchaber, A.; Khoshsima, M.; Teraoka, I.; Arnold, S. Protein detection by optical shift of a resonant microcavity. *Appl. Phys. Lett.* **2002**, *80*, 4057–4059. [CrossRef]

108. Chen, Y.; Fegadolli, W.S.; Jones, W.M.; Scherer, A.; Li, M. Ultrasensitive gas-phase chemical sensing based on functionalized photonic crystal nanobeam cavities. *ACS Nano* **2014**, *8*, 522–527. [CrossRef] [PubMed]

109. Duhr, S.; Braun, D. Why molecules move along a temperature gradient. *Proc. Natl. Acad. Sci. USA* **2006**, *103*, 19678–19682. [CrossRef] [PubMed]

110. Lin, S.; Crozier, K.B. Trapping-assisted sensing of particles and proteins using on-chip optical microcavities. *ACS Nano* **2013**, *7*, 1725–1730. [CrossRef] [PubMed]

111. Hu, J. Ultra-sensitive chemical vapor detection using micro-cavity photothermal spectroscopy. *Opt. Express* **2010**, *18*, 22174–22186. [CrossRef] [PubMed]

112. Lin, H.; Yi, Z.; Hu, J. Double resonance 1-D photonic crystal cavities for single-molecule mid-infrared photothermal spectroscopy: Theory and design. *Opt. Lett.* **2012**, *37*, 1304–1306. [CrossRef] [PubMed]

113. Ilchenko, V.S.; Paxton, A.H.; Kudryashov, A.V.; Poust, S.; Armani, A.M.; Shen, X.; Kovach, A.; Chen, D. Environmentally stable integrated ultra-high-Q optical cavities. *Proc. SPIE* **2018**. [CrossRef]

114. Grutter, K.E.; Davanço, M.I.; Balram, K.C.; Srinivasan, K. Invited Article: Tuning and stabilization of optomechanical crystal cavities through NEMS integration. *APL Photonics* **2018**, *3*, 100801. [CrossRef]

115. Lu, T.; Lee, H.; Chen, T.; Herchak, S.; Kim, J.-H.; Fraser, S.E.; Flagan, R.C.; Vahala, K. High sensitivity nanoparticle detection using optical microcavities. *Proc. Natl. Acad. Sci. USA* **2011**, *108*, 5976–5979. [CrossRef] [PubMed]

116. Ament, I.; Prasad, J.; Henkel, A.; Schmachtel, S.; Sonnichsen, C. Single unlabeled protein detection on individual plasmonic nanoparticles. *Nano Lett.* **2012**, *12*, 1092–1095. [CrossRef] [PubMed]

117. Aspelmeyer, M.; Kippenberg, T.J.; Marquardt, F. Cavity optomechanics. *Rev. Mod. Phys.* **2014**, *86*, 1391. [CrossRef]

118. Hu, Y.-W.; Xiao, Y.-F.; Liu, Y.-C.; Gong, Q. Optomechanical sensing with on-chip microcavities. *Front. Phys.* **2013**, *8*, 475–490. [CrossRef]

119. Pruessner, M.W.; Stievater, T.H.; Khurgin, J.B.; Rabinovich, W.S. Integrated waveguide-DBR microcavity opto-mechanical system. *Opt. Express* **2011**, *19*, 21904–21918. [CrossRef] [PubMed]

120. Zandi, K.; Wong, B.; Zou, J.; Kruzelecky, R.V.; Jamroz, W.; Peter, Y.-A. In-plane silicon-on-insulator optical MEMS accelerometer using waveguide fabry-perot microcavity with silicon/air bragg mirrors. In Proceedings of the 2010 IEEE 23rd International Conference on Micro Electro Mechanical Systems (MEMS), Hong Kong, China, 24–28 January 2010; pp. 839–842.

121. Jiang, X.; Lin, Q.; Rosenberg, J.; Vahala, K.; Painter, O. High-Q double-disk microcavities for cavity optomechanics. *Opt. Express* **2009**, *17*, 20911–20919. [CrossRef] [PubMed]

122. Anetsberger, G.; Arcizet, O.; Unterreithmeier, Q.P.; Rivière, R.; Schliesser, A.; Weig, E.M.; Kotthaus, J.P.; Kippenberg, T.J. Near-field cavity optomechanics with nanomechanical oscillators. *Nat. Phys.* **2009**, *5*, 909–914. [CrossRef]

123. Gavartin, E.; Verlot, P.; Kippenberg, T.J. A hybrid on-chip optomechanical transducer for ultrasensitive force measurements. *Nat. Nanotechnol.* **2012**, *7*, 509–514. [CrossRef] [PubMed]

124. Gavartin, E.; Braive, R.; Sagnes, I.; Arcizet, O.; Beveratos, A.; Kippenberg, T.J.; Robert-Philip, I. Optomechanical coupling in a two-dimensional photonic crystal defect cavity. *Phys. Rev. Lett.* **2011**, *106*, 203902. [CrossRef] [PubMed]

125. Safavi-Naeini, A.H.; Alegre, T.M.; Chan, J.; Eichenfield, M.; Winger, M.; Lin, Q.; Hill, J.T.; Chang, D.E.; Painter, O. Electromagnetically induced transparency and slow light with optomechanics. *Nature* **2011**, *472*, 69–73. [CrossRef] [PubMed]

126. Eichenfield, M.; Chan, J.; Camacho, R.M.; Vahala, K.J.; Painter, O. Optomechanical crystals. *Nature* **2009**, *462*, 78–82. [CrossRef] [PubMed]

127. Lee, C.; Thillaigovindan, J.; Chen, C.-C.; Chen, X.T.; Chao, Y.-T.; Tao, S.; Xiang, W.; Yu, A.; Feng, H.; Lo, G. Si nanophotonics based cantilever sensor. *Appl. Phys. Lett.* **2008**, *93*, 113113. [CrossRef]

128. Lee, C.; Radhakrishnan, R.; Chen, C.-C.; Li, J.; Thillaigovindan, J.; Balasubramanian, N. Design and modeling of a nanomechanical sensor using silicon photonic crystals. *J. Light. Technol.* **2008**, *26*, 839–846. [CrossRef]

129. Thompson, J.; Zwickl, B.; Jayich, A.; Marquardt, F.; Girvin, S.; Harris, J. Strong dispersive coupling of a high-finesse cavity to a micromechanical membrane. *Nature* **2008**, *452*, 72–75. [CrossRef] [PubMed]

130. Sankey, J.C.; Yang, C.; Zwickl, B.M.; Jayich, A.M.; Harris, J.G. Strong and tunable nonlinear optomechanical coupling in a low-loss system. *Nat. Phys.* **2010**, *6*, 707–712. [CrossRef]

131. Srinivasan, K.; Miao, H.; Rakher, M.T.; Davanco, M.; Aksyuk, V. Optomechanical transduction of an integrated silicon cantilever probe using a microdisk resonator. *Nano Lett.* **2011**, *11*, 791–797. [CrossRef] [PubMed]

132. Hryciw, A.C.; Barclay, P.E. Optical design of split-beam photonic crystal nanocavities. *Opt. Lett.* **2013**, *38*, 1612–1614. [CrossRef] [PubMed]

133. Lipfert, J.; Kerssemakers, J.W.; Jager, T.; Dekker, N.H. Magnetic torque tweezers: Measuring torsional stiffness in DNA and RecA-DNA filaments. *Nat. Methods* **2010**, *7*, 977–980. [CrossRef] [PubMed]

134. Weiss, T.; Nunnenkamp, A. Quantum limit of laser cooling in dispersively and dissipatively coupled optomechanical systems. *Phys. Rev. A* **2013**, *88*, 023850. [CrossRef]

135. Anetsberger, G.; Gavartin, E.; Arcizet, O.; Unterreithmeier, Q.P.; Weig, E.M.; Gorodetsky, M.L.; Kotthaus, J.P.; Kippenberg, T.J. Measuring nanomechanical motion with an imprecision below the standard quantum limit. *Phys. Rev. A* **2010**, *82*, 061804. [CrossRef]

136. Chan, J.; Eichenfield, M.; Camacho, R.; Painter, O. Optical and mechanical design of a "zipper" photonic crystal optomechanical cavity. *Opt. Express* **2009**, *17*, 3802–3817. [CrossRef] [PubMed]

137. Kippenberg, T.J.; Vahala, K.J. Cavity opto-mechanics. *Opt. Express* **2007**, *15*, 17172–17205. [CrossRef] [PubMed]

138. Fleming, W.J. Overview of automotive sensors. *IEEE Sens. J.* **2001**, *1*, 296–308. [CrossRef]

139. Li, H.-N.; Li, D.-S.; Song, G.-B. Recent applications of fiber optic sensors to health monitoring in civil engineering. *Eng. Struct.* **2004**, *26*, 1647–1657. [CrossRef]

140. Jolesz, F.A. MRI-guided focused ultrasound surgery. *Annu. Rev. Med.* **2009**, *60*, 417–430. [CrossRef] [PubMed]

141. Woo, X.Y.; Nagy, Z.K.; Tan, R.B.; Braatz, R.D. Adaptive concentration control of cooling and antisolvent crystallization with laser backscattering measurement. *Cryst. Growth Des.* **2008**, *9*, 182–191. [CrossRef]

142. Strouse, G. Standard platinum resistance thermometer calibrations from the Ar TP to the Ag FP. *NIST Spec. Publ.* **2008**, *250*, 81. [CrossRef]

143. Kim, G.-D.; Lee, H.-S.; Park, C.-H.; Lee, S.-S.; Lim, B.T.; Bae, H.K.; Lee, W.-G. Silicon photonic temperature sensor employing a ring resonator manufactured using a standard CMOS process. *Opt. Express* **2010**, *18*, 22215–22221. [CrossRef] [PubMed]

144. Rabiei, P. *Electro-Optic and Thermo-Optic Polymer Micro-Ring Resonators and Their Applications*; University of Southern California: Los Angeles, CA, USA, 2002.

145. Klimov, N.N.; Mittal, S.; Berger, M.; Ahmed, Z. On-chip silicon waveguide Bragg grating photonic temperature sensor. *Opt. Lett.* **2015**, *40*, 3934–3936. [CrossRef] [PubMed]

146. Kwon, M.-S.; Steier, W.H. Microring-resonator-based sensor measuring both the concentration and temperature of a solution. *Opt. Express* **2008**, *16*, 9372–9377. [CrossRef] [PubMed]

147. Kim, H.-T.; Yu, M. Cascaded ring resonator-based temperature sensor with simultaneously enhanced sensitivity and range. *Opt. Express* **2016**, *24*, 9501–9510. [CrossRef] [PubMed]

148. Klimov, N.N.; Purdy, T.; Ahmed, Z. Fabrication and characterization of on-chip integrated silicon photonic Bragg grating and photonic crystal cavity thermometers. In Proceedings of the 2015 TechConnect World Innovation Conference, Washington, DC, USA, 15–18 June 2015.

149. Klimov, N.N.; Purdy, T.P.; Ahmed, Z. On-Chip Integrated Silicon Photonic Thermometers. *Sens. Transducers J.* **2015**, *191*, 67–71.
150. Teng, J.; Dumon, P.; Bogaerts, W.; Zhang, H.; Jian, X.; Han, X.; Zhao, M.; Morthier, G.; Baets, R. Athermal Silicon-on-insulator ring resonators by overlaying a polymer cladding on narrowed waveguides. *Opt. Express* **2009**, *17*, 14627–14633. [CrossRef] [PubMed]
151. Desiatov, B.; Goykhman, I.; Levy, U. Parabolic tapered photonic crystal cavity in silicon. *Appl. Phys. Lett.* **2012**, *100*, 041112. [CrossRef]
152. Zhang, Y.; Han, S.; Zhang, S.; Liu, P.; Shi, Y. High-Q and High-Sensitivity Photonic Crystal Cavity Sensor. *IEEE Photonics J.* **2015**, *7*, 1–6. [CrossRef]
153. Ahn, B.-H.; Kang, J.-H.; Kim, M.-K.; Song, J.-H.; Min, B.; Kim, K.-S.; Lee, Y.-H. One-dimensional parabolic-beam photonic crystal laser. *Opt. Express* **2010**, *18*, 5654–5660. [CrossRef] [PubMed]
154. Karnutsch, C.; Smith, C.L.; Graham, A.; Tomljenovic-Hanic, S.; McPhedran, R.; Eggleton, B.J.; O'Faolain, L.; Krauss, T.F.; Xiao, S.; Mortensen, N.A. Temperature stabilization of optofluidic photonic crystal cavities. *Appl. Phys. Lett.* **2009**, *94*, 231114. [CrossRef]
155. Hagino, H.; Takahashi, Y.; Tanaka, Y.; Asano, T.; Noda, S. Effects of fluctuation in air hole radii and positions on optical characteristics in photonic crystal heterostructure nanocavities. *Phys. Rev. B* **2009**, *79*, 085112. [CrossRef]
156. Lim, A.E.-J.; Liow, T.-Y.; Song, J.-F.; Yu, M.-B.; Li, C.; Tu, X.-G.; Chen, K.-K.; Tern, R.P.-C.; Huang, Y.; Luo, X.-S. Path to Silicon Photonics Commercialization: The Foundry Model Discussion. In *Silicon Photonics III*; Springer: Berlin, Germany, 2016; pp. 191–215.
157. O'Brien, P.; Carrol, L.; Eason, C.; Lee, J.S. Packaging of silicon photonic devices. In *Silicon Photonics III*; Springer: Berlin, Germany, 2016; pp. 217–236.

micromachines

MDPI

Article

Design of a Micromachined Z-axis Tunneling Magnetoresistive Accelerometer with Electrostatic Force Feedback

Bo Yang [1,2,*], Binlong Wang [1,2], Hongyu Yan [1,2] and Xiaoyong Gao [1,2]

[1] School of Instrument Science and Engineering, Southeast University, Nanjing 210096, China; 220162760@seu.edu.cn (B.W.); 230189689@seu.edu.cn (H.Y.); 220173252@scu.edu.cn (X.G.)
[2] Key Laboratory of Micro-Inertial Instrument and Advanced Navigation Technology, Ministry of Education, Nanjing 210096, China
* Correspondence: 101011019@seu.edu.cn; Tel.: +86-25-8379-3559

Received: 22 January 2019; Accepted: 21 February 2019; Published: 25 February 2019

check for updates

Abstract: This paper presents the design, simulation, fabrication and experiments of a micromachined z-axis tunneling magnetoresistive accelerometer with electrostatic force feedback. The tunneling magnetoresistive accelerometer consists of two upper differential tunneling magnetoresistive sensors, a middle plane main structure with permanent magnetic films and lower electrostatic feedback electrodes. A pair of lever-driven differential proof masses in the middle plane main structure is used for sensitiveness to acceleration and closed-loop feedback control. The tunneling magnetoresistive effect with high sensitivity is adopted to measure magnetic field variation caused by input acceleration. The structural mode and mass ratio between inner and outer proof masses are optimized by the Ansys simulation. Simultaneously, the magnetic field characteristic simulation is implemented to analyze the effect of the location of tunneling magnetoresistive sensors, magnetic field intensity, and the dimension of permanent magnetic film on magnetic field sensitivity, which is beneficial for the achievement of maximum sensitivity. The micromachined z-axis tunneling magnetoresistive accelerometer fabricated by the standard deep dry silicon on glass (DDSOG) process has a device dimension of 6400 μm (length) × 6400 μm (width) × 120 μm (height). The experimental results demonstrate the prototype has a maximal sensitivity of 8.85 mV/g along the z-axis sensitive direction under the gap of 1 mm. Simultaneously, Allan variance analysis illustrate that a noise floor of 86.2 μg/Hz$^{0.5}$ is implemented in the z-axis tunneling magnetoresistive accelerometer.

Keywords: accelerometer; tunnel magnetoresistive effect; electrostatic force feedback

1. Introduction

Numerous applications require high-performance accelerometers including individual navigation and positioning, earthquake early warning systems, the attitude adjustment of micro/nano satellites, oil and gas exploration, and the motion control of micro-autonomous systems [1–3]. Various techniques, such as capacitive technology [1], optical principle [4,5], resonance [6–8], tunneling [9], piezoelectric [10] and piezoresistive [11] effect, are studied to detect acceleration signals. Among them, the tunneling effect is a potential high precision acceleration measurement method. Traditional tunnel current micro electro mechanical systems (MEMS) accelerometers perform high-resolution detection of displacement based on quantum tunneling effects, which enables high resolution acceleration detection. Compared with other accelerometers, tunneling current accelerometers have the advantages of high sensitivity and wide bandwidth. Many research institutes have implemented extensive research on tunneling tip processing technology, thin film deposition technology, feedback control technology and tunneling current testing technology [12]. The traditional tunneling MEMS accelerometer can obtain extremely high resolution, however, its dynamic range is extremely narrow. The main reason is that the nm gap between the tunneling tip and the substrate needs to be controlled in order to achieve the tunneling effect, which results in limited dynamic range. At the same time, the current fabrication tolerance of micromachining process makes the nm gap hard to process, which makes the traditional tunneling current accelerometer difficult to implement.

Inspired by the successful application in other commercial products, a new tunneling magnetoresistance technology with ultra-high sensitivity, which surpasses the shortcomings of traditional tunneling current effects, has rapidly penetrated the field of micro-inertial devices in recent years [13]. Phan et al. at the Eindhoven University of Technology present a polydimethylsiloxane (PDMS) based biaxial accelerometer by using the magnetoresistive (MR) detection method [14]. The proof mass of the sensor is a mushroom-shaped polymer magnet, whose minute movement under lateral acceleration is precisely sensed by a set of MR sensors. A sensitivity of 0.32 mV/(V.g) and a noise density of 35 µg/Hz$^{0.5}$ at 5 Hz was obtained in the device. Literature from Pacesetter Inc proposed a magnetoresistive-based position sensor for use in an implantable electrical device [15]. The sensor includes a magnetoresistive sensor made from giant magnetoresistive (GMR) materials and a magnet positioned on a flexible cantilevered beam. The relative movement of the magnet, which originates from the movement of the patient, is detected by the magnetoresistive sensor. The realization of the device is not given in the literature. Ali Alaoui at Crocus Technology SA proposed a MLU (Magnetic Logic Unit) based biaxial accelerometer using a magnetic tunneling junction [16]. An ultra-sensitive displacement sensing device is presented by Olivas et al. of the National Aeronautics and Space Administration (NASA) for use in accelerometers, pressure gauges, temperature transducers [17]. The device comprises a sputter deposited, multilayer, magnetoresistive field sensor with a variable electrical resistance based on an imposed magnetic field. The scheme implements a magnetoresistive structure and gap control using micromachining technology, which already has the basic elements of a miniature magnetoresistive accelerometer. Similarly, a cantilever-based precision force detection MEMS device with magnetoresistance technology was used to measure acceleration information at the Sony Precision Engineering Center in Singapore [18]. In addition, David P. Fries et al. in the University of South Florida has presented a gyroscope based on giant magnetoresistance effect [19]. Currently, most of the above accelerometers are based on the traditional magnetoresistance or giant magnetoresistance (GMR) effects, which limit the further improvement of accelerometer accuracy due to the restriction of displacement sensitivity. The literature [20] has proposed a small tunnel magnetoresistive accelerometer based on 3D printing, however the accelerometer with a large volume and weight is fabricated by traditional macro processing technology. Therefore, the structure is not a true micromechanical accelerometer, and the support cantilever beam of proof mass with three-dimensional structure is also difficult to achieve through micromachining processing technology. Simultaneously, the accelerometer can only measure the input acceleration through open loop detection due to the lack of feedback torque mechanism.

This paper presents a micromachined z-axis tunneling magnetoresistive accelerometer with electrostatic force feedback. The tunneling magnetoresistive accelerometer consists of two upper differential tunneling magnetoresistive sensors, a middle plane main structure with permanent magnetic films and lower electrostatic feedback electrodes. The proof mass with permanent magnet film on the plane main structure is driven to move by the input acceleration, then the caused magnetic field change is measured by tunneling magnetoresistive sensors. Finally, the electrostatic feedback electrodes are used to adjust the proof mass to the equilibrium position, thereby closed loop detection is implemented. In Section 2, the structure principle of the micromachined z-axis tunneling magnetoresistive accelerometer is briefly presented. Then the simulation analysis as well as the measurement and control circuit are given in Sections 3 and 4. We demonstration the experimental results in Section 5. Concluding remarks are given in the last section.

2. Structure Principle

The cantilever beam structure with proof mass is one of the core components of the micromechanical accelerometer and can directly convert input acceleration into stress or displacement. It has the advantages of a simple structure, high sensitivity and high conversion efficiency. The cantilever beam structure with proof mass has to be innovatively designed in order to be compatible with tunneling magnetoresistive effects for acceleration measure. A structural schematic of the micromachined z-axis tunneling magnetoresistive accelerometer, which consists of upper, middle and lower layers, is shown in Figure 1.

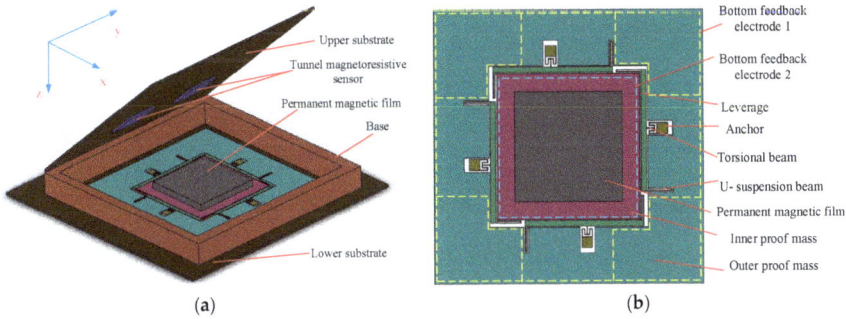

(a)　　　　　　　　　　　　　　　(b)

Figure 1. Structural schematic of the micromachined z-axis tunnel magnetoresistive accelerometer: (**a**) The structure layout of tunnel magnetoresistive accelerometer; (**b**) The plane main structure of tunnel magnetoresistive accelerometer.

Two tunnel magnetoresistive sensors with opposite sensitive direction (y-axis in the Figure 1a) in the upper substrate which are symmetrically arranged on the sides of the centerline are used to measure the variations of the surrounding magnetic field caused by the input acceleration. The middle layer is the plane main structure of tunnel magnetoresistive accelerometer shown in Figure 1b. The plane main structure is composed of four lever mechanisms, the permanent magnetic film, outer and inner proof mass. The permanent magnetic film is arranged on the inner proof mass. The outer proof mass is connected to the inner proof mass by four leverages. When the acceleration is input along the z-axis, the inertia forces at both ends of the leverage will be unbalanced due to the difference in mass between the outer proof mass and inner proof mass with permanent magnetic film, which cause the inner and outer proof masses to move in opposite directions along the z-axis. Since the gap between the permanent magnetic film and the tunnel magnetoresistive sensor varies due to the input acceleration, there will be a redistribution of the surrounding magnetic field in the tunneling magnetoresistive sensor. Therefore, the input acceleration can be measured indirectly by the output of the tunneling magnetoresistive sensor. The tunneling magnetoresistance effect with high sensitivity is directly generated by the multilayer nano-film layer in the tunneling magnetoresistive sensors, which can be easily achieved by a multi-thin-film deposition process. Therefore, the micromachining requirement for the nano-gap between the tunnel tip and the plane in the traditional tunnel effect is avoided. Simultaneously, the gap between the permanent magnetic film and the tunnel magnetoresistive sensor can be further amplified, which will facilitate the realization of a large dynamic range. Furthermore, two feedback electrodes on the lower substrate, which are arranged under the inner and outer proof masses, are used to achieve an electrostatic force feedback and closed loop control in subsequent operations.

The entire structural model is simplified, as shown in Figure 2. The leverage equations are

$$k_t\delta = (m_1 + m_2)a \tag{1}$$

$$m_2 a l_2 = m_1 a l_1 + k_{t\theta} \theta \tag{2}$$

$$z_1 = l_1 \theta - \delta - \frac{m_1 a}{k_o} \tag{3}$$

$$z_2 = l_2 \theta + \delta + \frac{m_2 a}{k_i} \tag{4}$$

where k_t and $k_{t\theta}$ are the stiffness along the z-axis and torsional stiffness of torsional beam respectively, k_o and k_i is stiffness of U-suspension beam, l_1 and l_2 are the arm length of leverage, z_1 and z_2 are the displacement of outer and inner proof masses, respectively.

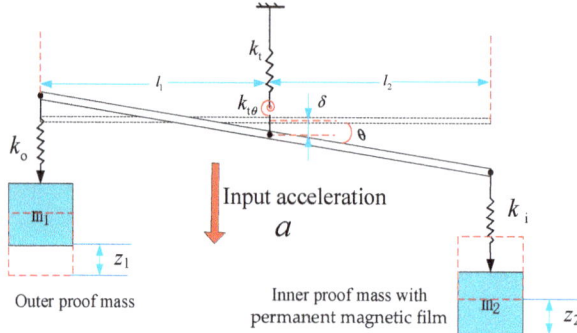

Figure 2. Simplified structural model.

The weighted displacement z is

$$z = \frac{m_1 z_1 + m_2 z_2}{m_1 + m_2} = \left(\frac{m_2^2 l_2^2 - m_1^2 l_1^2}{k_{t\theta}(m_1 + m_2)} + \frac{m_2 - m_1}{k_t} + \frac{m_2^2}{k_i(m_1 + m_2)} - \frac{m_1^2}{k_o(m_1 + m_2)} \right) a \tag{5}$$

Suppose $k_i = k_o$ and $l_1 = l_2 = l$, then

$$z = \frac{G_m a}{\omega_n^2} \tag{6}$$

where G_m is mass ratio, ω_n is modal natural frequency of accelerometer along the z-axis, k_e is the equivalent stiffness and

$$G_m = \frac{m_2 - m_1}{m_1 + m_2}, \quad \omega_n^2 = \frac{k_e}{m_1 + m_2}, \quad k_e = \frac{1}{\frac{l^2}{k_{t\theta}} + \frac{1}{k_t} + \frac{1}{k_i}} \tag{7}$$

The magnetic field distribution along the y-axis due to a rectangular permanent magnetic film can be expressed approximately as $B_y (x, y, z)$ [21]. Only the magnetic field distribution along the y-axis is given because the sensitive direction of two tunneling magnetoresistive sensors is the y-axis.

$$B_y(x, y, z) = \frac{\mu_0 M}{4\pi} \ln \frac{F_2(-y, x, -z) F_2(y, x, z)}{F_2(y, x, -z) F_2(-y, x, z)} \tag{8}$$

where

$$F_2(x, y, z) = \frac{\sqrt{(x + a)^2 + (y - b)^2 + (z + c)^2} + b - y}{\sqrt{(x + a)^2 + (y + b)^2 + (z + c)^2} - b - y} \tag{9}$$

M is the moment density. a, b and c are half of length along the x-axis, half of width along the y-axis and half of thickness along the z-axis in the rectangular permanent magnetic film, respectively.

The magnetic field distribution along the y direction caused by the displacement movement in the z-axis is simplified as

$$B_{yz}(x,y,z) = k_{Bz}\Delta z + B_y(x_0, y_0, z_0) \tag{10}$$

where $k_{Bz} = \left. \frac{\partial B_y(x,y,z)}{\partial z} \right|_{(x_0,y_0,z_0)}$.

The outputs of the tunnel magnetoresistive sensor are

$$V_z \approx k_v B_{yz}(x,y,z) \approx \frac{G_m k_v k_{Bz}}{\omega_n^2} a + k_v B_y(x_0, y_0, z_0) \tag{11}$$

where k_v is equivalent transform coefficient of tunneling magnetoresistive sensor from magnetic field to voltage. The structure parameters of the tunneling magnetoresistive accelerometer are shown in Table 1.

Table 1. Structure parameters.

Parameter	Value	Parameter	Value
Inner proof mass (kg)	2.14×10^{-5}	Thickness of main structure (μm)	120
Outer proof mass (kg)	4.68×10^{-6}	U-suspension beam (length × width (μm))	465 × 15
Equivalent stiffness k_e (N/m)	53.75	Torsional beam (length × width (μm))	519 × 15
Mode frequency ω_n (rad/s)	1435.61	Leverage (length × width (μm))	3600 × 150
Length 2a (μm)	3000	Gap d_1 (between proof mass and feedback electrode (μm))	10
Width 2b (μm)	3000	Feedback electrode area (mm²)	15.36
Thickness 2c (μm)	500	Gap d_2 (between tunnel magnetoresistive sensor and proof mass (μm))	1000
Inner proof mass (length × width (μm))	4000 × 4000	Moment density M (mT)	250
Outer proof mass (length × width (mm))	6400 × 6400	Maximum equivalent transform coefficien k_v (mV/mT)	3000

3. Simulation Analysis

The main structure of the tunneling magnetoresistive accelerometer is simulated by the Ansys in order to verify the structural principle. The modal simulation results are shown in Figure 3. The first mode with a frequency of 228.6 Hz is the sense mode of acceleration along the z-axis. The inner proof mass is pushed to move downward along the z-axis by inertial force, and promote simultaneously the outer proof mass to move upward by the leverage, since the mass of the inner proof mass is greater than that of the outer proof mass. Two interference modes shown in Figure 3b,c are off-plane rotation movement of inner proof mass respectively. The interference mode shown in Figure 3d is the in-plane translational movement of inner proof mass in the fourth mode. Other disturbance modes are shown in Table 2. Simultaneously, the simulation results demonstrate that the mechanical structure of the tunnel magnetoresistive accelerometer has a mechanical sensitivity of 3.05 μm/g in z-axis sense direction.

Table 2. The first six mode frequencies of the tunnel magnetoresistive accelerometer.

Modal	1	2	3	4	5	6
Frequency (Hz)	228.6	567.7	568.4	1051.7	1052.5	1516.2

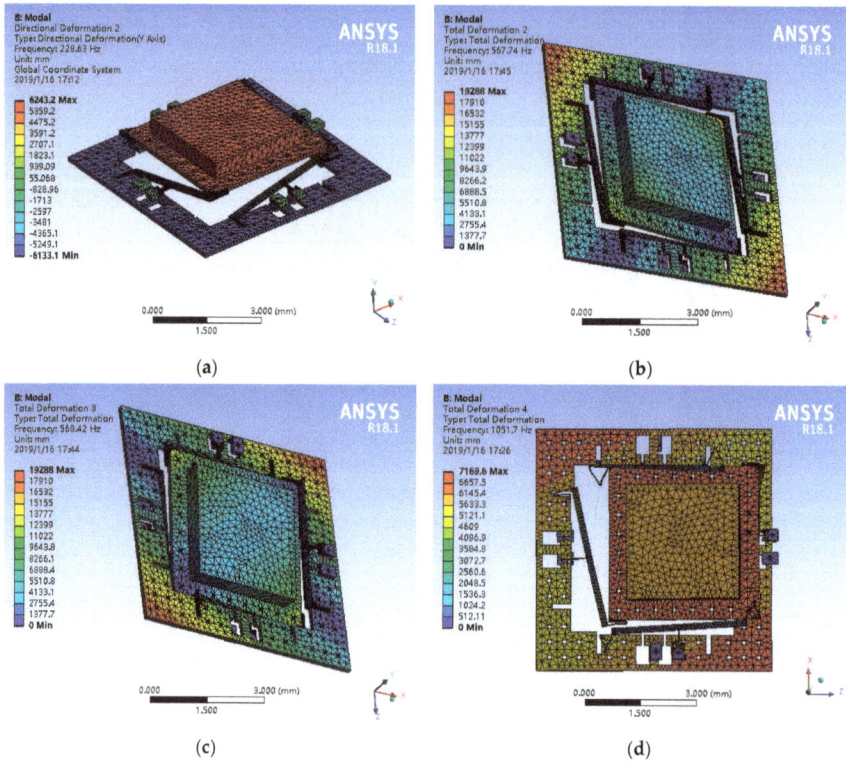

(a)

(b)

(c)

(d)

Figure 3. The selected modes of the micromachined z-axis tunneling magnetoresistive accelerometer. (**a**) Linear movement of proof mass along z direction in first mode; (**b**) The off-plane rotation movement of inner proof mass in second mode; (**c**) The off-plane rotation movement of inner proof mass in the third mode; (**d**) The in-plane translational movement of inner proof mass in the fourth mode.

The effect of mass ratio and first-order modal frequency on mechanical sensitivity is shown in Figure 4. As can be seen from Figure 4a,b, the mechanical sensitivity is clearly proportional to the mass ratio at the same first-order modal frequency and inversely proportional to the first-order modal frequency at the same mass ratio. Simultaneously, the first-order modal frequency is proportional to the mass ratio under the same mechanical sensitivity conditions, shown in Figure 4c. That is, in order to maintain the same mechanical sensitivity, the mass ratio must be increased accordingly when the first-order mode is added. Actually, the increase of the mass ratio is beneficial to enhance the mechanical sensitivity. In the limit case, if $m_1 = 0$, the maximum mass ratio can be obtained, which can implement the maximum mechanical sensitivity. However, a compromise design has been made in order to be compatible with the electrostatic torque devices in the structure. Reverse electrostatic forces for closed loop feedback control are implemented by the electrostatic torque devices which are constituted by a pair of differential masses (m_1 and m_2) with a pair of differential electrodes and lever mechanisms. In summary, the above simulation results are in good agreement with the above theory, which confirms the correctness of theoretical analysis.

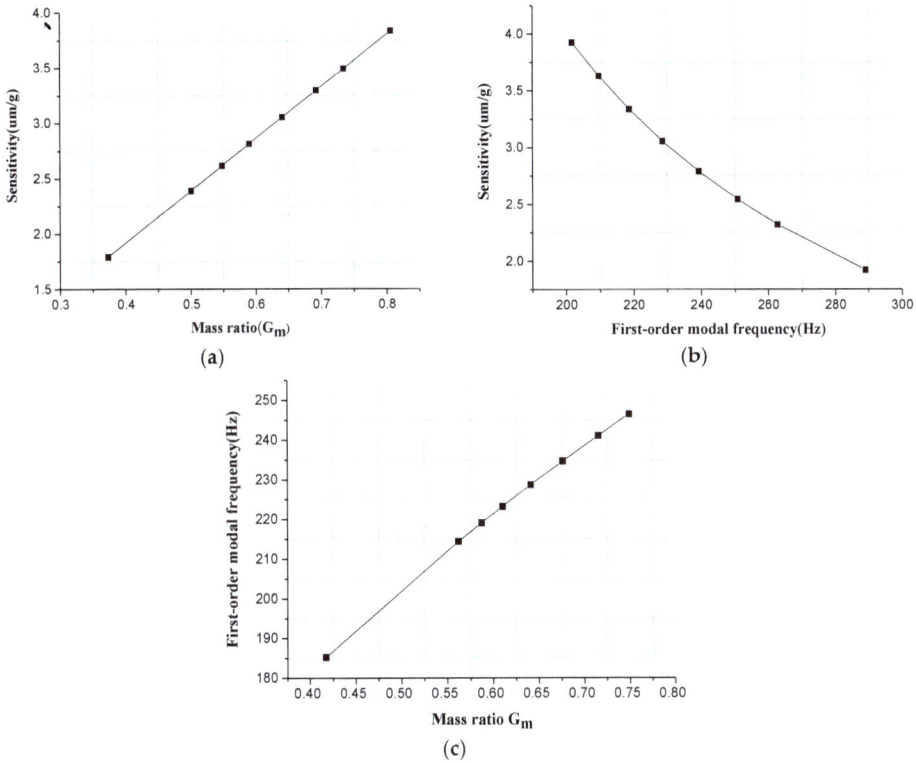

Figure 4. The effect of mass ratio and first-order modal frequency on mechanical sensitivity (**a**) The relationship between mechanical sensitivity and mass ratio in the same first-order modal frequencies; (**b**) The relationship between mechanical sensitivity and first-order modal frequencies in the same mass ratio; (**c**) The relationship between mass ratio and first-order modal frequency in the same mechanical sensitivity.

Furthermore, a finite element simulation based on a solid model and a numerical simulation are implemented to analyze the magnetic field characteristics of the entire accelerometer. The physical structure is constructed using Comsol software according to the parameters shown in Table 1. The finite element simulation of magnetic field distribution for permanent magnetic film is shown in Figure 5.

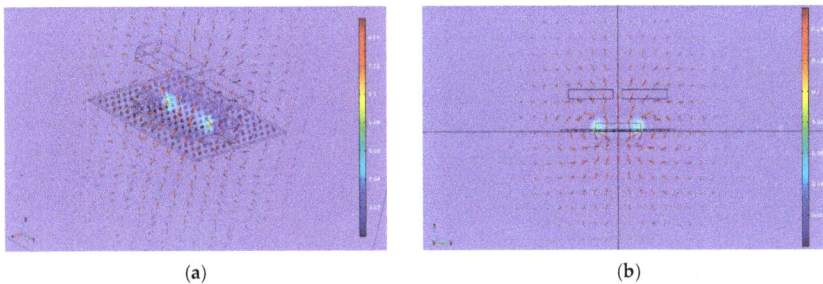

Figure 5. Finite element simulation of magnetic field distribution for permanent magnetic film. (**a**) Three dimensional magnetic field distribution; (**b**) Planar magnetic field distribution in the middle section along the y-direction.

The magnetic field simulation results demonstrate that the magnetic field generated by the miniature rectangular permanent magnetic film is basically similar to the macro magnetic field. And the closer to the permanent magnetic film, the greater the magnetic field strength, as shown in Figure 5a. Simultaneously, the planar magnetic field distribution in the middle section shown in Figure 5b indicates that two tunnel magnetoresistive sensors located directly above the boundary of the permanent magnetic film have opposite magnetic field directions. The sensitive direction of the tunneling magnetoresistive sensor is the y direction. The magnetic field change caused by the displacement in the z direction still has a component in the y direction since the magnetic field direction on the tunnel magnetoresistive sensor is not orthogonal to the y direction, which is the sensitive mechanism of the tunnel magnetoresistive accelerometer.

The magnetic field data under different conditions are extracted for further quantitative analysis. The magnetic field distribution in the y direction is mainly considered since the sensitive direction of the tunnel magnetoresistive sensor is the y axis. Figure 6 shows the magnetic field characteristic in different conditions.

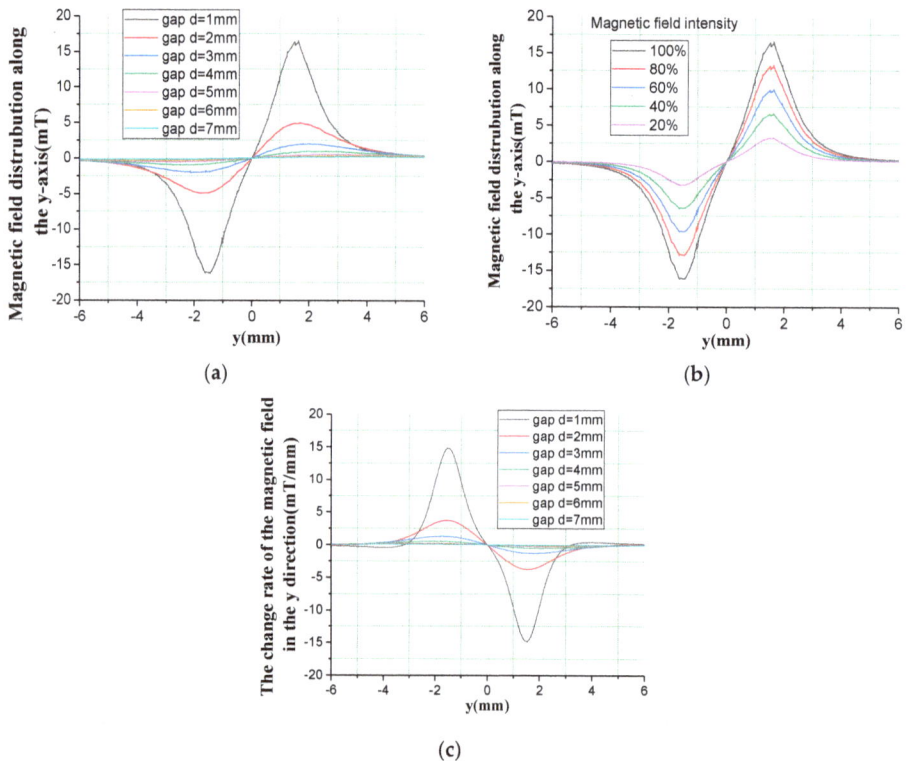

(a)

(b)

(c)

Figure 6. Magnetic field characteristic in different conditions (**a**) Magnetic field distribution along the y-axis in the tunnel magnetoresistive sensor versus different gaps; (**b**) Magnetic field distribution along the y-axis in the tunnel magnetoresistive sensor versus different magnetic field intensity of permanent magnetic film (d = 1 mm); (**c**) The change rate of the magnetic field in the y direction due to a displacement variation in the z direction versus different gaps.

Magnetic field intensity along the y-axis in the tunneling magnetoresistive sensor shown in Figure 6a is inversely proportional to the gap between the tunnel magnetoresistive sensor and the permanent magnetic film. The maximum magnetic field intensity is decreased by 77.1 times when the gap is increased from 1 mm to 7 mm. The maximum magnetic field intensity in the tunnel magnetoresistive sensor reached 16.57 mT under a gap of 1 mm. The entire magnetic field intensity cannot approach the saturated area or exceed the measurable range of the tunneling magnetoresistive sensor, otherwise it will cause sensitivity attenuation, saturation and lose the measurement capability in the tunneling magnetoresistive sensor. This will limit the optional gap and shift position of y between the permanent magnetic film and tunneling magnetoresistive sensor. Therefore, the normal magnetic field signal detection can be achieved in the suitable gap and shift position of y by the tunneling magnetoresistive sensor. Figure 6b shows the magnetic field distribution along the y-axis in the tunnel magnetoresistive sensor versus different magnetic field intensities of permanent magnetic film (d = 1 mm). The maximum magnetic field intensity in the tunneling magnetoresistive sensor can significantly reduce by decreasing the magnetic field strength of the magnetic film under the same gap. When the magnetic field strength of the permanent magnetic film is reduced to 20 %, the maximum magnetic field intensity in the tunneling magnetoresistive sensor is decreased from 16.58 mT to 3.31 mT.

Further, a numerical simulation is differentiated to abstract the change rate of the magnetic field due to the displacement variation in the z directions. Apparently, the change rate of the magnetic field in the y direction due to a displacement variation in the z direction decreases as the gap increases, shown in Figure 6c. The maximum change rate of the magnetic field is reduced from 14.8 mT/mm to 0.079 mT/mm when the gap is increased from 1 mm to 7 mm. Simultaneously, the change rate of the magnetic field due to a displacement variation along the z direction in different horizontal shift between the tunnel magnetoresistive sensor and permanent magnetic film is shown in Figure 6c. The change rate of the magnetic field is essentially zero when horizontal shift y = 0, which indicates that the minimal sensitivity in the tunneling magnetoresistive sensor appears at this point. The main reason for this is that the direction of the magnetic field is orthogonal to the sensitive direction of the tunnel magnetoresistive sensor at this point, which is basically consistent with the finite element simulation of Figure 5b. At the same time, when the horizontal shift is equal to 1.5 mm, the change rate of the magnetic field in the y direction has approximately the largest absolute value, which demonstrates that the tunneling magnetoresistive sensor has the greatest sensitivity at the boundary of the permanent magnetic film. This is also the theoretical basis for the layout optimization of the tunneling magnetoresistive sensor.

The structure dimension of permanent magnetic film not only affects the distribution of the magnetic field, but also affects the mechanical sensitivity. Figure 7 shows the influences of structure dimension of permanent magnetic films on characteristics of the tunneling magnetoresistive accelerometer. Obviously, the variation in the structure dimension of the permanent magnet film causes the maximum change rate of magnetic field to appear at different positions, shown in Figure 7a. However, all the maximum change rates of magnetic field appear almost around the boundary of the permanent magnetic film. At the same time, the mechanical sensitivity of the displacement increases by 2.62 times as the structure dimension of permanent magnetic film widens from 2.5 mm × 2.5 mm to 4 mm × 4 mm, as shown in Figure 7b. Combining the above two effects, the change rate of the magnetic field in the y direction due to input acceleration in the z direction is proportional to the change of structure dimension of the permanent magnetic film, as shown in Figure 7c. As the structure dimension of the permanent magnetic film is enlarged from 2.5 mm × 2.5 mm to 4 mm × 4 mm, the magnetic field sensitivity in the y direction due to input acceleration in the z direction increases by 17.62 times.

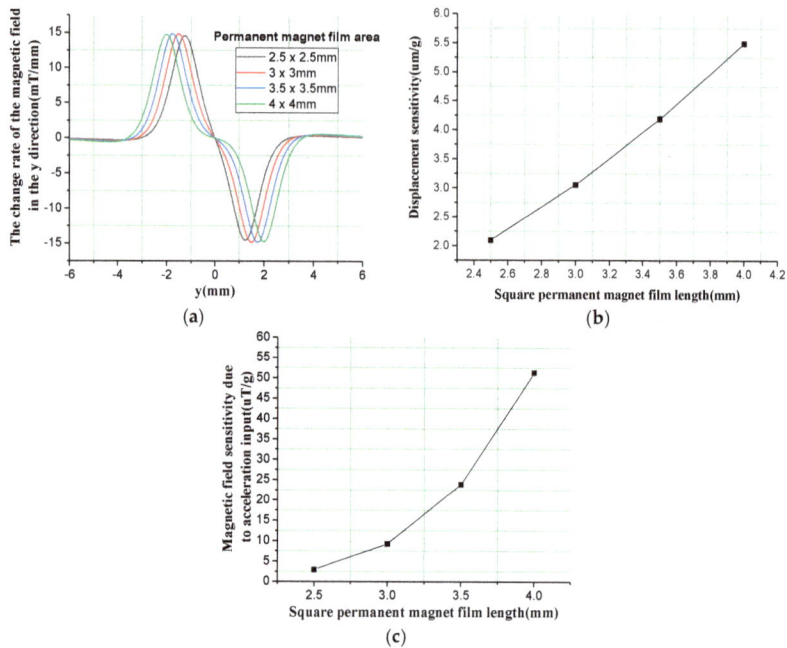

(a)

(b)

(c)

Figure 7. Influence of structure dimension of permanent magnet film on characteristics of the tunneling magnetoresistive accelerometer. (**a**) The change rate of the magnetic field in the y direction due to a displacement variation in the z direction versus different structure dimension of permanent magnet film; (**b**) Mechanical sensitivity versus different structure dimension of permanent magnet film; (**c**) The magnetic field sensitivity in the y direction due to input acceleration in the z direction versus different structure dimension of permanent magnet film (y = 2.5 mm).

4. Measurement and Control Circuit

The scheme of the measurement and control circuit is shown in Figure 8. The displacement of the permanent magnetic film due to input acceleration causes the resistance change of the tunnel magnetoresistive sensor. Two resistance bridges are formed by eight tunneling magnetoresistive sensors arranged on the two boundaries of the permanent magnetic film. An interface amplifier circuit consisting of three operational amplifiers is used to measure resistance changes. Then, the signal is filtered by a low pass filter (LPF) and control signals are generated by the proportional-integral (PI) controller. Finally, the output signal of the PI controller is divided into two paths, which are respectively applied to the two feedback electrodes. The feedback correction force is generated by the electrostatic force mechanism to realize the closed-loop detection of the tunneling magnetoresistive accelerometer.

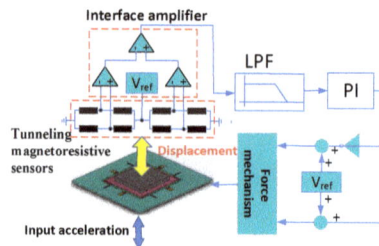

Figure 8. The scheme of measurement and control circuit.

5. Experiment

The micromachined z-axis tunneling magnetoresistive accelerometer is fabricated by a standard deep dry silicon on glass (DDSOG) process in order to verify the theoretical analysis and evaluate the characteristics. Three masks are used to implement the fabrication of the micromachined z-axis tunnel magnetoresistive accelerometer. The first mask is utilized to pattern and expose the bonding anchors in a monocrystalline wafer with 200 μm thickness by lithography, and steps of bonding anchors with 10 μm height are etched by the deep reactive ion etching (DRIE). The electrode wires and pads with a Cr/Ti/Au stack layer are established by the sputtering process with the second mask in a Pyrex glass substrate with 500 μm thickness. Then the silicon wafer and the Pyrex glass wafer are linked together by the electrostatic anodic bonding process, and the silicon wafer is thinned to 120 μm thickness by a wet etching process with KOH solution. Subsequently, the mechanical structure is lithographically patterned and etched to release by the DRIE with the third mask. Finally, two commercial tunneling magnetoresistive sensors and a permanent magnetic film are patterned and installed by the micro-assembly process. The performance of the prototype is sensitive to the distance between the permanent magnetic film and the tunnel magnetoresistive sensor. We adjust the gap between the permanent magnetic film and the tunnel magnetoresistive sensor through the base. The base is realized by 3D printing technology, and its precision can be controlled within several tens of μm. The tunneling magnetoresistive sensors adopts the commercial linear sensor of TMR9001 in Multi-Dimension Technology [22]. The optical micrograph of micromachined z-axis tunnel magnetoresistive accelerometer is shown in Figure 9. The fabricated plane main structure of tunnel magnetoresistive accelerometer has a dimension of 6400 μm (length) × 6400 μm (width) × 120 μm (height) with a permanent magnet film of 3000 μm (length) × 3000 μm (width) × 500 μm (height).

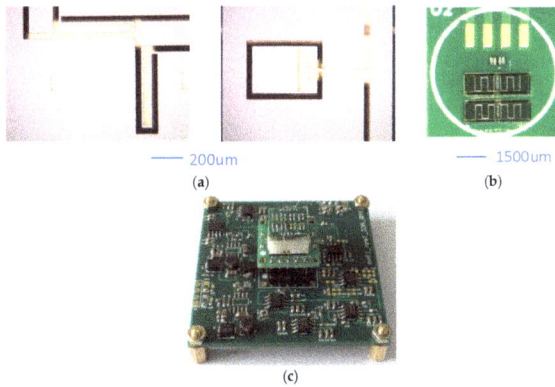

Figure 9. The optical micrograph of micromachined z-axis tunnel magnetoresistive accelerometer. (a) The plane main structure of tunnel magnetoresistive accelerometer; (b) Commercial tunnel magnetoresistive sensor; (c) Micro-assembled overall structure.

In order to fully evaluate the performance of tunnel magnetoresistive accelerometers, system experiments under various conditions are implemented. We test the acceleration input and output response characteristics under different gaps in the tunneling magnetoresistive accelerometer, shown in Figure 10. The experiment results demonstrate that the signal sensitivity is inversely proportional to gap variation. When the gap is reduced from 2 mm to 1 mm, the signal sensitivity is increased by 7.37 times from 0.559 mV/g to 4.12 mV/g. This confirms that the decrease of the gap between the tunneling magnetoresistive sensors and permanent magnetic film can significantly increase the change rate of the magnetic field due to a displacement variation along the z direction, which is consistent with the previous simulation analysis.

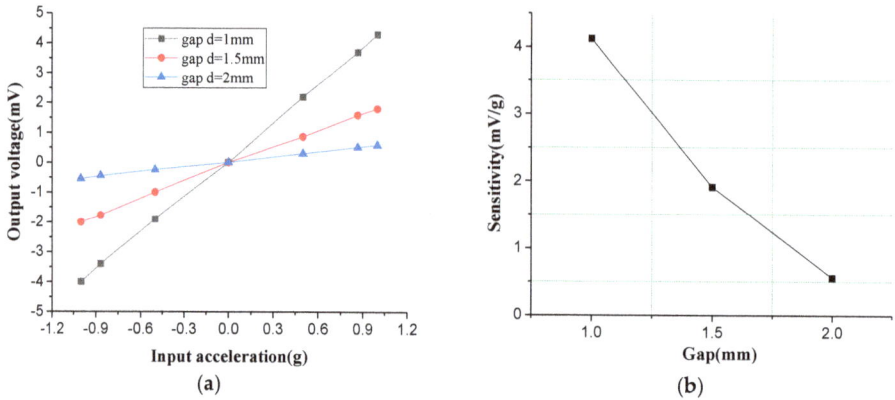

Figure 10. Acceleration input and output response characteristics under different gaps in the tunneling magnetoresistive accelerometer. (**a**) Input acceleration versus output voltage under different gaps; (**b**) The sensitivity versus gap.

Acceleration input and output response characteristics experiments under different dimensions of permanent magnetic film are implemented to analyze the influence on the performance of the tunneling magnetoresistive accelerometer, shown in Figure 11. The experiment results illustrate that the signal sensitivity is proportional to the dimension of permanent magnetic film. When the dimension of the permanent magnetic film is widened from 3 mm × 3 mm to 4 mm × 4 mm, the sensitivity is increased by 6.81 times from 1.30 mV/g to 8.85 mV/g. This demonstrates that the area amplification of the magnetic field can significantly improve the mechanical displacement sensitivity and the magnetic field change rate due to a displacement variation along the z direction, and ultimately improve the signal sensitivity of the tunneling magnetoresistive accelerometer, which is consistent with the aforementioned simulation analysis.

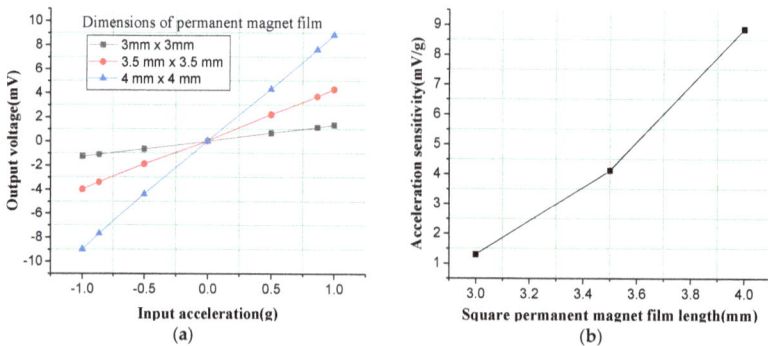

Figure 11. Acceleration input and output response characteristics under different dimensions of permanent magnetic film in the tunneling magnetoresistive accelerometer. (**a**) Input acceleration versus output voltage under different dimensions of permanent magnetic film; (**b**) The sensitivity versus dimensions of permanent magnet film.

Finally, the stability and noise performance measures of the z-axis tunneling magnetoresistive accelerometer are implemented to evaluate the performance of the prototype. In order to optimize system performance, the gap and shift of the tunneling magnetoresistive sensor are slightly adjusted for maximum sensitivity along z-axis sensitive direction. The output voltage noise spectrum of prototype is shown in Figure 12.

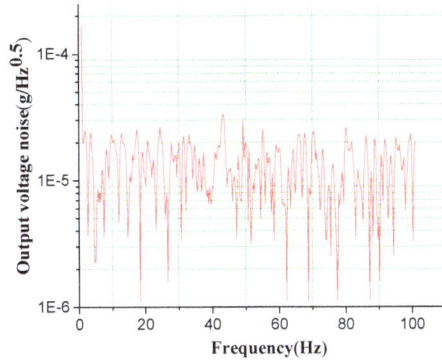

Figure 12. Output voltage noise spectrum of z-axis tunneling magnetoresistive accelerometer.

Experimental results demonstrate that the tunneling magnetoresistive accelerometer has a noise floor of 86.2 µg/Hz$^{0.5}$ along z-axis sensitive direction. The noise comes from mechanical thermal noise of proof mass, resistance thermal noise of tunnel magnetoresistive sensor and circuit noise of the interface amplifier. The noisy force generator of mechanical thermal noise corresponds to an equivalent input acceleration $a_n = (4k_BT\xi)^{0.5}/(9.8m)$ [23]. Suppose the Boltzman's constant $k_B = 1.38 \times 10^{-23}$ J/K, the absolute temperature $T = 300$ K, the viscous damping coefficient $\xi = 0.5$ N/m/s (high air damping), $m = 2.61 \times 10^{-5}$ kg, the equivalent acceleration noise of the mechanical thermal noise of the proof mass is 0.36 µg/Hz$^{0.5}$. Obviously, the main noise is dominated by the resistance thermal noise of tunnel magnetoresistive sensor and circuit noise of the interface amplifier, and the prototype of tunneling magnetoresistive accelerometer has a large potential for improvement. Simultaneously, a drift stability curve of the z-axis tunneling magnetoresistive accelerometer based on Allan variance is shown in Figure 13 [24]. Allan variance analysis illustrates that the tunneling magnetoresistive accelerometer has a bias stability of 482 µg along z-axis sensitive direction. In summary, the above experimental results prove that the scheme of z-axis tunneling magnetoresistive accelerometer is feasible and effective and achieves a considerable performance.

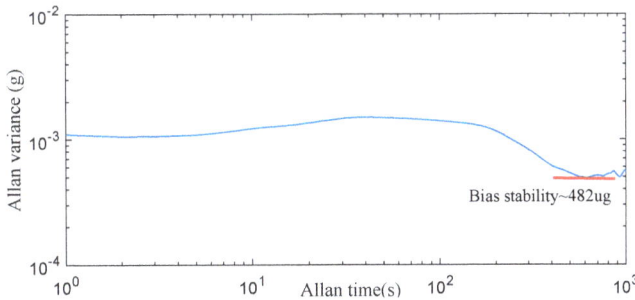

Figure 13. Drift stability curve of z-axis tunneling magnetoresistive accelerometer based on Allan variance.

6. Conclusions

The design, simulation, fabrication and experiments of a micromachined z-axis tunneling magnetoresistive accelerometer with electrostatic force feedback are proposed in the paper. The tunneling magnetoresistive accelerometer adopted the tunneling magnetoresistive effect with high sensitivity to measure magnetic field variation caused by input acceleration. A pair of lever-driven differential proof masses in the middle plane main structure is used for sensitiveness to acceleration and closed-loop feedback control. Compared with the traditional tunnel effect, the tunneling magnetoresistance effect with high sensitivity is directly generated by the multi-layer nano-film layer in the tunneling magnetoresistive sensors, which can be easily achieved with a multi-thin film deposition process. This avoids the micromachining requirement for a nano-gap between the tunnel tip and the plane. We constructed the finite element model of the mechanical structure and optimized the structural mode and mass ratio between inner and outer proof mass by the Ansys simulation. Simultaneously, we establish a magnetic field simulation model to analyze the magnetic field distribution and magnetic field change rate around the tunnel magnetoresistive sensor, as well as the effect of the location of tunneling magnetoresistive sensors, magnetic field intensity, and the dimension of permanent magnetic film on magnetic field sensitivity, which is beneficial to the achievement of maximum sensitivity. A standard deep dry silicon on glass (DDSOG) process is used to fabricate the micromachined z-axis tunneling magnetoresistive accelerometer which has a device dimension of 6400 µm (length) × 6400 µm (width) × 120 µm (height). The experimental results demonstrate the prototype has a maximal sensitivity of 8.85 mv/g along the z-axis sensitive direction under a gap of 1 mm. Simultaneously, Allan variance analysis illustrate that a noise floor of 86.2 µg/Hz$^{0.5}$ is implemented in the z-axis tunneling magnetoresistive accelerometer, which proves that the scheme of z-axis tunneling magnetoresistive accelerometer is feasible and effective, and has great development potential in the future.

Author Contributions: B.Y. conceived and designed theoretically the tunnel magnetoresistive accelerometer, and wrote the paper; B.W. was responsible for the design, simulation and partial experiment of tunnel magnetoresistive accelerometer; H.Y. completed the magnetic field simulation and partial experiment of tunnel magnetoresistive accelerometer; X.G. implemented partial structure simulation and partial experiment of tunnel magnetoresistive accelerometer.

Funding: This research was funded by the National Natural Science Foundation of China (Grant No. 61874025 and 61571126), Equipment pre-research field foundation (Grant No. 6140517010316JW06001), the Fundamental Research Funds for the Central Universities (Grant No. 2242018k1G017) and the Eleventh Peak Talents Programme Foundation in the Six New Industry Areas.

Conflicts of Interest: The authors declare no conflict of interest.

References

1. Aydemir, A.; Terzioglu, Y.; Akin, T. A new design and a fabrication approach to realize a high performance three axes capacitive MEMS accelerometer. *Sens. Actuators A* **2016**, *244*, 324–333. [CrossRef]
2. Zwahlen, P.; Balmain, D.; Habibi, S.; Etter, P. Open-loop and Closed-loop high-end accelerometer platforms for high demanding applications. In Proceedings of the 2016 IEEE/ION Position, Location and Navigation Symposium, Savannah, GA, USA, 11–14 April 2016; pp. 932–937.
3. Sadeghi, M.-M.; Peterson, R.-L.; Najafi, K. Hair-based sensors for micro-autonomous systems. In Proceedings of the SPIE, Baltimore, MD, USA, 3 May 2012; Volume 8373, p. 83731L-1.
4. Gerberding, O.; Cervantes, F.-G.; Melcher, J. Optomechanical reference accelerometer. *Metrologia* **2015**, *52*, 654–665. [CrossRef]
5. Cervantes, F.-G.; Kumanchik, L.; Pratt, J. High sensitivity optomechanical reference accelerometer over 10 kHz. *Appl. Phys. Lett.* **2014**, *104*, 221111. [CrossRef]
6. Hopkills, R.; Miola, J.; Setterlund, R. The silicon oscillating accelerometer: A High-Performance MEMS accelerometer for Precision Navigation and Strategic Guidance Applications. In Proceedings of the 2005 National Technical Meeting of the Institute of Navigation, San Diego, CA, USA, 24–26 January 2005; pp. 970–979.
7. Zotov, S.-A.; Simon, B.-R.; Trusov, A.-A.; Shkel, A.-M. High Quality Factor Resonant MEMS Accelerometer with Continuous Thermal Compensation. *IEEE Sens. J.* **2015**, *15*, 5045–5052. [CrossRef]
8. Comi, C.; Corigliano, A.; Langfelder, G. A Resonant Microaccelerometer with High Sensitivity Operating in an Oscillating Circuit. *J. Microelectromech. Syst.* **2010**, *19*, 1140–1152. [CrossRef]
9. Bose, S.; Raychowdhury, A.; Jatolia, M. Design of PID controller for ultra-sensitive Nano-g resolution MEMS tunneling accelerometer. In Proceedings of the 2014 IEEE International Conference on Control System, Computing and Engineering (ICCSCE), Batu Ferringhi, Malaysia, 28–30 November 2014; pp. 658–662.

10. Xu, M.-H.; Feng, Y.-J.; Zhou, H.; Sheng, J.-N. Noise analysis of the triaxial piezoelectric micro-accelerometer. In Proceedings of the 2017 Symposium on Piezoelectricity, Acoustic Waves, and Device Applications (SPAWDA), Chengdu, China, 27–30 October 2017; pp. 288–292.

11. Messina, M.; Njuguna, J.; Palas, C. Computational analysis and optimization of a MEMS-based piezoresistive accelerometer for head injuries monitoring. In Proceedings of the 2017 IEEE SENSORS, Glasgow, UK, 29 October–1 November 2017; pp. 1–3.

12. Kumar, V.; Guo, X.-B.; Pourkamali, S. Single-mask field emsission based tunable MEMS tunneling accelerometer. In Proceedings of the 2015 IEEE 15th International Conference on Nanotechnology (IEEE-NANO) Rome, Italy 27–30 July 2015; pp. 1171–1174.

13. Phan, K.-L.; Mauritz, A.; Homburg, F.-G.-A. A novel elastomer-based magnetoresistive accelerometer. *Sens. Actuators A* **2008**, *145–146*, 109–115. [CrossRef]

14. Phan, K.-L. Methods to correct for creep in elastomer-based sensors. In Proceedings of the 2008 IEEE Sensors, Lecce, Italy, 26–29 October 2008; pp. 1119–1122.

15. McNeil, K.-R.; Shankar, B.; Vogel, A.-B.; Gibson, S. Magnetoresistive-Based Position Sensor for Use in an Implantable Electrical Device. U.S. Patent 6,430,440, 6 August 2012.

16. Alaoui, A. MLU Based Accelerometer Using a Magnetic Tunnel Junction. U.S. Patent US20170160308A1, 8 June 2017.

17. Olivas, J.-D.; Lairson, B.-M.; Ramesham, R. Ultra-Sensitive Magnetoresistive Displacement Sensing Device. U.S. Patent 6,507,187, 14 January 2003.

18. Takada, A. Force Sensing MEMS Device for Sensing an Oscillating Force. U.S. Patent 6,722,206, 20 April 2004.

19. Fries, D.-P. Giant Magnetoresistance Based Gyroscope. U.S. Patent 7,113,104, 26 September 2006.

20. Yang, B.; Wang, B.-L.; Gao, X.-Y. Research on a small tunnel magnetoresistive accelerometer based on 3D printing. *Microsyst. Technol.* **2018**, 1–12. Available online: https://link.springer.com/article/10.1007/s00542-018-4218-2 (accessed on 11 February 2019). [CrossRef]

21. Camacho, J.-M.; Sosa, V. Alternative method to calculate the magnetic field of permanent magnets with azimuthal symmetry. *Revista Mexicana de Física E* **2013**, *59*, 8–17.

22. The Date Sheet of TMR9001. Available online: http://www.dowaytech.com/1884.html (accessed on 11 February 2019).

23. Lemkin, M.; Boser, B.E. A three-axis micromachined accelerometer with a CMOS position-sense interface and digital Offset-trim electronis. *IEEE J. Solid-State Circuits* **1999**, *34*, 456–468. [CrossRef]

24. D'Alessandro, A.; Vitale, G.; Scudero, S.; D'Anna, R.; Costanza, A.; Fagiolini, A.; Greco, L. Characterization of MEMS accelerometer self-noise by means of PSD and AllanVariance analysis. In Proceedings of the 2017 7th IEEE International Workshop on Advances in Sensors and Interfaces (IWASI), Vieste, Italy, 15–16 June 2017; pp. 160–164.

micromachines

MDPI

Article

The Effect of the Anisotropy of Single Crystal Silicon on the Frequency Split of Vibrating Ring Gyroscopes

Zhengcheng Qin [1,2], Yang Gao [3], Jia Jia [1,2] , Xukai Ding [1,2] , Libin Huang [1,2] and Hongsheng Li [1,2,*]

[1] School of Instrument Science and Engineering, Southeast University, Nanjing 210096, China; 230189281@seu.edu.cn (Z.Q.); 230169207@seu.edu.cn (J.J.); dingxukai@126.com (X.D.); huanglibin@seu.edu.cn (L.H.)
[2] Key Laboratory of Micro-Inertial Instruments and Advanced Navigation Technology, Ministry of Education, Nanjing 210096, China
[3] Artificial Intelligence Institute of Industrial Technology, Nanjing Institute of Technology, Nanjing 211167, China; ygao@njit.edu.cn
[*] Correspondence: hsli@seu.edu.cn; Tel.: +86-25-8379-5920

Received: 10 January 2019; Accepted: 11 February 2019; Published: 14 February 2019

check for updates

Abstract: This paper analyzes the effect of the anisotropy of single crystal silicon on the frequency split of the vibrating ring gyroscope, operated in the $n = 2$ wineglass mode. Firstly, the elastic properties including elastic matrices and orthotropic elasticity values of (100) and (111) silicon wafers were calculated using the direction cosines of transformed coordinate systems. The (111) wafer was found to be in-plane isotropic. Then, the frequency splits of the $n = 2$ mode ring gyroscopes of two wafers were simulated using the calculated elastic properties. The simulation results show that the frequency split of the (100) ring gyroscope is far larger than that of the (111) ring gyroscope. Finally, experimental verifications were carried out on the micro-gyroscopes fabricated using deep dry silicon on glass technology. The experimental results are sufficiently in agreement with those of the simulation. Although the single crystal silicon is anisotropic, all the results show that compared with the (100) ring gyroscope, the frequency split of the ring gyroscope fabricated using the (111) wafer is less affected by the crystal direction, which demonstrates that the (111) wafer is more suitable for use in silicon ring gyroscopes as it is possible to get a lower frequency split.

Keywords: single crystal silicon; anisotropy; vibrating ring gyroscope; frequency split

---

1. Introduction

With the development of the microelectromechanical systems (MEMS) technology, MEMS inertial sensors have been widely adopted into many fields, such as aerospace, vehicle navigation, and consumer electronic products including smartphones, tablets, and wearable sensors [1]. One of the most common MEMS sensors is the vibrating gyroscope. Compared with the traditional gyroscope, the MEMS gyroscope has many advantages, including a smaller volume, lower power consumption, wider measurement range, and higher reliability. The vibrating ring gyroscope (VRG) is a kind of MEMS gyroscope that has the following merits over other similar types. Firstly, the VRG has better mechanical sensitivity characteristics as well as being less sensitive to environmental interferences such as shock and temperature variations. This is because of its symmetric structure and the equal resonant frequencies of the drive and sense modes [2]. In addition, the VRG has a wider bandwidth and full-scale range [3]. Moreover, the VRG can be used to measure the angle directly, which helps avoid the continuous accumulation of the test error during the process of the integration by measuring the angle rate [4].

At present, the high respect ratio VRG can be fabricated using single crystal silicon (SCS) because of the mature bulk silicon micromachining technique. The SCS is a very significant MEMS material, and it is often used to fabricate the MEMS resonator and substrate due to its excellent mechanical and electrical properties. However, the anisotropy of the SCS makes the elastic properties of the SCS vary with respect to the crystal orientations [5], which may influence the effective stiffness of the drive and sense modes of the VRG. As a result, resonant frequencies of the $n = 2$ wineglass mode of the VRG are not always equal. The frequency mismatch can cause many problems, such as the angle-dependent bias and a decrease in sensitivity thus hindering the performance of the VRG [6]. So it is essential to analyze the relationship between elastic properties and crystal directions of the SCS during the process of designing the VRG. In order to avoid this trouble, some researchers prefer the $n = 3$ mode rather than the $n = 2$ mode, because the $n = 3$ mode is inherently identical, which helps to eliminate the frequency mismatch induced by the anisotropic elastic properties of the SCS [7]. However, both the angular gain and the effective mass of the $n = 3$ mode are smaller than those of the $n = 2$ mode, and the sensitivity of the $n = 3$ mode is lower [8,9]. Besides, resonant frequencies of the $n = 3$ mode are higher, and the electrode arrangement is more complicated, which raises the difficulty of the circuit design and realization. For the reasons mentioned above, the $n = 2$ mode of the SCS VRG is usually the ideal choice.

There are some theories about the relationship between the frequency split and the anisotropy of the SCS circular ring. For example, Hamilton's principle was used to derive equations of the in-plane motion of the SCS circular ring, and the frequency split was explained by the conservation of averaged mechanical energy [10]. In addition, in-plane and out-of-plane resonant modes of the circular ring were solved by Lagrange's equations, and the effect of anisotropy was considered in the strain energy formulation [5]. However, these theories derived from the perspective of the single ring were not concretely combined with a real SCS VRG or confirmed by experiments. The effects of the anisotropy of SCS on resonant frequencies of other resonators such as the cantilever and the disk were reported. For instance, the influence of crystallographic orientation on resonant frequencies of (100)-oriented resonators was investigated by a micromachined array consisting of 36 cantilevers in Reference [11]. In Reference [12], frequency splits of (100) and (111) disk resonators with different modes were examined and compared experimentally. Besides, an analytical formulation of frequency splits of SCS disk resonators was represented by taking material anisotropy into account [13]. Though some researchers prefer the (111) SCS to fabricate $n = 2$ mode ring and disk gyroscopes because of the uniform material properties of this material, the reason for this was not analyzed and explained systematically in these articles [14,15]. Therefore, there is still a lack of research about how the anisotropy influences the frequency characteristics of the SCS VRG.

This paper aims to analyze the influence of the anisotropy of the SCS on the frequency split of the SCS VRG. The elastic properties of (100) and (111) wafers are analyzed, and the ring gyroscopes of these wafers are simulated, fabricated, and tested to prove that the (111) VRG has a tiny frequency split, so the (111) wafer is more suitable than the (100) wafer to make VRGs. This paper is organized as follows. Section 2 calculates the elastic matrices and orthotropic elasticity values of (100) and (111) wafers. In Section 3, (100) and (111) VRGs are simulated by finite element analysis. VRGs of different wafers are fabricated and tested in Section 4 to verify the theory and the simulation. Section 5 is the conclusion of this paper.

2. Anisotropic Elastic Properties of SCS

The linear proportional relationship between the strain ε and the stress σ of solids such as silicon is proved by the Hooke's law,

$$\varepsilon_{ij} = S_{ijkl}\sigma_{kl}, \sigma_{ij} = C_{ijkl}\varepsilon_{kl} \tag{1}$$

where S and C are the compliance tensor and the stiffness tensor, respectively; $i,j,k,l = x,y,z$ in the rectangular coordinate. The fourth rank stiffness or compliance tensor can be simplified as a 6×6 matrix with 21 independent constants due to the symmetry of the crystal. Meanwhile, subscripts with four elements in Equation (1) can be simplified to two elements: $xx \rightarrow 1$, $yy \rightarrow 2$, $zz \rightarrow 3$, yz and $zy \rightarrow 4$, xz and $zx \rightarrow 5$, xy and $yx \rightarrow 6$ [16].

The anisotropy of the SCS is reflected in the variety of physical properties related to crystal orientations. Each crystal plane or crystal orientation can be identified by Miller indices and written in the form of (ijk) or $[ijk]$, respectively [17]. Three axes X ([100]), Y ([010]), Z ([001]) of the default coordinate system based on the right-hand rule lay along three baselines of the crystal cubic, so corresponding default stiffness and compliance matrices along the crystal orientations according to Hooke's law can be represented as

$$C = \begin{bmatrix} c_{11} & c_{12} & c_{12} & 0 & 0 & 0 \\ c_{12} & c_{11} & c_{12} & 0 & 0 & 0 \\ c_{12} & c_{12} & c_{11} & 0 & 0 & 0 \\ 0 & 0 & 0 & c_{44} & 0 & 0 \\ 0 & 0 & 0 & 0 & c_{44} & 0 \\ 0 & 0 & 0 & 0 & 0 & c_{44} \end{bmatrix} = S^{-1} = \begin{bmatrix} s_{11} & s_{12} & s_{12} & 0 & 0 & 0 \\ s_{12} & s_{11} & s_{12} & 0 & 0 & 0 \\ s_{12} & s_{12} & s_{11} & 0 & 0 & 0 \\ 0 & 0 & 0 & s_{44} & 0 & 0 \\ 0 & 0 & 0 & 0 & s_{44} & 0 \\ 0 & 0 & 0 & 0 & 0 & s_{44} \end{bmatrix}^{-1} \tag{2}$$

where $c_{11} = 1.657 \times 10^{11}$ Pa, $c_{12} = 0.639 \times 10^{11}$ Pa, $c_{44} = 0.796 \times 10^{11}$ Pa, $s_{11} = 0.768 \times 10^{-11}$ Pa^{-1}, $s_{12} = -0.214 \times 10^{-11}$ Pa^{-1}, $s_{44} = 1.26 \times 10^{-11}$ Pa^{-1} at room temperature [18]. Next, anisotropic elasticity analyses of (100) and (111) silicon wafers are based on these two matrices.

Though the top surface of the (100) silicon wafer is perpendicular to the [100] direction, the actual coordinate system of the structure fabricated on the (100) silicon wafer is not consistent with the default coordinate system. The xyz coordinate system of the (100) wafer can be obtained by rotating the default coordinate system anti-clockwise through 45 degrees about the Z-axis [19]. So orthogonal axes (x, y, z) keep the same directions with three yellow dashed lines in Figure 1a. Figure 1b indicates that the primary flat of the (100) wafer is parallel to the [110] direction that is exactly the x-axis. For the (111) silicon wafer, the [111] orientation is vertical to the crystal surface of this wafer. Coordinate axes (x', y', z') of the (111) wafer are represented by three red dot-dashed lines in Figure 1a successively, because coordinate axes of the (111) wafer shown in Figure 1a can be converted by rotating the default coordinate system twice. At first, the default coordinate system is rotated clockwise 135 degrees about the Z-axis, then through a clockwise angle of 54.7 degrees about the transformed Y-axis (i.e., y'-axis). The x'-axis in Figure 1b is parallel to the primary flat of the (111) wafer and perpendicular to the y'-axis which is the [1$\bar{1}$0] direction.

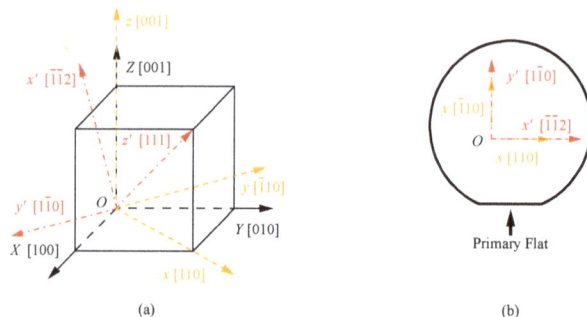

Figure 1. Coordinate systems and overhead views of the (100) and (111) silicon wafers: (**a**) Coordinate systems of two wafers, (**b**) Overhead views of two wafers.

After getting the coordinate systems of the (100) and (111) wafers, corresponding stiffness matrices and compliance matrices can be figured out using the method developed by Bond [16]. Direction cosines of each rotated axis should be firstly given, which are three cosines of angles between the rotated axis and three default axes (i.e., [100], [010], and [001]). Three terms l, m, and n are used to define direction cosines of an axis. Therefore, direction cosines of three axes of any rotated coordinate system can build a 3×3 matrix R to describe the transformation from the default coordinate system. Matrices Rs of coordinate systems of (100) and (111) wafers are presented as

$$R_{100} = \begin{bmatrix} \frac{1}{\sqrt{2}} & \frac{1}{\sqrt{2}} & 0 \\ \frac{-1}{\sqrt{2}} & \frac{1}{\sqrt{2}} & 0 \\ 0 & 0 & 1 \end{bmatrix}, \tag{3}$$

$$R_{111} = \begin{bmatrix} \frac{-1}{\sqrt{6}} & \frac{-1}{\sqrt{6}} & \frac{2}{\sqrt{6}} \\ \frac{1}{\sqrt{2}} & \frac{-1}{\sqrt{2}} & 0 \\ \frac{1}{\sqrt{3}} & \frac{1}{\sqrt{3}} & \frac{1}{\sqrt{3}} \end{bmatrix}. \tag{4}$$

Then, stiffness matrices of (100) and (111) wafers can be obtained according to the method in Reference [16]. Because the compliance matrix is the inverse of the stiffness matrix, only the stiffness matrices of (100) and (111) wafers are calculated and shown as

$$C_{100} = \begin{bmatrix} 1.944 & 0.352 & 0.639 & 0 & 0 & 0 \\ 0.352 & 1.944 & 0.639 & 0 & 0 & 0 \\ 0.639 & 0.639 & 1.657 & 0 & 0 & 0 \\ 0 & 0 & 0 & 0.796 & 0 & 0 \\ 0 & 0 & 0 & 0 & 0.796 & 0 \\ 0 & 0 & 0 & 0 & 0 & 0.509 \end{bmatrix} \times 10^{11} \, (\text{Pa}), \tag{5}$$

$$C_{111} = \begin{bmatrix} 1.9440 & 0.5433 & 0.4477 & 0 & -0.1353 & 0 \\ 0.5433 & 1.9440 & 0.4477 & 0 & 0.1353 & 0 \\ 0.4477 & 0.4477 & 2.0397 & 0 & 0 & 0 \\ 0 & 0 & 0 & 0.6047 & 0 & 0.1353 \\ -0.1353 & 0.1353 & 0 & 0 & 0.6047 & 0 \\ 0 & 0 & 0 & 0.1353 & 0 & 0.7003 \end{bmatrix} \times 10^{11} \, (\text{Pa}). \tag{6}$$

Obtaining correct elastic properties of a wafer is the premise of achieving the correct modal simulation of a structure fabricated on this wafer. In the practical simulation process, apart from inputting the stiffness matrix or the compliance matrix, employing orthotropic elasticity values of the wafer can be used to simulate resonant frequencies of the VRG [19].

Orthotropic elasticity values which contain Young's modulus E, Poisson ratio v, and the shear modulus G also depend on the direction cosines of coordinate axes of the wafer. These values can be given as

$$\frac{1}{E} = s_{11} - 2(s_{11} - s_{12} - \frac{1}{2}s_{44})(l^2m^2 + m^2n^2 + l^2n^2), \tag{7}$$

$$v = -\frac{s_{12} + (s_{11} - s_{12} - \frac{1}{2}s_{44})(l_1^2l_2^2 + m_1^2m_2^2 + n_1^2n_2^2)}{s_{11} - 2(s_{11} - s_{12} - \frac{1}{2}s_{44})(l_1^2m_1^2 + m_1^2n_1^2 + l_1^2n_1^2)}, \tag{8}$$

$$\frac{1}{G} = s_{44} + 4(s_{11} - s_{12} - \frac{1}{2}s_{44})(l_1^2l_2^2 + m_1^2m_2^2 + n_1^2n_2^2) \tag{9}$$

where (l_1, m_1, n_1) and (l_2, m_2, n_2) are two groups of direction cosines of two orthogonal directions with respect to the crystal axes [20,21].

In particular, direction cosines of any directions in the plane of the (111) wafer meet the conditions that $l + m + n = 0$ and $l^2 + m^2 + n^2 = 1$. Therefore, the conclusion can be derived that $l^2m^2 + l^2n^2 + m^2n^2 = 1/4$, which means Young's modulus in the plane of the (111) wafer is invariable. Further, another conclusion that $l_1^2l_2^2 + m_1^2m_2^2 + n_1^2n_2^2 = 1/6$ can be proved by using an additional condition that $l_1l_2 + m_1m_2 + n_1n_2 = 0$. As a result, the Poisson ratio as well as the shear modulus in the (111) plane are also irrelevant to the crystal directions. Therefore, the in-plane elasticity of the (111) wafer can be regarded as isotropic. It might be predicted that the frequency splits of the simulation and test of the (111) VRG would be very tiny, even zero. According to the above methodology, the orthotropic elasticity values of the (100) and (111) wafers are calculated and listed in Table 1, and these values could be used to simulate resonant frequencies of a VRG.

Table 1. Orthotropic elasticity values of (100) and (111) silicon wafers.

Orthotropic Elasticity Values	(100) Wafer	(111) Wafer
E_x, E_y, E_z (GPa)	169, 169, 130	169, 169, 188
v_{yz}, v_{zx}, v_{xy}	0.36, 0.28, 0.064	0.16, 0.18, 0.26
G_{yz}, G_{zx}, G_{xy} (GPa)	79.4, 79.4, 50.9	57.8, 57.8, 66.9

3. Working Principle and Modal Simulation of VRG

3.1. Working Principle

The working principle of the VRG is based on the wineglass mode. The vibrating pattern of the $n = 2$ mode is elliptical with four nodes, and two orthogonal modes are 45 degrees apart from each other. When the gyroscope is being driven at the $n = 2$ mode, once it is under the rotation, the generated Coriolis acceleration will force the energy to transfer between two orthogonal modes [22]. The angle of rotation of the VRG can be measured by two methods. The first method, called the whole angle pattern, utilizes the inertia of the vibrating modes to measure the angle of rotation directly [23]. In the rotation, the rotation of the vibrating mode lags behind that of the case, and the lagging angle is called the precession angle [24]. The precession angle is obtained by the product of the angle of rotation and the angular gain A_g. It is noticed that the angular gain is a fixed value that only depends on the geometry and the vibrating mode of a VRG, and has nothing to do with the lifetime, etc. [25]. The angular gain A_g defined as γ/nM is also the scale factor of the VRG used as a rate-integrating gyroscope [3]. Here, γ is the Coriolis mass, and M is the effective mass. The other method, named the force-to-rebalance pattern, can be used to measure the angle rate. In this pattern, the amplitude of the sense mode is driven to zero by control electrodes while the vibration of the VRG is only aligned with the drive mode, and then the angle rate can be figured out because the force produced by the control electrodes is proportional to the angle rate [22].

3.2. Modal Simulation

The VGR simulated and fabricated in this paper is based on the structure that has been designed and reported in Reference [26]. The VRG shown in Figure 2 consists of three elements: a ring, eight support beams, and an anchor.

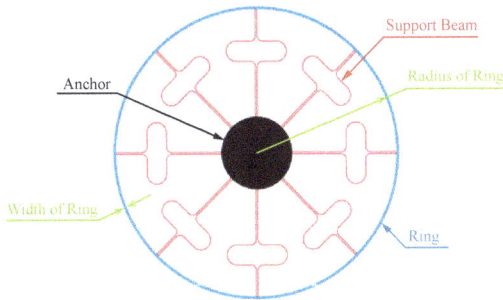

Figure 2. The schematic of the vibrating ring gyroscope (VRG).

Resonant frequencies of the $n = 2$ mode of the circular ring structure are crucial to those of the VRG, so it is necessary to simulate the effects of structural parameters of the circular ring on resonant frequencies and the frequency split of the $n = 2$ mode by COMSOL (v5.4). Since the thickness of the circular ring almost does not affect the resonant frequencies of the $n = 2$ mode, only the width and the radius are considered. In order to simulate the resonant frequencies of the circular ring of the (100) wafer, the **Anisotropic** option and the **Voigt** option are chosen in the lists of the **Solid model** section and the **Material data ordering** section, respectively, and the matrix in Equation (5) is typed in the table of **Elasticity matrix** in the **Settings** window for **Linear Elastic Material** in the interface of COMSOL. When it comes to the (111) wafer, the procedure is the same, but the matrix is changed to Equation (6).

3.2.1. Width of Circular Ring

The width plays an important role in the resonant frequencies of the circular ring because it contributes to the mass and the stiffness of the ring. The radius and thickness of the ring are fixed at 2500 μm and 120 μm, respectively, while the width distribution is from 10 μm to 100 μm, and the interval is 5 μm. Simulation results of different widths of (100) and (111) circular rings are shown in Figure 3.

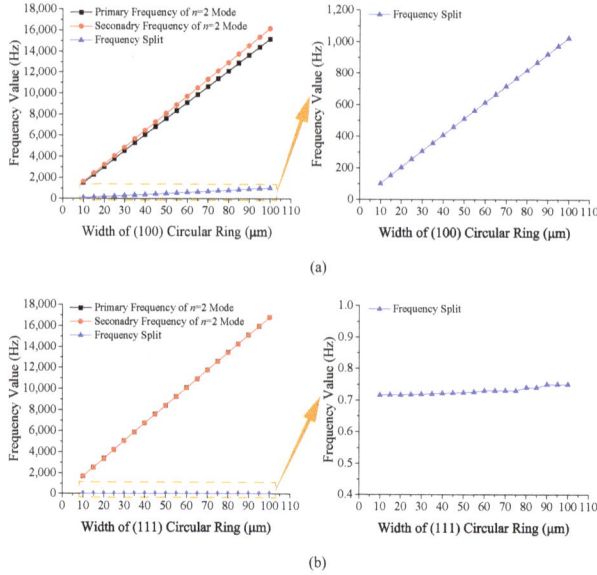

Figure 3. Resonant frequencies and frequency splits of rings with different widths: (**a**) (100) ring, (**b**) (111) ring.

It can be seen that the resonant frequencies of two types of rings increase linearly with the increase in width. The frequency split of the $n = 2$ mode of the (100) ring caused by the anisotropy is large, and it increases linearly as the width increases. However, the primary frequency of the $n = 2$ mode of the (111) circular ring is always almost the same as the secondary frequency, which means the frequency split of the (111) ring is very tiny and irrelevant to the width.

3.2.2. Radius of Circular Ring

Radius is the other factor that significantly impacts the frequency characters of the circular ring. In this case, the radius ranges from 2000 μm to 4000 μm with 21 points while the width remains at 60 μm, and the thickness remains at 120 μm. The resonant frequencies and frequency splits of two kinds of rings with different radii are shown in Figure 4.

As shown in Figure 4, the resonant frequencies of the (100) and (111) circular rings decrease as parabolic curves when the radius increases. It can be seen that the difference between the two resonant frequencies of the (100) ring drops obviously with the increment of the radius. However, the two curves of resonant frequencies of the (111) ring almost overlap, which means the frequency split is quite small.

According to the simulation above, the resonant frequencies of the (111) circular ring are always larger than those of the (100) circular ring at the same size, which indicates that the global in-plane stiffness of the (111) wafer is larger. Besides, the simulation results show that the frequency split of the (100) ring caused by the anisotropy is obvious, and it has a strong relationship with the geometry of the (100) ring while the (111) wafer is in-plane isotropic.

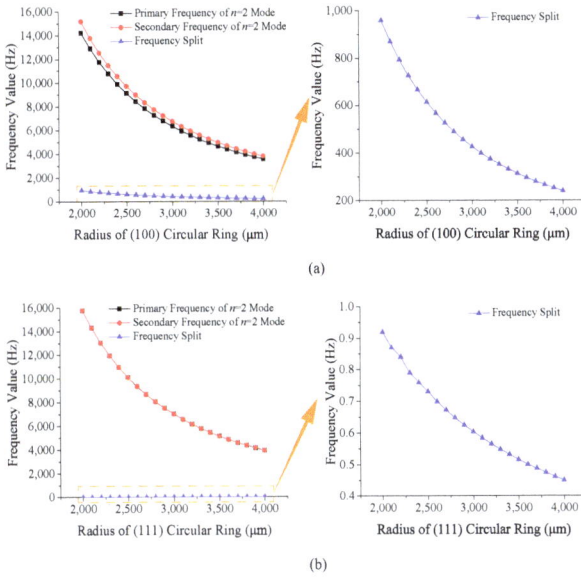

Figure 4. Resonant frequencies and frequency splits of rings with different radii: (**a**) (100) ring, (**b**) (111) ring.

In order to further study effects of different elastic properties of (100) and (111) wafers on the resonant frequencies of the VRG, a simulation of a VRG with specific structural parameters was performed. The resonant frequencies of the VRG are set at around 10 kHz, so the final sizes of the VRG are determined after taking into account the simulation results of the circular rings and some important structural parameters as shown in Table 2. According to the simulation using stiffness matrices, modal shapes and resonant frequencies of (100) and (111) VRGs are shown in Figure 5.

Table 2. Partial structure parameters of VRG.

Structure Parameter	Value (μm)
Radius of Ring	3000
Width of Ring	80
Radius of Anchor	750
Width of Support Beam	20
Thickness of Ring	120

According to Figure 5, the primary frequency f_1 of the $n = 2$ mode of the (100) VRG is 9739.3 Hz, and the secondary frequency f_2 is 10,144.1 Hz, so the frequency split $\Delta f_{(100)}$ between the two resonant frequencies is 404.8 Hz. The lower resonant frequency f_1 of the (111) VRG is 10,625.6Hz while the higher resonant frequency f_2 is 10,627.4 Hz, and thus the frequency split $\Delta f_{(111)}$ is 1.8 Hz. In summary, the (111) VRG has a few higher $n = 2$ mode resonant frequencies, and its frequency split is much smaller than that of the (100) VRG, which is consistent with the theory of elastic properties of the two silicon wafers and the simulation results of the circular ring. Then, it was verified by fabrication and the test.

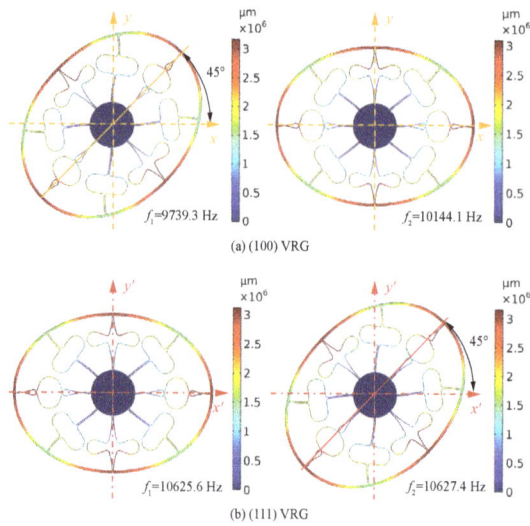

Figure 5. Modal shapes and resonant frequencies of (100) and (111) VRGs.

4. Fabrication and Test

The fabrication technology of the VRG is deep dry silicon on glass (DDSOG) process, which is a kind of mature bulk micromachining [27]. The silicon-glass bonding of this technology has the advantages of being a simple process, with small parasitic capacitance, and less structural stress.

The fabrication process is shown in Figure 6: (a) the photoresist is laid and photo-etched on the silicon wafer; (b) the bonding area of the silicon is formed by dry etching; (c) the metal is deposited in small grooves which are processed in the glass by wet etching; (d) the silicon wafer is bonded to the glass substrate by anodic bonding; (e) the silicon wafer is thinned and polished; (f) the structure of the VRG, of which the dimensions are shown in Table 2, is deeply etched by dry etching and then released.

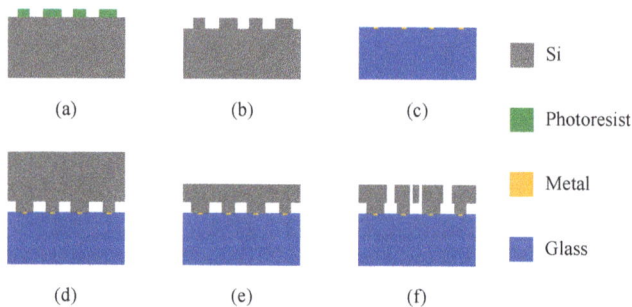

Figure 6. The fabrication process of VRG: (a) Photo-etching, (b) Bonding area etching, (c) Metal deposition, (d) Anodic bonding, (e) Thinning and polishing, (f) Dry etching and structure release.

After the fabrication, the structure of one VRG under the microscope is shown in the middle of Figure 7. The VRGs tested on the probe station in the picture were excited by applying an 8 V DC and a 6 V AC. A signal generator was used to produce sinusoidal signals with different frequencies, then resonant frequencies (f_1, f_2) of $n = 2$ mode of each VRG were obtained through the spectrum analyzer shown in Figure 7 by acquiring the peak value of second harmonic frequencies [28].

Besides, the amplitude-frequency curve which is displayed on the screen of the spectrum analyzer can be recorded by the soft disk. Figure 8 shows the tested second harmonic frequency data of four (100) VRG samples (#1, #2, #3, #4) while the frequency data of four (111) VRG samples (#1′, #2′, #3′, #4′) are shown in Figure 9. It should be noticed that the frequency corresponding to the peak value of each curve is twice the resonant frequency in Figures 8 and 9, so the difference between the two frequencies of one VRG is also double. The specific resonant frequency data of the (100) and (111) VRGs read from the signal generator as well as calculation results are represented in Table 3, where $|\Delta f|$ is the absolute value of f_1-f_2 while $\overline{|\Delta f|}$ and $\sigma(|\Delta f|)$ are the average and standard deviation values of the absolute values.

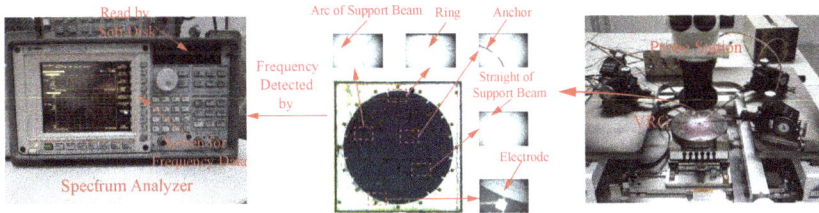

Figure 7. Microscope image of the VRG on the probe station.

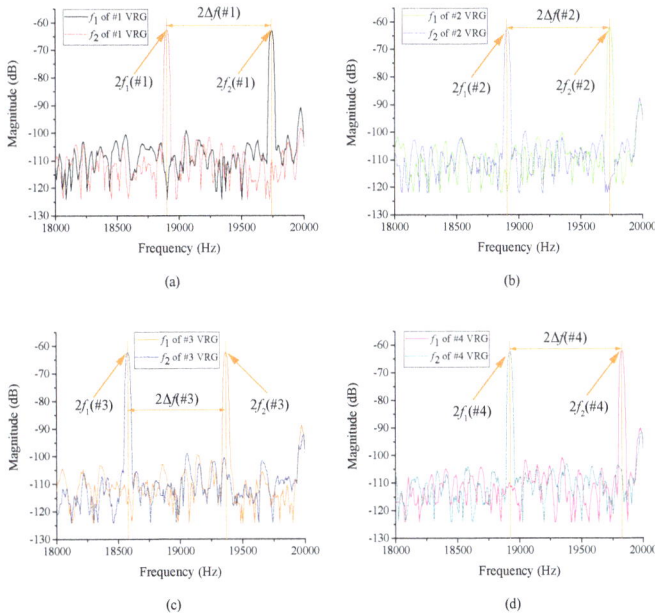

Figure 8. Second harmonic frequency data of four (100) VRG samples: (**a**) #1 VRG, (**b**) #2 VRG, (**c**) #3 VRG, (**d**) #4 VRG.

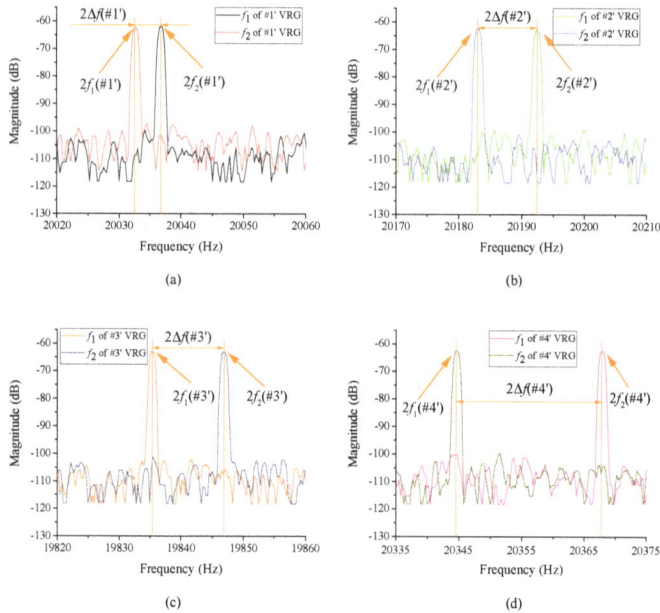

Figure 9. Second harmonic frequency data of four (111) VRG samples: (**a**) #1′ VRG, (**b**) #2′ VRG, (**c**) #3′ VRG, (**d**) #4′ VRG.

According to test results, practical resonant frequencies and frequency splits of the (100) and (111) VRGs are close to the simulation results. Existing differences between the test data and simulation results are mainly caused by over-etching in the fabrication. Over-etching usually reduces the widths of support beams and the circular ring, which leads to a decline in the resonant frequencies. Besides, fabricated (100) VRGs generally have very large frequency splits, and the average value is 419.6 Hz while frequency splits of fabricated (111) VRGs are around 5.9 Hz. It indicates that the frequency splits of the (100) VRGs are strongly affected by the anisotropy of the (100) wafer while the small frequency splits of the (111) VRGs benefit from the in-plane isotropy; only from this point of view, is the (111) silicon wafer the preferred material for the VRG.

Table 3. Frequency test data of (100) and (111) VRGs.

	(100) VRG				(111) VRG			
	#1	#2	#3	#4	#1′	#2′	#3′	#4′
f_1 (Hz)	9871.1	9865.7	9685.4	9913.2	10,018.4	10,096.2	9917.7	10,183.4
f_2 (Hz)	9451.3	9453.6	9290.5	9461.7	10,016.3	10,091.5	9923.4	10,172.3
$|\Delta f|$ (Hz)	419.8	412.1	394.9	451.5	2.1	4.7	5.7	11.1
$|\overline{\Delta f}|$ (Hz)		419.6				5.9		
$\sigma(|\Delta f|)$ (Hz)		23.7				3.8		

5. Conclusions

In the field of MEMS, SCS is widely used to fabricate high respect ratio VRGs along with being used in bulk silicon micromachining techniques; this is because of its outstanding mechanical and electrical properties. However, the anisotropy of SCS causes the elastic properties of the resonator to vary depending on the crystal directions, which influences the frequency characteristic of the SCS VRG. This paper presents how the anisotropy of SCS affects the frequency splits of VRGs, which can provide a method to calculate the elastic properties of various SCS wafers. Furthermore, this paper offers a basis with which to choose a suitable material to achieve VRGs of which the drive and sense modes are well matched.

In this paper, the coordinate systems of (100) and (111) SCS wafers are obtained by rotating the default coordinate system. Then, the elastic matrices and orthotropic elasticity values of these wafers are calculated using the direction cosines of transformed coordinate systems. It is found that Young's modulus, Poisson ratio, and the shear modulus of the (111) wafer are irrelevant to the crystal directions, so the frequency split of the (111) VRG should, in theory, be near zero. Next, resonant frequencies of the $n = 2$ mode of (100) and (111) circular rings with different sizes are simulated, and the simulation results show that the frequency split of $n = 2$ mode of the (100) ring is significantly influenced by the geometry, while the frequency split of the (111) ring is always very tiny as a result of the in-plane isotropy. After that, modal simulations of (100) and (111) VRGs are carried out, and frequency tests of fabricated VRGs are performed to detect the resonant frequencies on the probe station. The test results match the simulation results of the VRGs well to prove that (100) VRGs have much larger frequency splits than (111) VRGs, so the resonant frequencies of the $n = 2$ mode of (100) VRGs are more sensitive to the crystal direction, and (111) wafer is a suitable choice for the fabrication of VRGs with restrained frequency splits.

This paper analyzes the effect of the anisotropy of SCS on the frequency splits of VRGs, which can offer a reference for analyzing the elastic properties of VRGs fabricated using various SCS wafers. Future work will include the analysis of how the anisotropy of SCS influences other characteristics of the VRG, such as the quality factor.

Author Contributions: Conceptualization: Z.Q., L.H. and H.L.; Formal analysis: Z.Q.; Investigation: Z.Q.; Software: Y.G.; Validation: J.J.; Writing—review & editing: Z.Q., X.D.

Funding: This research received no external funding.

Conflicts of Interest: The authors declare no conflict of interest.

References

1. Shaeffer, D.K. MEMS inertial sensors: A tutorial overview. *IEEE Commun. Mag.* **2013**, *51*, 100–109. [CrossRef]
2. Kou, Z.; Liu, J.; Cao, H.; Han, Z.; Sun, Y.; Shi, Y.; Ren, S.; Zhang, Y. Investigation, modeling, and experiment of an MEMS S-springs vibrating ring gyroscope. *J. Micro/Nanolithography MEMS MOEMS* **2018**, *17*, 015001. [CrossRef]
3. Cho, J.Y. High-Performance Micromachined Vibratory Rate-and Rate-Integrating Gyroscopes. Ph.D. Thesis, University of Michigan, Ann Arbor, MI, USA, 2012.
4. Shkel, A.M.; Acar, C.; Painter, C. Two types of micromachined vibratory gyroscopes. In Proceedings of the Sensors, 2005 IEEE, Irvine, CA, USA, 30 October–3 November 2005; pp. 531–536.
5. Eley, R.; Fox, C.; McWilliam, S. Anisotropy effects on the vibration of circular rings made from crystalline silicon. *J. Sound Vib.* **1999**, *228*, 11–35. [CrossRef]
6. Nitzan, S.H.; Taheri-Tehrani, P.; Defoort, M.; Sonmezoglu, S.; Horsley, D.A. Countering the effects of nonlinearity in rate-integrating gyroscopes. *IEEE Sens. J.* **2016**, *16*, 3556–3563. [CrossRef]
7. Senkal, D.; Askari, S.; Ahamed, M.; Ng, E.; Hong, V.; Yang, Y.; Ahn, C.H.; Kenny, T.; Shkel, A. 100K Q-factor toroidal ring gyroscope implemented in wafer-level epitaxial silicon encapsulation process. In Proceedings of the 2014 IEEE 27th International Conference on Micro Electro Mechanical Systems (MEMS), San Francisco, CA, USA, 26–30 January 2014; pp. 24–27.
8. Ren, J.; Liu, C.Y.; Li, M.H.; Chen, C.C.; Chen, C.Y.; Li, C.S.; Li, S.S. A mode-matching 130-kHz ring-coupled gyroscope with 225 ppm initial driving/sensing mode frequency splitting. In Proceedings of the Transducers-2015 18th International Conference on Solid-State Sensors, Actuators and Microsystems (TRANSDUCERS), Anchorage, AK, USA, 21–25 June 2015; pp. 1057–1060.
9. Ahn, C.H.; Ng, E.J.; Hong, V.A.; Yang, Y.; Lee, B.J.; Flader, I.; Kenny, T.W. Mode-matching of wineglass mode disk resonator gyroscope in (100) single crystal silicon. *J. Microelectromech. Syst.* **2015**, *24*, 343–350. [CrossRef]

10. Chang, C.O.; Chang, G.E.; Chou, C.S.; Chien, W.T.C.; Chen, P.C. In-plane free vibration of a single-crystal silicon ring. *Int. J. Solids Struct.* **2008**, *45*, 6114–6132. [CrossRef]

11. Wang, D.F.; Ono, T.; Esashi, M. Crystallographic influence on nanomechanics of (100)-oriented silicon resonators. *Appl. Phys. Lett.* **2003**, *83*, 3189–3191. [CrossRef]

12. Ghaffari, S.; Ahn, C.H.; Ng, E.J.; Wang, S.; Kenny, T.W. Crystallographic effects in modeling fundamental behavior of MEMS silicon resonators. *Microelectron. J.* **2013**, *44*, 586–591. [CrossRef]

13. Wei, X.; Seshia, A.A. Analytical formulation of modal frequency split in the elliptical mode of SCS micromechanical disk resonators. *J. Micromech. Microeng.* **2014**, *24*, 025011. [CrossRef]

14. He, G.; Najafi, K. A single-crystal silicon vibrating ring gyroscope. In Proceedings of the Fifteenth IEEE International Conference on Micro Electro Mechanical Systems, Las Vegas, NV, USA, 20–24 January 2002; Volume 2024, p. 718721.

15. Johari, H.; Ayazi, F. High-frequency capacitive disk gyroscopes in (100) and (111) silicon. In Proceedings of the IEEE 20th International Conference on Micro Electro Mechanical Systems, Hyogo, Japan, 21–25 January 2007; pp. 47–50.

16. Auld, B.A. *Acoustic Fields and Waves in Solids*; Wiley: Hoboken, NJ, USA, 1973.

17. Gad-el Hak, M. *The MEMS Handbook*; CRC Press: Boca Raton, FL, USA, 2001.

18. Wortman, J.; Evans, R. Young's modulus, shear modulus, and Poisson's ratio in silicon and germanium. *J. Appl. Phys.* **1965**, *36*, 153–156. [CrossRef]

19. Hopcroft, M.A.; Nix, W.D.; Kenny, T.W. What is the Young's Modulus of Silicon? *J. Microelectromech. Syst.* **2010**, *19*, 229–238. [CrossRef]

20. Brantley, W. Calculated elastic constants for stress problems associated with semiconductor devices. *J. Appl. Phys.* **1973**, *44*, 534–535. [CrossRef]

21. Zhang, L.; Barrett, R.; Cloetens, P.; Detlefs, C.; Sanchez del Rio, M. Anisotropic elasticity of silicon and its application to the modelling of X-ray optics. *J. Synchrotron Radiat.* **2014**, *21*, 507–517. [CrossRef] [PubMed]

22. Putty, M.W. A Micromachined Vibrating Ring Gyroscope. Ph.D. Thesis, University of Michigan, Ann Arbor, MI, USA, 1995.

23. Gallacher, B.J. Principles of a micro-rate integrating ring gyroscope. *IEEE Trans. Aerosp. Electron. Syst.* **2012**, *48*, 658–672. [CrossRef]

24. Challoner, A.D.; Howard, H.G.; Liu, J.Y. Boeing disc resonator gyroscope. In Proceedings of the Position, Location and Navigation Symposium-PLANS 2014, Monterey, CA, USA, 5–8 May 2014; pp. 504–514.

25. Ayazi, F. A High Aspect-Ratio High-Performance Polysilicon Vibrating Ring Gyroscope. Ph.D. Thesis, University of Michigan, Ann Arbor, MI, USA, 2000.

26. Yoon, S.; Park, U.; Rhim, J.; Yang, S.S. Tactical grade MEMS vibrating ring gyroscope with high shock reliability. *Microelectron. Eng.* **2015**, *142*, 22–29. [CrossRef]

27. Gao, Y.; Huang, L.; Ding, X.; Li, H. Design and Implementation of a Dual-Mass MEMS Gyroscope with High Shock Resistance. *Sensors* **2018**, *18*, 1037. [CrossRef] [PubMed]

28. Xu, L.; Li, H.; Liu, J.; Ni, Y.; Huang, L. Research on nonlinear dynamics of drive mode in z-axis silicon microgyroscope. *J. Sens.* **2014**, *2014*, 801618. [CrossRef]

micromachines

MDPI

Article

Evolution of Si Crystallographic Planes-Etching of Square and Circle Patterns in 25 wt % TMAH

Milče M. Smiljanić [1,*], Žarko Lazić [1], Branislav Radjenović [2], Marija Radmilović-Radjenović [2] and Vesna Jović [1]

[1] Institute of Chemistry, Technology and Metallurgy-Centre of Microelectronic Technologies (IHTM-CMT), University of Belgrade, Njegoševa 12, 11000 Belgrade, Serbia; zlazic@nanosys.ihtm.bg.ac.rs (Z.L.); vjovic@nanosys.ihtm.bg.ac.rs (V.J.)
[2] Institute of Physics, University of Belgrade, Pregrevica 118, 11080 Belgrade, Serbia; bradjeno@ipb.ac.rs (B.R.); marija@ipb.ac.rs (M.R.-R.)
*   Correspondence: smilce@nanosys.ihtm.bg.ac.rs; Tel.: +381-112-630-757

Received: 20 December 2018; Accepted: 26 January 2019; Published: 31 January 2019

check for updates

Abstract: Squares and circles are basic patterns for most mask designs of silicon microdevices. Evolution of etched Si crystallographic planes defined by square and circle patterns in the masking layer is presented and analyzed in this paper. The sides of square patterns in the masking layer are designed along predetermined <n10> crystallographic directions. Etching of a (100) silicon substrate is performed in 25 wt % tetramethylammonium hydroxide (TMAH) water solution at the temperature of 80 °C. Additionally, this paper presents three-dimensional (3D) simulations of the profile evolution during silicon etching of designed patterns based on the level-set method. We analyzed etching of designed patterns in the shape of square and circle islands. The crystallographic planes that appear during etching of 3D structures in the experiment and simulated etching profiles are determined. A good agreement between dominant crystallographic planes through experiments and simulations is obtained. The etch rates of dominant exposed crystallographic planes are also analytically calculated.

Keywords: tetramethylammonium hydroxide (TMAH); wet etching; silicon; 3D simulation; level-set method

1. Introduction

Anisotropic wet etching of a (100) silicon substrate in 25 wt % tetramethylammonium hydroxide (TMAH) water solution is a well-known process [1–11]. Most of the results have been obtained through the analysis of the etching square or rectangular patterns in the masking layer with sides along <110> crystallographic direction. During etching of these mesa structures, a severe convex corner undercutting appears. The convex corner compensation techniques for the TMAH water solutions etching are developed for the patterns with sides along <110> crystallographic directions [12–22]. Additionally, etched circular patterns were compared for pure and surfactant added 25 wt % TMAH water solution [23]. When a (100) silicon substrate is etched through the aperture in the masking layer that only has concave angles, the structure shape will be defined after a sufficiently long etching time with the slowest etching crystallographic planes {111}. Appearance of {111} is inevitable and cannot be eliminated.

Previous studies of silicon wet etching processes [1–32] were conducted using various etching solutions of KOH and TMAH at different temperatures and silicon wafers of various crystallographic orientations. Etching of a (100) silicon in the KOH water solution using circular mask patterns, as well as square, rectangular and octagonal mask patterns with sides along different crystallographic directions was explored in References [2,26–32]. Etching of circular and square patterns with sides along <100> crystallographic directions was analyzed in References [26–28]. Etching of squares with sides along <210> and <310> crystallographic directions was studied to develop triangular corner compensation structure [31]. Etching of octagonal patterns with sides along <210>, <310> and <410> crystallographic directions was discussed in References [28,29] for KOH, and in Reference [9] for TMAH water solutions. Additionally, etching of squares [32] and rectangles [30] was analyzed as their side directions change angle with <110> crystallographic direction. A special case study considered changing of the obtuse angle with an increment of 2° [24]. In this paper, we explore etching of square patterns that are designed along predetermined <n10> crystallographic directions ($0 \leq n < 10$) and circle patterns in the masking layer on a (100) silicon. Squares (rectangles) and circles are basic patterns for a great majority of mask designs in micromachining of membranes, bosses, convex corner compensations, etc. We analyze etching of square and circle islands in the masking layer using both experiments and simulations, and provide a comprehensive insight into the evolution of patterns for different crystallographic directions. Etched silicon structures are limited by various planes during etching in 25 wt % TMAH water solution at a temperature of 80 °C. Because of the differences in etch rates, some planes will appear, while others will disappear during etching [1–24]. We performed our experiments for the etching depths of up to 300 μm.

Our aim is to observe and analyze the appearance of various crystallographic planes and verify agreement of simulation with experimental results. Knowing evolution of crystallographic planes during etching is necessary for the successful mask design of silicon microdevices. Simulated etching profiles are obtained by the level-set (LS) method [33–40]. This method for evolving interfaces belongs to the geometric type of methods, and it is specially designed for profiles that can develop sharp corners, change of topology and undergo orders of magnitude changes in speed. All simulations are performed using a three-dimensional (3D) anisotropic etching simulator based on the sparse field method for solving the level-set equations, described in our previous publications [13,34–37]. Pictures of the simulated etching profiles are rendered with the Paraview visualization package [41]. The scanning electron microscopy (SEM) micrographs for several subsequent etch depths and simulated etching profiles are presented to demonstrate evolution of all exposed crystallographic planes. Additionally, we derive relation between parameters to calculate etch rates of the exposed planes. The etch rates are determined by measuring change of the side *a* of square island in the masking layer with time.

2. Experimental Setup

We used phosphorus-doped {100} oriented 3" silicon wafers (Wacker, Burghausen, Germany) with mirror-like single side polished surfaces and 1–5 Ωcm resistivity. Anisotropic etching was performed in pure TMAH 25 wt % water solution (Merck, Darmstadt, Germany). The etching temperature was 80 °C. Wafers were standard Piranha and RCA cleaned and covered with SiO_2 thermally grown at 1100 °C in an oxygen ambient saturated with water vapour (at least 1 μm thick). Cleaning of Si wafers before oxidation was accomplished by using freshly-prepared mixture of concentrated sulfuric acid (H_2SO_4, 95–98%) and hydrogen peroxide (H_2O_2, 30%). Mixture had volume ratio 3:1 (H_2SO_4:H_2O_2). RCA processing steps used a mixture of ammonium hydroxide (NH_4OH, 29%), hydrogen peroxide and water (1:1:5) and a mixture of hydrochloric acid (HCl, 37%), hydrogen peroxide and water (1:1:6). SiO_2 was etched in buffered hydrofluoric acid (BHF) in a photolithographic process in order to define square patterns along predetermined crystallographic directions and circle patterns. BHF solution consisted of 7 parts by volume of ammonium fluoride (NH_4F, 40%) and 7 parts by volume of hydrofluoric acid (HF, about 50%.) Again, wafers were subjected to standard cleaning procedure and were dipped before etching for 30 s in HF (10%) to remove native SiO_2 followed by rinsing in deionized water. Etching of the whole 3" wafer was carried out in a thermostated glass vessel containing around 0.8 dm^3 of the solution with electronic temperature controller stabilizing temperature within ± 0.5 °C. The vessel was on the top of a hot plate and closed with a Teflon lid that included a water-cooled condenser to minimize evaporation during etching. The wafer was oriented vertically in a Teflon basket inside the glass vessel. Throughout the process, the solution was electromagnetically stirred with a velocity of 300 rpm. After reaching the desired depth, the wafer was rinsed in deionized water and dried with nitrogen.

3. Discussion and Results

3.1. Square Islands in the Masking Layer with Sides Along <n10> Crystallographic Directions

Various 3D shapes were obtained by etching of square islands in the masking layer with sides along <n10> crystallographic directions. At the beginning of etching, sides of 3D structure aligned to <n10> direction belonging to {n11} crystallographic planes. The convex corners were undercut by fast etching crystallographic planes. As etching continued, for almost all square patterns, fast etching crystallographic planes dominated. Convex corners of squares differed from case to case, and their shapes changed with etching depth. Some of them were defined with rugged, others were defined with smooth crystallographic planes.

In the case of square island with sides along <110> directions, the slowest etching planes {111} defined sidewalls of silicon structure. There was a severe convex corner undercutting. Convex corner is defined by the rugged planes of {m0n} families and smooth planes of {311} and {211} families [13,24,42], as shown in Figure 1a. Observing etched shapes obtained in the experiments, we conjectured that the rugged planes of {m0n} families belong to the {301} or {401} families because the average angles γ_{m0n} between {m0n} planes and (100) plane are close to 71° and 75°. We also noticed in SEM micrographs that most of the {m0n} planes consist of smooth upper and rugged lower part, as shown in Figure 2a. In the simulated etching profiles there are two different planes instead of one from the {m0n} family. Average angles of intersections of these {m0n} planes and (100) silicon plane are close to 75° and 55.8°. We assume that they belong to the {401} and {203} families, where the {401} plane is closer to the masking layer and the {203} plane to the etched bottom. The appearance of two planes in the simulation is likely to be related to the smooth upper and rugged lower parts of the {m0n} planes in the SEM micrographs. Additionally, at the bottom of the {m0n} planes, there are two almost imperceptible symmetrical planes, as shown in Figure 2a. We assume that they belong to the {331} family. As etching continues, the plane of {111} family on the sidewall of structure disappears and two smooth planes of {211} family from the nearby convex corner define new convex corner, as shown in Figure 1a. Two smooth planes of {211} family are slightly undercut by planes that are hard to determine. Further etching will not change the silicon structure defined by the new convex corners and undercut convex corners.

In the case of square with sides along <100> directions, etching planes of the {100} family define sidewalls of the silicon structure, as shown in Figure 1b. As in the previous case, there is severe undercutting of the convex corner but only by the smooth planes of {311} family, as shown in Figure 1b. As etching continues, the {100} plane on the sidewall of structure disappears and two planes of {311} family from the nearby convex corners adjoin. At their cross section, new fast etching rugged planes from presumed {301} (or {401}–{203}) and {331} families appear, as shown in Figure 1b. The new shape of the silicon structure with eight convex corners sustains during the further etching. Similar shape was obtained in Reference [9] when etching with an octagonal mask in the pure TMAH 25 wt % water solution. The authors observed that the planes of {311} family are inclined only to the surface, while the planes of {331} family are at the bottom.

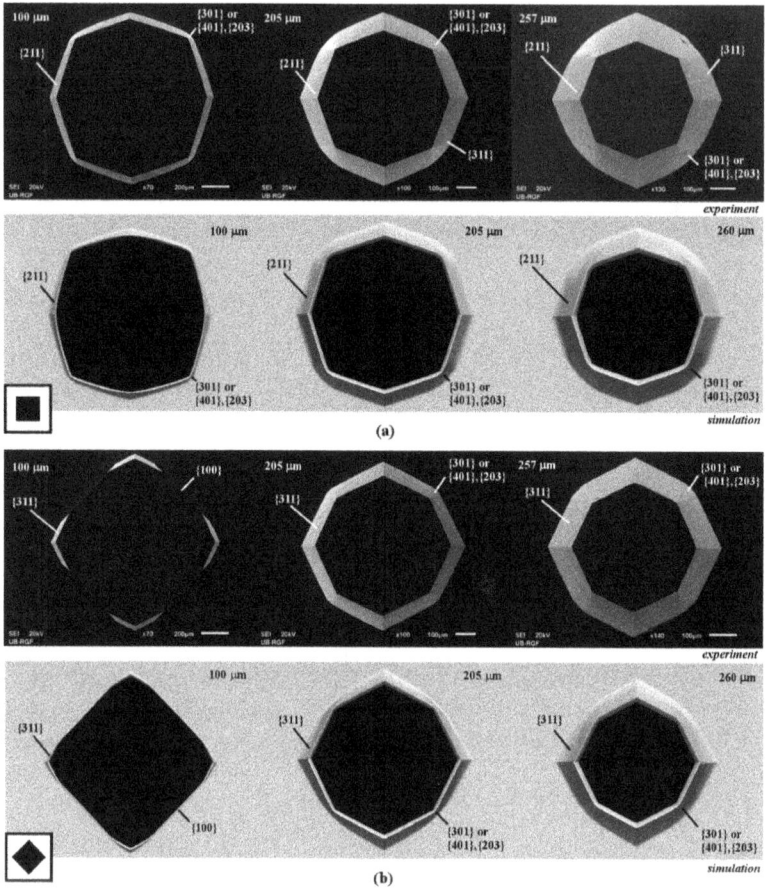

Figure 1. Shematic mask patterns, scanning electron microscopy (SEM) micrographs and simulated etching profiles of the etched square island with sides along: (**a**) <110> directions; (**b**) <100> directions. In the experiment the depths of etching were 100, 205 and 257 μm. In the simulation the depths of etching were 100, 205 and 260 μm.

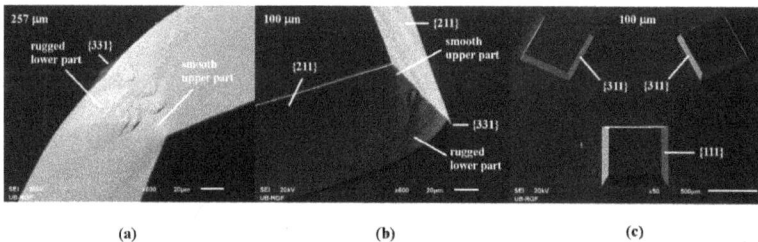

Figure 2. SEM micrographs: (**a**) Enlarged detail of the smooth upper and rugged lower part of undercut convex corner of etched square island with sides along <110> directions; (**b**) Convex corners undercut asymmetrically by the smooth upper and rugged lower part in the case of etched square island with sides along <210> directions; (**c**) Etched silicon structures from square island with sides along <110> directions with convex corner compensation and two symmetrical square islands with sides along <310> directions in the masking layer.

3D shapes obtained by etching of square islands with sides along <210> and <310> crystallographic directions are very similar, as shown in Figure 3. If the sides are along <210> directions, at the beginning of etching obtained shape is a truncated pyramid with sides defined by the planes of {211} family. In Reference [9], the planes of {211} family are inclined to the surface, and the planes of {221} family are at the bottom. These planes were observed at the etched octagonal mask shape with sides along <210> crystallographic directions in pure TMAH 25 wt % water solution. Convex corner undercutting appears, as shown in Figure 3a. Convex corner is defined by the rugged planes of {301} (or {401}–{203}) and {331} families and the smooth planes of {211} family, as shown in Figure 2b. This asymmetrical undercutting is less destructive than in the case of square patterns with sides along <110> and <100> directions due to smaller differences of etch rates [13]. After a sufficiently long etching time, the shape will be changed into truncated pyramid defined by {311} planes. Convex corner's shape will not be changed. If the sides are along <310> directions, the pyramidal shape with sides defined by the planes of {311} family is formed from the beginning of etching. In Reference [9], the planes from {311} family are inclined to the surface, and planes of {331} family are at the bottom in the case of octagone sides along <310> crystallographic directions. As in the previous case, there is also asymmetrical undercutting of the convex corner by the rugged planes of {301} family (or {401}–{203} families), as shown in Figure 3b. Additionally, at the bottom, there is almost imperceptible plane from the presumed {331} family. This 3D shape will not be changed during further etching.

Figure 3. Schematic mask patterns, SEM micrographs and simulated etching profiles of the etched square island with sides along: (**a**) <210> directions; (**b**) <310> directions. In the experiment the depths of etching were 100, 205 and 257 µm. In the simulation the depths of etching were 100, 205 and 260 µm.

We observed that the 3D shapes obtained in the cases of etching of square islands with sides along <410>, <510> and <610> crystallographic directions are also very similar, as shown in Figures 4 and 5a. In the case of the square with sides along <410> directions, at the beginning of etching pyramidal shape with sides defined by planes of {411} family is obtained. Same planes were also noticed in Reference [9] for the octagone with sides along <410> crystallographic directions. Convex corner undercutting appears again, as shown in Figure 4a. Convex corner is defined by the smooth planes of {311} and {411} families. Neither planes of {301} nor the planes of {401}–{203} families appear. As etching continues, the {411} plane on the sidewall of pyramid disappears and smooth {311} plane from the nearby convex corner starts to dominate, as shown in Figure 4a. The shape will be changed into truncated pyramid after a sufficiently long etching time as in the case of square island with sides along <310> crystallographic directions. Evolution of the etched silicon structure will be the same in the cases of square islands with sides along <510> and <610> crystallographic directions, as shown in Figures 4b and 5a. At the beginning, the pyramid sidewalls will be the planes of {511} and {611} families, respectively. Convex corners are defined by the smooth planes of {311} and {511} or {311} and {611} families, as shown in Figures 4b and 5a. As in the previous case, the {311} planes dominate after a sufficiently long etching time. Again, pyramidal shapes are obtained as in the case of square island with sides along <310> crystallographic directions. It can be noticed that planes of {411}, {511} and {611} families are not as smooth as planes of {211} and {311} families. It looks like that they consist of consecutive facets of negligible areas.

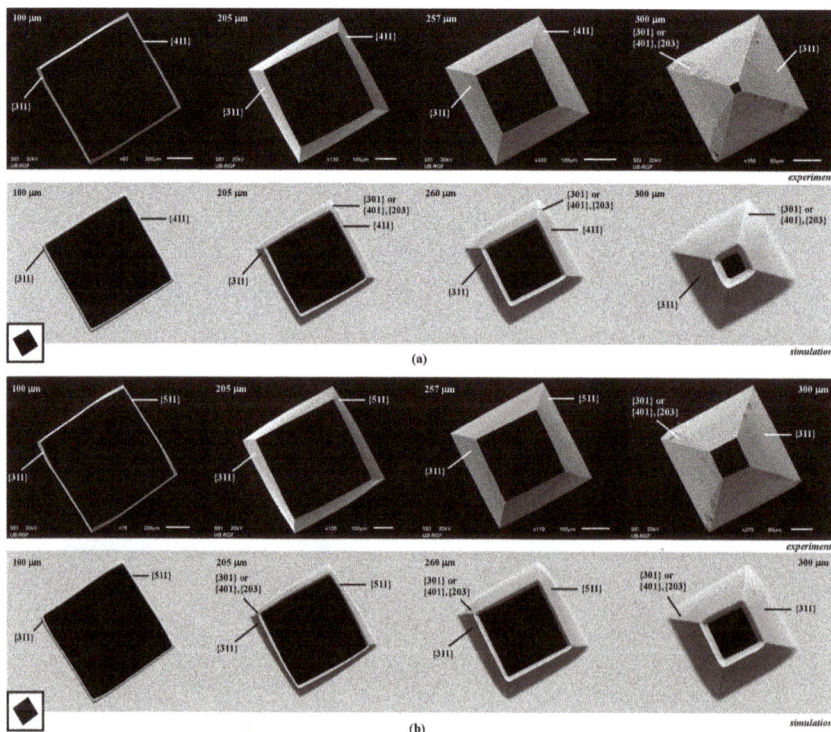

Figure 4. Schematic mask patterns, SEM micrographs and simulated etching profiles of the etched square island with sides along: (**a**) <410> directions; (**b**) <510> directions. In the experiment the depths of etching were 100, 205, 257 and 300 μm. In the simulation the depths of etching were 100, 205, 260 and 300 μm.

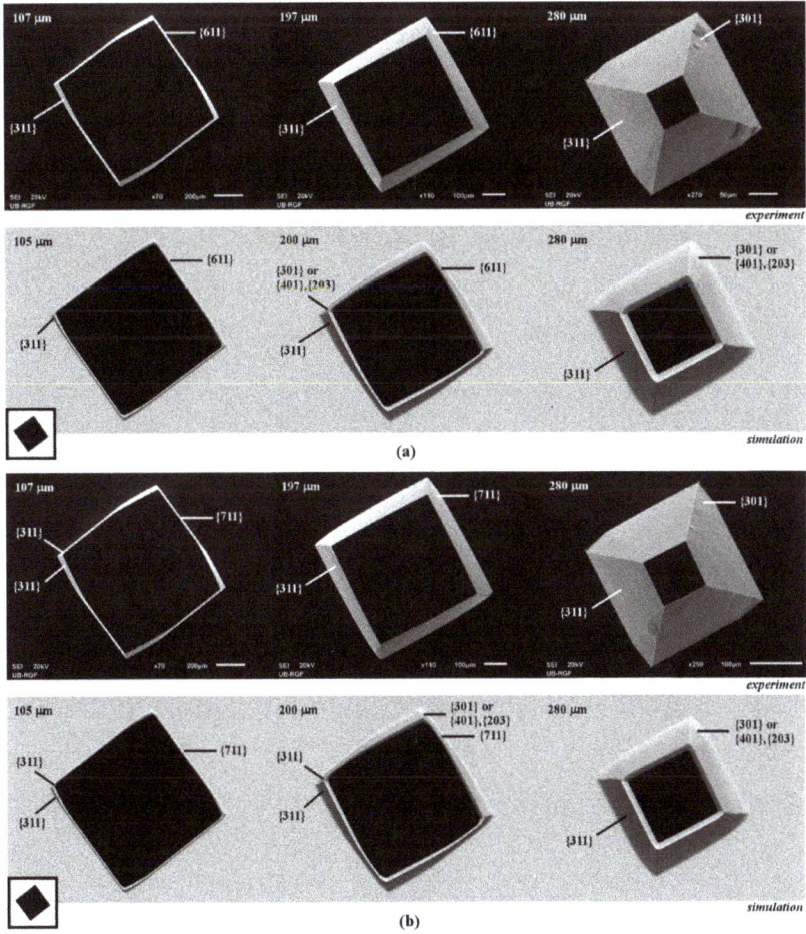

Figure 5. Schematic mask patterns, SEM micrographs and simulated etching profiles of the etched square island with sides along: (**a**) <610> directions; (**b**) <710> directions. In the experiment the depths of etching were 107, 197 and 280 μm. In the simulation the depths of etching were 105, 200 and 280 μm.

Other similarities can be noticed when observing the 3D shapes obtained by etching of the square islands with sides along <710>, <810> and <910> crystallographic directions, as shown in Figures 5b and 6. In the case of the square with sides along <710> directions, the pyramidal shape with sides defined by the planes of {711} family is obtained at the beginning of etching, as shown in Figure 5b. Asymmetrical convex corner's undercut is done by two smooth {311} planes, as in the case of square island with sides along <100> directions, as shown in Figure 1b. At the cross section of the {711} plane and {311} plane with a smaller area, the rugged planes from presumed {301} (or {401}–{203}) and {331} families appear. As etching continues, the {711} and 'smaller' {311} planes on the pyramid sidewall disappear and smooth {311} plane from the nearby convex corner starts to dominate, as shown in Figure 5b. Evolution of the similar shapes is observed in the cases of square islands with sides along <810> and <910> crystallographic directions, as shown in Figure 6. At the beginning of etching, the pyramid sidewalls will be the planes of the {811} and {911} families, respectively. Convex corners are defined by two smooth planes of {311} family in both cases, as shown in Figure 6. As in the previous case, one of the {311} planes takes over after a sufficiently long etching time. In all three cases, etched silicon structures become pyramids defined by {311} planes with convex corners undercut asymmetrically by the rugged planes of {301} (or {401}–{203}) and {331} families, as shown in Figures 5b and 6. It can be noticed that planes of {711}, {811} and {911} families also consist of consecutive facets of negligible areas.

The appearance of the planes directly under the masking layer can be noticed in all figures of simulated etching profiles [13]. These planes have smaller surface areas than the dominant ones and form shapes resembling ship prows. The planes obtained in simulation are more round and the edges of the convex corners tend to soften, so it is difficult to determine the orientation of small etched surfaces. There is a good agreement between dominant crystallographic planes obtained through experiments and simulations.

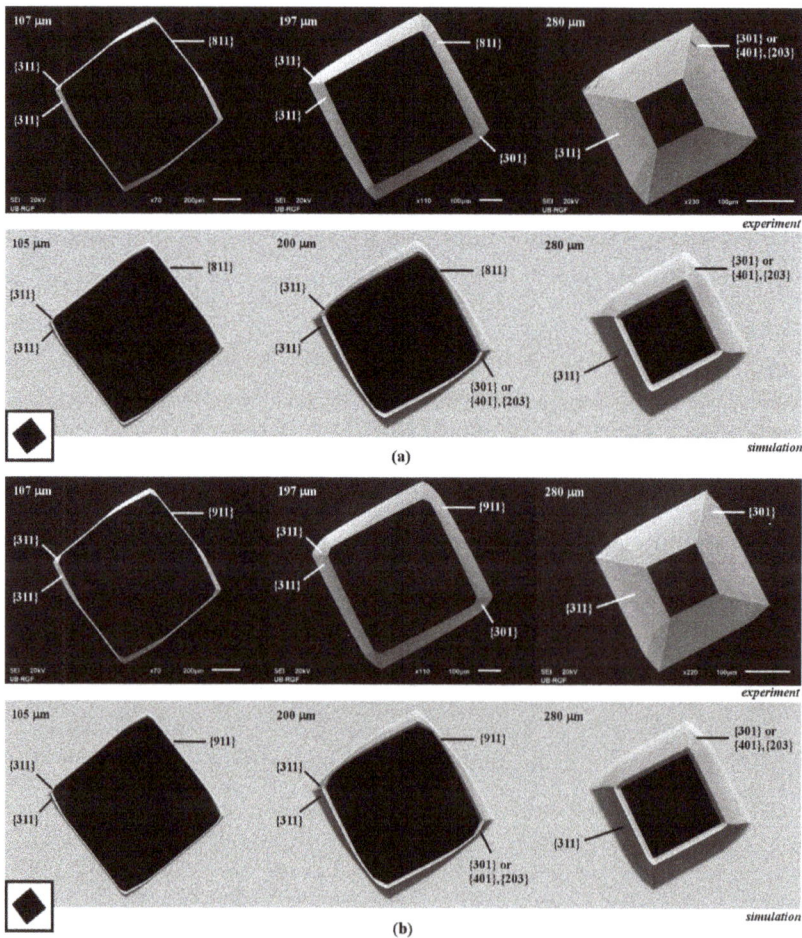

Figure 6. Schematic mask patterns, SEM micrographs and simulated etching profiles of the etched square island with sides along: (**a**) <810> directions; (**b**) <910> directions. In the experiment the depths of etching were 107, 197 and 280 μm. In the simulation the depths of etching were 105, 200 and 280 μm.

The most important result is the case of square island with sides along <310> crystallographic directions. In this case, the 3D shape will not be changed during etching. The convex corner compensation is not necessary as undercutting of {301} planes is not so severe. Together with square island with sides along <110> crystallographic directions (where convex corner compensation is applied), it could be used for future designs of sensors and actuators, as shown in Figure 2c. After a sufficiently long etching time, all square islands with sides along <n10> crystallographic directions, where n>1, become silicon truncated pyramids defined by {311} planes with convex corners undercut asymmetrically by the rugged planes of {301} (or {401}–{203}) and {331} families.

3.2. Square Apertures in the Masking Layer with Sides Along <n10> Crystallographic Cirections

3D shapes obtained by etching of square apertures in the masking layer with sides along <n10> crystallographic directions also have some similarities. All obtained cavities have a bottom that is determined by {100} plane of etched silicon substrate. Sides of squares aperture aligned to the <n10> direction allow developing of {n11} crystallographic planes at the beginning of etching. The concave corners are defined by the slowest etching planes, as shown in Figure 7.

The two most studied cavities were fabricated by etching square apertures in the masking layer with sides along <110> and <100> crystallographic directions. In the first case, sidewalls of a cavity are defined by the slowest etching planes {111}. In the second case, sidewalls of a cavity are defined by the planes of {100} and {111} families. After a sufficiently long etching time, the cavity will be changed into an inverse pyramid with the sides that are defined by {111} planes, as in the first case.

In all other cases of cavities obtained by etching of the square apertures with sides along <n10> crystallographic directions (1 < n < 10), the initial right concave corners are turned into three new concave corners, as shown in Figure 7. The first new concave corner is defined by the slowest etching planes {111} and {n11}. Appropriate <110> and <n10> crystallographic directions form angle smaller than 45° (in fact, concave angle in the masking layer larger than 145°). The second concave corner is defined by planes of {n11} and {100} families. The third concave corner is defined by the planes of {100} and {111} families. Appropriate <n10> crystallographic direction form the concave angle with <110> direction in the masking layer larger than concave angle formed by <110> and <100> crystallographic directions. After a sufficiently long etching time, cavities will have shape as in the case of etching square aperture in the masking layer with sides along <100> direction. Further etching will produce inverse pyramid with sides that are defined by {111} planes.

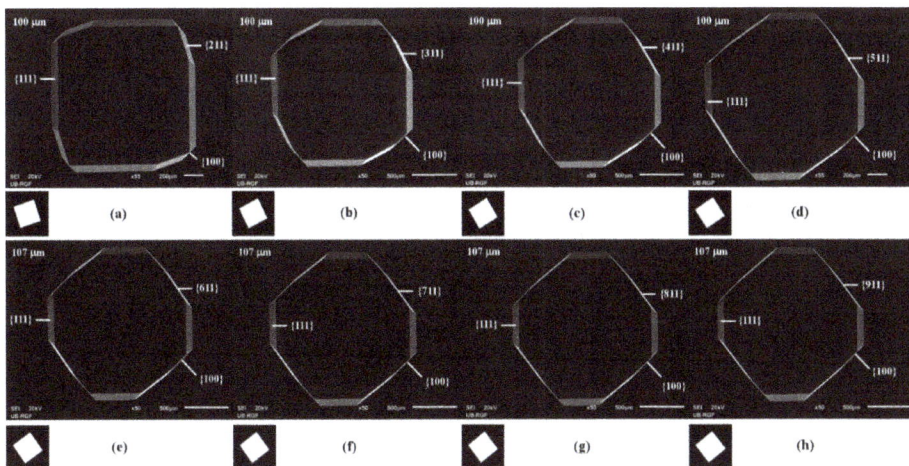

Figure 7. Schematic mask patterns, SEM micrographs of the etched square apertures: (**a**) <210>; (**b**) <310>; (**c**) <410>; (**d**) <510>; (**e**) <610>; (**f**) <710>; (**g**) <810>; (**h**) <910>. The etching depths were 100 and 107 μm.

3.3. Circle Island and Aperture in the Masking Layer

The 3D shape obtained by etching of circle island in masking layer is similar to the shapes obtained in the case of square islands with sides along <110> and <100> directions after a sufficiently long etching time, as shown in Figure 8a. At the beginning of etching, the first type of convex corner is defined by the rugged planes of {301} family (or {401}–{203} families) and the smooth planes of {311} families, as shown in Figure 8a. The second type of convex corner is defined by two smooth planes of {211} family. As etching continues, the plane of {211} family on the sidewall of pyramid disappears and smooth plane of {311} family from the nearby convex corner takes over. Further etching will not change the silicon structure defined by these convex corners.

The cavity obtained by etching of the circle aperture in the masking layer is similar to the cavity obtained by etching of the square aperture with sides along <100> crystallographic directions. At the beginning of the etching, the transition of plane {100} to {111} at the concave corner is not so abrupt as the transition at concave corner of the square aperture, as can be observed from Figure 8b. As in other cases of apertures, after a sufficiently long etching time, the cavity will become an inverse pyramid with the sides defined by the slowest etching planes of {111} family.

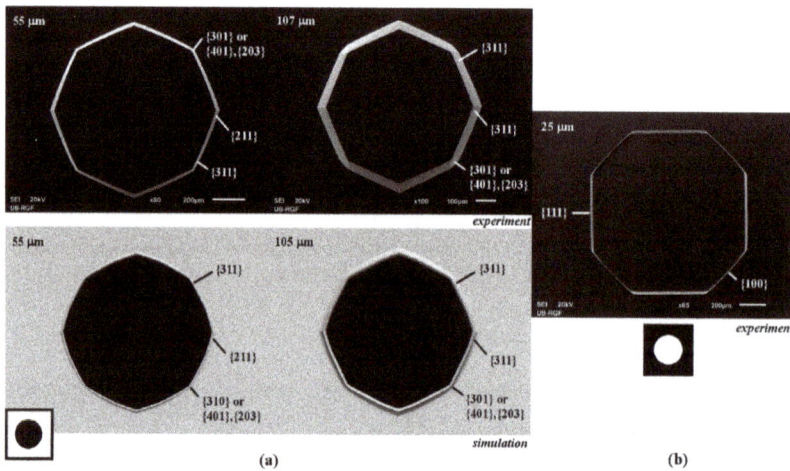

Figure 8. Schematic mask patterns, SEM micrographs and simulated etching profiles of the etched circle: (**a**) island; (**b**) aperture. In the experiment the depths of etching for island are 55 and 107 μm. In the experiment the etching depth for aperture was 25 μm. In the simulation the depths of etching were 55 and 105 μm.

3.4. Etch Rates of Exposed Planes

Table 1 includes the etch rates for all dominant exposed crystallographic planes during etching that have not been considered in our previous work [12,13]. The etch rate of {100} plane is 0.46 μm/min. The etch rates are determined indirectly by measuring change of the square side a_{n11} of the island in the masking layer with time for both the experiment and the simulated etching profile [9], as shown in Figure 9:

$$r_{n11} = (\Delta a_{n11}/\Delta t) \sin \gamma_{n11} \tag{1}$$

where

$$\Delta t = t_2 - t_1, \Delta a = |a_2 - a_1|$$

and where t_1 and t_2 are two subsequent moments of the etching, a_1 and a_2 are the sides of the square in the <n10> direction for the moments t_1 and t_2 and γ_{n11} is the angle between {n11} and (100) planes. All angles used in (1) are given in Table 1. Table 1 gives insight into derivations of average angles γ_{n11} and etch rates which are measured in the experiments and simulations. The second column are the theoretical angles, while the third column represents the input etch rates for simulation or the numerically interpolated etch rates [13,43]. The calculated etch rates for planes in the experiments are in good agreement with the values obtained by the authors of Reference [8]. The etch rates measured in the simulations change in time and they are dependent on the surface area of crystallographic plane. After some time and sufficient surface area, the etch rate will be close to its input or numerically interpolated value. It can be concluded that all differences between 3D shapes obtained in the experiments and simulations are due to the differences of their etch rates.

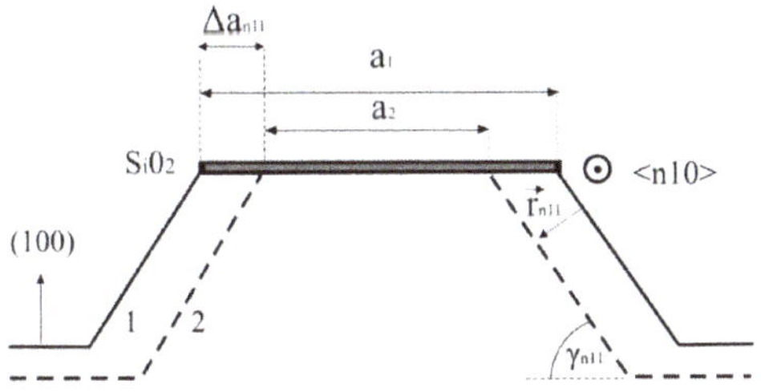

Figure 9. Schematic picture of cross section of the etched square island with sides along <n10> directions. a_1 and a_2 are the sides of the square in the <n10> direction for the moments t_1 and t_2. r_{n11} is etch rate. γ_{n11} is the angle between {n11} and (100) planes.

Table 1. The etch rates of various crystallographic Si planes in the 25 wt % TMAH water solution at the temperature of 80 °C obtained by numerical interpolation, experiment and simulation, and average angles γ_{n11} between {n11} planes and (100) plane in the experiment and simulation obtained from normal cross-sections and their theoretical values. We used a combination of Paraview and Gimp software tools and SEM micrographs, microscope photographs and depth mesaurments to determine angles, as in References [9,26].

Plane {n11}	γ_{n11} theo (°)	Etch Rate r_{n11} input (μm/min)	γ_{n11} exp (°)	Etch Rate r_{n11} exp (μm/min)	γ_{n11} sim (°)	Etch Rate r_{n11} sim (μm/min)
{111}	54.7	0.02	54.2	0.02	54.7	0.03
{211}	65.9	0.87	66.7	0.87	65.3	0.81
{311}	72.5	0.93	74.2	0.93	69.8	0.88
{411}	76.4	0.82	78.7	0.85	76.9	0.79
{511}	78.9	0.75	80.9	0.81	79.4	0.74
{611}	80.7	0.71	81	0.73	80.8	0.70
{711}	82	0.67	83.1	0.69	82.1	0.66
{811}	82.9	0.65	83.1	0.66	82.9	0.64
{911}	83.7	0.63	84.1	0.63	83.8	0.62

4. Conclusions

In this paper we studied silicon etching of square and circle patterns in the masking layer when 25 wt % TMAH water solution is used at the temperature of 80 °C. Almost all crystallographic planes that appear during etching are determined. Etch rates of dominant exposed planes are calculated using the derived relation. Good agreement of experimental and simulation results has been presented. In this way, we confirm that the simulations based on the level-set model can help cost reduction when designing silicon microdevices. Analyzed behavior of the crystallographic planes that appeared during etching described in this paper contributes to a better understanding of anisotropic etching in 25 wt % TMAH water solution at the temperature of 80 °C. The 3D shape of square island in the masking layer with sides along <310> crystallographic directions will not be changed during etching and no convex corner compensation is needed. This mechanism provides advantages for controllable designs of complex silicon structures not only using the most common directions <110> and <100>. In addition, other observed effects can be used in future designs of various silicon microdevices.

Author Contributions: M.M.S., Z.L. and V.J. performed the experiments; B.R. and M.R.-R. developed the simulation program.

Funding: This work has been partially funded by the Ministry of Education, Science and Technological Development of the Republic of Serbia within the framework of the project TR32008 and O171036.

Conflicts of Interest: The authors declare no conflicts of interest.

References

1. Lindroos, V.; Tilli, M.; Lehto, A.; Motooka, T. *Handbook of Silicon Based MEMS Materials and Technologies*; William Andrew: Norwich, NY, USA, 2010.
2. Frühauf, J. *Shape and Functional Elements of the Bulk Silicon Microtechnique*; Springer: Berlin, Germany, 2005.
3. Shikida, M.; Sato, K.; Tokoro, K.; Uchikawa, D. Differences in anisotropic etching properties of KOH and TMAH solutions. *Sens. Actuators A* **2000**, *80*, 179–188. [CrossRef]
4. Sato, K.; Shikida, M.; Yamashiro, T.; Asaumi, K.; Iriye, Y.; Yamamoto, M. Anisotropic etching rates of single-crystal silicon for TMAH water solution as a function of crystallographic orientation. *Sens. Actuators A* **1999**, *73*, 131–137. [CrossRef]
5. Resnik, D.; Vrtacnik, D.; Aljancic, U.; Amon, S. Wet etching of silicon structures bounded by (311) sidewalls. *Microelectron. Eng.* **2000**, *51–52*, 555–566. [CrossRef]
6. Resnik, D.; Vrtacnik, D.; Amon, S. Morphological study of {311} crystal planes anisotropically etched in (100) silicon: Role of etchants and etching parameters. *J. Micromech. Microeng.* **2000**, *10*, 430–439. [CrossRef]
7. Yang, H.; Bao, M.; Shen, S.; Li, X.; Zhang, D.; Wu, G. A novel technique for measuring etch rate distribution of Si. *Sens. Actuators A* **2000**, *79*, 136–140. [CrossRef]
8. Landsberger, L.M.; Naseh, S.; Kahrizi, M.; Paranjape, M. On Hillocks Generated During Anisotropic Etching of Si in TMAH. *J. Microelectromech. Syst.* **1996**, *5*, 106–116. [CrossRef]
9. Zubel, I.; Barycka, I.; Kotowska, K.; Kramkowska, M. Silicon anisotropic etching in alkaline solution IV: The effect of organic and inorganic agents on silicon anisotropic etching process. *Sens. Actuators A* **2001**, *87*, 163–171. [CrossRef]
10. Trieu, H.K.; Mokwa, W. A generalized model describing corner undercutting by the experimental analysis of TMAH/IPA. *J. Micromech. Microeng.* **1998**, *8*, 80–83. [CrossRef]
11. Sarro, P.M.; Brida, D.; Vlist, W.V.D.; Brida, S. Effect of surfactant on surface quality of silicon microstructures etched in saturated TMAHW solutions. *Sens. Actuators A* **2000**, *85*, 340–345. [CrossRef]
12. Smiljanić, M.M.; Jović, V.; Lazić, Ž. Maskless convex corner compensation technique on a (100) silicon substrate in a 25 wt. % TMAH water solution. *J. Micromech. Microeng.* **2012**, *22*, 115011. [CrossRef]
13. Smiljanic, M.M.; Radjenović, B.; Radmilović-Radjenović, M.; Lazić, Ž.; Jović, V. Simulation and experimental study of maskless convex corner compensation in TMAH water solution. *J. Micromech. Microeng.* **2014**, *24*, 115003. [CrossRef]
14. Mukhiya, R.; Bagolini, A.; Margesin, B.; Zen, M.; Kal, S. <100> bar corner compensation for CMOS compatible anisotropic TMAH etching. *J. Micromech. Microeng.* **2006**, *16*, 2458–2462. [CrossRef]
15. Bagolini, A.; Faes, A.; Decarli, M. Influence of Etching Potential on Convex Corner Anisotropic Etching in TMAH Solution. *J. Microelectromech. Syst.* **2010**, *19*, 1254–1259. [CrossRef]
16. Mukhiya, R.; Bagolini, A.; Bhattacharyya, T.K.; Lorenzelli, L.; Zen, M. Experimental study and analysis of corner compensation structures for CMOS compatible bulk micromachining using 25 wt % TMAH. *Microelectron. J.* **2011**, *42*, 127–134. [CrossRef]
17. Merlos, A.; Acero, M.C.; Bao, M.H.; Bausells, J.; Esteve, J. A study of the undercutting characteristics in the TMAH-IPA system. *J. Micromech. Microeng.* **1992**, *2*, 181–183. [CrossRef]
18. Merlos, A.; Acero, M.C.; Bao, M.H.; Bausells, J.; Esteve, J. TMAH/IPA anisotropic etching characteristics. *Sens. Actuators A* **1993**, *37–38*, 737–743. [CrossRef]
19. Pal, P.; Sato, K.; Shikida, M.; Gosalvez, M.A. Study of corner compensating structures and fabrication of various shape of MEMS structures in pure and surfactant added TMAH. *Sens. Actuators A* **2009**, *154*, 192–203. [CrossRef]

20. Pal, P.; Sato, K.; Chandra, S. Fabrication techniques of convex corners in a (100)-silicon wafer using bulk micromachining: A review. *J. Micromech. Microeng.* **2007**, *17*, R111–R133. [CrossRef]
21. Powell, O.; Harrison, H.B. Anisotropic etching of {100} and {110} planes in (100) silicon. *J. Micromech. Microeng.* **2001**, *11*, 217–220. [CrossRef]
22. Pal, P.; Sato, K. A comprehensive review on convex and concave corners in silicon bulk micromachining based on anisotropic wet chemical etching. *Micro Nano Syst. Lett.* **2015**, *3*, 1–42. [CrossRef]
23. Pal, P.; Sato, K.; Gosalvez, M.A.; Shikida, M. Study of rounded concave and sharp edge convex corners undercutting in CMOS compatible anisotropic etchants. *J Micromech. Microeng.* **2007**, *17*, 2299–2307. [CrossRef]
24. Pal, P.; Haldar, S.; Singh, S.S.; Ashok, A.; Yan, X.; Sato, K. A detailed investigation and explanation to the appearance of different undercut profiles in KOH and TMAH. *J Micromech. Microeng.* **2014**, *24*, 095026. [CrossRef]
25. Zubel, I.; Kramkowska, M. Development of etch hilloks on different Si (hkl) planes in silicon anisotropic etching. *Surf. Sci.* **2008**, *602*, 1712–1721. [CrossRef]
26. Barycka, I.; Zubel, I. Silicon anisotroping etching in KOH-isopropanol etchant. *Sens. Actuators A* **1995**, *48*, 229–238. [CrossRef]
27. Zubel, I.; Barycka, I. Silicon anisotropic etching in alkaline solutions I. The geometric description of figures developed under etching Si (100) in various solutions. *Sens. Actuators A* **1998**, *70*, 250–259. [CrossRef]
28. Zubel, I. The influence of atomic configuration of (hkl) planes on adsorption processes associated with anisotropic etching of silicon. *Sens. Actuators A* **2001**, *94*, 76–86. [CrossRef]
29. Zubel, I. Silicon anisotropic etching in alkaline solutions III. On the possibility of spatial structures forming in the course of Si (100) anisotropic etching in KOH and KOH+IPA solutions. *Sens. Actuators A* **2000**, *84*, 116–125. [CrossRef]
30. Gosalvez, M.A.; Nieminen, R.; Kilpinen, P.; Haimi, E.; Lindroos, V.K. Anisotropic wet chemical etching of crystalline silicon: Atomistic Monte-Carlo simulations and experiments. *Appl. Surf. Sci.* **2001**, *178*, 7–26. [CrossRef]
31. Puers, B.; Sansen, W. Compensation structures for convex corner micromachining in silicon. *Sens. Actuators A* **1990**, *21–23*, 1036–1041. [CrossRef]
32. Monterio, T.S.; Kastytis, P.; Goncalves, L.M.; Minas, G.; Cardoso, S. Dynamic Wet Etching of Silicon through Isopropanol Alcohol Evaporation. *Micromachines* **2015**, *10*, 1534–1545. [CrossRef]
33. Osher, S.; Sethian, J.A. Fronts Propagating with Curvature Dependent Speed: Algorithms Based on Hamilton-Jacobi Formulations. *J. Comp. Phys.* **1988**, *79*, 12–49. [CrossRef]
34. Radjenović, B.; Lee, J.K.; Radmilović-Radjenović, M. Sparse field level set method for non-convex Hamiltonians in 3D plasma etching profile simulations. *Comput. Phys. Commun.* **2006**, *174*, 127–132. [CrossRef]
35. Radjenović, B.; Radmilović-Radjenović, M.; Mitrić, M. Non-convex Hamiltonians in 3D level set simulations of the wet etching of silicon. *Appl. Phys. Lett.* **2006**, *89*, 213102. [CrossRef]
36. Radjenović, B.; Radmilović-Radjenović, M. 3D simulations of the profile evolution during anisotropic wet etching of silicon. *Thin Solid Films* **2009**, *517*, 4233–4237. [CrossRef]
37. Radjenović, B.; Radmilović-Radjenović, M.; Mitrić, M. Level Set Approach to Anisotropic Wet Etching of Silicon. *Sensors* **2010**, *10*, 4950–4967. [CrossRef] [PubMed]
38. Montoliu, C.; Ferrando, N.; Gosalvez, M.A.; Cerda, J.; Colom, R.J. Level set implementation for the simulation of anisotropic etching: Application to complex MEMS micromachining. *J. Micromech. Microeng.* **2013**, *23*, 075017. [CrossRef]
39. Montoliu, C.; Ferrando, N.; Gosalvez, M.A.; Cerda, J.; Colom, R.J. Implementation and evaluation of the Level Set method—Towards efficient and accurate simulation of wet etching for microengineering applications. *Comput. Phys. Commun.* **2013**, *184*, 2299–2309. [CrossRef]
40. Yu, J.C.; Zhou, Z.F.; Su, J.L.; Xia, C.F.; Zhang, X.W.; Wu, Z.Z.; Huang, Q.A. Three-Dimensional Simulation of DRIE Process Based on the Narrow Band Level Set and Monte Carlo Method. *Micromachines* **2018**, *2*, 74. [CrossRef]
41. Available online: http://www.paraview.org (accessed on 26 January 2019).

Micromachines **2019**, *10*, 102

42. Shikida, M.; Nanbara, K.; Koizumi, T.; Sasaki, H.; Odagaki, M.; Sato, K.; Ando, M.; Furuta, S.; Asaumi, K. A model explaining mask-corner undercut phenomena in anisotropic silicon etching: A saddle point in the etching-rate diagram. *Sens. Actuators A* **2002**, *97–98*, 758–763. [CrossRef]

43. Hubbard, T.J. MEMS Design-Geometry of Silicon Micromachining. Ph.D. Thesis, California Institute of Technology, Pasadena, CA, USA, 1994.

micromachines

MDPI

Article

A Novel Fabricating Process of Catalytic Gas Sensor Based on Droplet Generating Technology

Liqun Wu [1], Ting Zhang [1], Hongcheng Wang [1,*], Chengxin Tang [2] and Linan Zhang [1]

[1] School of Mechanical Engineering, Hangzhou Dianzi University, Hangzhou 310018, China;
wuliqun@hdu.edu.cn (L.W.); 172010042@hdu.edu.cn (T.Z.); zhanglinan@hdu.edu.cn (L.Z.)

[2] School of Media and Design, Hangzhou Dianzi University, Hangzhou 310018, China; tcx@hdu.edu.cn

* Correspondence: wanghc@hdu.edu.cn; Tel: +86-571-8691-9052

Received: 25 November 2018; Accepted: 15 January 2019; Published: 20 January 2019

check for
updates

Abstract: Catalytic gas sensors are widely used for measuring concentrations of combustible gases to prevent explosive accidents in industrial and domestic environments. The typical structure of the sensitive element of the sensor consists of carrier and catalyst materials, which are in and around a platinum coil. However, the size of the platinum coil is micron-grade and typically has a cylindrical shape. It is extremely difficult to control the amount of carrier and catalyst materials and to fulfill the inner cavity of the coil, which adds to the irreproducibility and uncertainty of the sensor performance. To solve this problem, this paper presents a new method which uses a drop-on-demand droplet generator to add the carrier and catalytic materials into the platinum coil and fabricate the micropellistor. The materials in this article include finely dispersed Al_2O_3 suspension and platinum palladium (Pd-Pt) catalyst. The size of the micropellistor with carrier material can be controlled by the number of the suspension droplets, while the amount of Pd-Pt catalyst can be controlled by the number of catalyst droplets. A bridge circuit is used to obtain the output signal of the gas sensors. The original signals of the micropellistor at 140 mV and 80 mV remain after aging treatment. The sensitivity and power consumption of the pellistor are 32 mV/% CH_4 and 120 mW, respectively.

Keywords: gas sensor; micropellistor; microdroplet; pulse inertia force; methane

1. Introduction

Catalytic combustion type gas detectors that operate on catalytic oxidation of combustible gases are widely used for detecting gas concentration and maintaining it at below the lower explosion limit (LEL) [1]. It is an effective means to prevent explosive accidents in industrial and domestic environments [2,3]. The combustive gases include hydrogen (H_2), methane (CH_4), carbon monoxide (CO), organic vapors, etc. As is shown in Figure 1, the widely applicable structure of the sensitive element (called a pellistor) in the catalytic combustion type gas detectors consists of a porous structure and a platinum coil [4,5]. Therefore, the pellistor is a kind of solid phase gas sensor [6] and has a much higher sensitivity, though much effort has been put into developing a silicon microheater potentially with a high-temperature and low-power consumption [7,8]. The porous structure, constructed around the platinum coil, is called a carrier. On the inner surface is catalyst which has a catalytic effect during the detecting process. The platinum coil can heat the catalyst to a sufficiently high temperature, at which any flammable gas molecules present can produce flameless combustion [9,10] and release combustion heat. Besides that, the Pt coil serves not only as a catalyst heater, but also as a resistance thermometer.

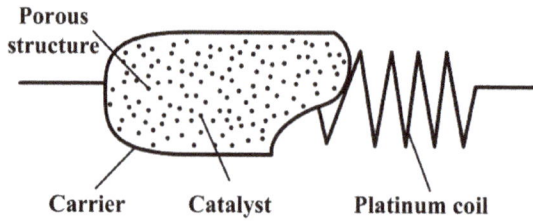

Figure 1. Structure of the micropellistor in a catalytic gas sensor.

The surface of the carrier is a porous structure, the precursor of which is nano-scale particle suspension (typically nano-scale Al_2O_3 suspension [11,12]) and can form the porous structure after evaporation of the solvent. Wang et al. [5] researched performances of sensors with different carriers, including Al_2O_3, n-Al_2O_3, and n-Ce-Al_2O_3. The carrier and catalytic materials were coated onto the platinum coil via the sometric impregnation method.

The catalytic material is immersed into the inner-surface of the porous structure to form a large amount of catalyst activated points, which determine the sensitivity of the sensor. The greater the number of activated points the catalyst is on, the higher the sensitivity of the sensor [13].

However, the size of the platinum coil is micron-grade and typically cylindrically shaped. Therefore, coating the platinum coil with carrier and catalytic materials appeared to be an extremely critical step in practice. The cumbersome dip and drop technique [14], a thick film processing step, was adopted to coat the catalytic materials. Unfortunately, it was very difficult to control in the case of viscous species and to fulfill the inner cavity of the coil, which adds to the irreproducibility and uncertainty of the fabrication process and causes unstable performance. To solve this problem, this paper proposes a new method which uses a drop-on-demand droplet generator to introduce the carrier and catalytic materials into the platinum coil and fabricate the micropellistor.

2. Materials and Methods

2.1. Materials

The platinum coil was wound with coil diameter of 250 μm and a length of 800 μm, as shown in Figure 2. The diameter of the platinum wire was about 25 μm. The catalytic materials used included finely dispersed Al_2O_3 suspension and platinum palladium (Pd-Pt) catalyst.

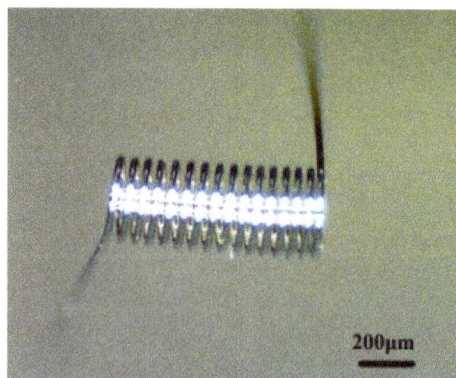

Figure 2. Structure of platinum coil.

The catalytic materials used included finely dispersed Al_2O_3 suspension and platinum palladium catalyst, which were ejected into the platinum coil successively. The catalytic materials needed to exhibit appropriate viscosity and stability, so the catalytic materials—especially the Al_2O_3 suspension—required homogeneous mixing before being used, and the optimum viscosity of the catalytic materials was below 100 cp.

2.2. Methods

2.2.1. Droplet Generating Method Based on Pulse Inertia Force

One of the main tasks of this work is to propose a droplet generator to eject catalytic materials into the platinum coil. Droplet generating technologies mainly include two classes: continuous mode and drop-on-demand (DOD) mode. In recent years, the DOD mode has gradually replaced the continuous mode because of its promising better manufacturing control and better control of ejection time, position, and volume by a function generator. The DOD mode droplet generators include piezoelectrically [15], thermally, pneumatically, electrostatically, pulse electromagnetic force [16], and membrane-piston [17] actuated generators. The piezoelectric droplet generator has an ink containing chamber with piezoelectric elements on one or two chamber external surfaces. Displacement of piezoelectric elements can change the volume of the chamber, generate pressure waves, and eject a droplet from an orifice while the elements are applied with a pulse voltage signal. However, it is difficult for the above DOD droplet generators to be applied in ejecting the catalytic materials in this article, because the catalytic materials may corrode the nozzle, and most of the nozzles are made of metallic materials.

As is shown in Figure 3, an apparatus for producing catalytic material droplets in an air environment was fashioned. The nozzle filled with catalytic materials was clamped by connector B, which was fixed to the bottom face of a lead zirconate titanate (PZT) stack actuator (PAL200VS25, NanoMotions, Shanghai, China) through connector A, while the upper face was fixed to a three-dimensional adjustable frame and kept stationary through the connector. There was an approximate linearity between applied voltage amplitude and the displacement of the bottom face of the actuator. Therefore, the actuator instantaneously caused a larger displacement and consequently provided a greater pulse inertia force [18] for the nozzle and catalytic materials inside when applying a higher pulse driving voltage. When the pulse inertia force was large enough and exceeded the viscous force, a droplet of liquid was ejected from the micro-nozzle drop by drop in the direction of inertial force. Pulse inertia force had no influence on the catalytic material, while the piezoelectric micro-dispenser pushed/squeezed the catalytic material, and the thermal micro-dispenser heated material to above 300 °C to form micro-droplets. Both micro-dispensers may have changed the properties of catalytic material.

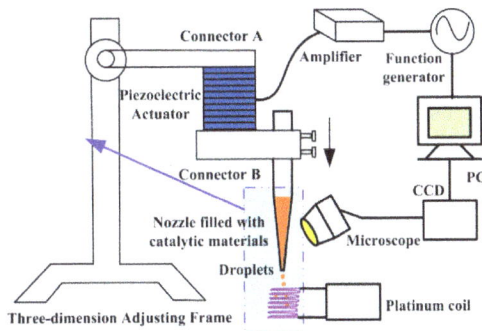

Figure 3. Schematic of the droplet generating device for coating catalytic materials.

Glass material was chosen to make the tapered glass capillary because of several advantages, such as good chemical resistance, smooth surface, ease of manufacture and observation, and low cost. The raw material was borosilicate glass capillary (Beijing Zhengtianyi Scientific and Trading Co., Ltd.,

Beijing, China). The dimensions of the glass capillary were 1.0 mm, 0.6 mm, and 100 mm in external diameter, internal diameter, and length, respectively, as is shown in Figure 4a. A glass heating process was adopted to fabricate the micro-nozzle without complicated micro-fabrication technology and can be divided into two steps: (1) pulling a capillary to form a micro-nozzle with a straight outlet, as is shown in Figure 4b, and (2) forging the straight outlet to form a shrinkage one, as is shown in Figure 4c. The micro-nozzles with different outlet diameters were obtained by varying the control parameters (the outlet diameter in this article means the inner diameter of the nozzle tip). The fabricated micro-nozzle with an outlet diameter of 100 µm is shown in Figure 4d. The raw material of the micro-nozzle was borosilicate glass which, having good chemical inertness, allowed for no chemical reactions to occur between the catalytic materials and nozzle; conversely, most of the nozzles of micro-dispensers on the market are made of metal material.

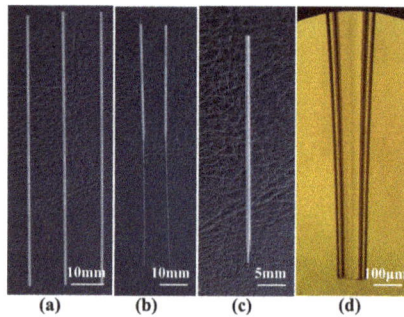

Figure 4. Fabrication of the borosilicate glass micro-nozzle: (**a**) borosilicate glass pipe; (**b**) micro-nozzle after being pulled; (**c**) micro-nozzle after being cut; (**d**) micrograph of micro-nozzle after being cut.

2.2.2. Manufacturing Process of the Pellistor

As is shown in Figure 5, the fabrication process of the pellistor using the droplet generator proposed above included the following:

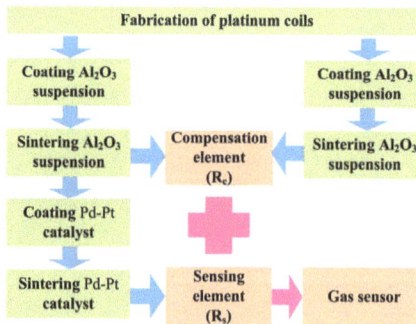

Figure 5. Manufacturing process of the catalytic gas sensor.

(1) Adding Al_2O_3 suspension

As is shown in Figure 6, the platinum coil cylinder axis and micro-nozzle were kept vertical. The distance between the coil and the micro-nozzle was less than 2 mm to avoid forming satellites. The micro-nozzle was made to inhale a certain amount of Al_2O_3 suspension by a negative-pressure apparatus. The PZT function generator and amplifier was then started to excite enough inertia force for the Al_2O_3 suspension which was ejected droplet by droplet from of the nozzle orifice. The driving voltage and frequency were in the range of 0–80 V and 1–256 Hz, respectively. The droplet size was controlled by the driving voltage signal and orifice diameter of the micro-nozzle. The ejected Al$_2$O$_3$ suspension covered the entirety of the platinum coil and avoided creating a hole defect inside of the carrier.

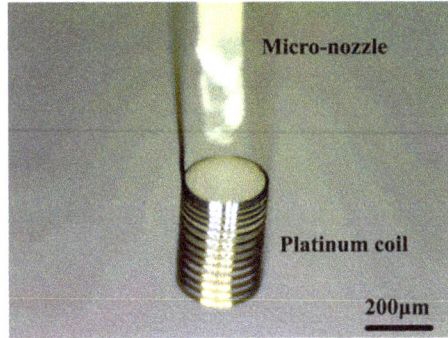

Figure 6. Relative position between the platinum coil and micro-nozzle.

(2)   Formation of the carrier

The added Al$_2$O$_3$ suspension was sintered to form a porous Al$_2$O$_3$ matrix (γ-Al$_2$O$_3$ layer) by self-heating of the underlying platinum coil. The porous Al$_2$O$_3$ matrix was called the carrier. The resulting alumina structure established a perfect thermo-mechanical contact to the platinum coil in order to form an outer surface with sufficient temperature for catalysis and to conduct heat, which was developed by the catalytic combustion of the present gas, to the coil resistor, acting as a temperature sensor.

The sintering temperature was set at 750 °C in the experiment and the temperature holding time was twenty minutes. The temperature was controlled by the parameter of the voltage applied to the platinum coil and can be calculated by Equation (1):

$$\frac{R - R_0}{T - T_0} = R_0 \times k \tag{1}$$

where k is the temperature coefficient of resistance of platinum, a constant of 0.0026/°C. T_0 is the room temperature, and R_0 is the resistance of platinum at room temperature. The platinum coil with porous Al$_2$O$_3$ matrix can be used as a compensation element.

(3)   Adding catalytic material

The micro-nozzle was made to draw in a certain amount of catalytic material by a negative-pressure apparatus, just like in step (1). The ejected catalyst soaked into and adhered to the porous structure as soon as it made contact with the matrix. The amount of added catalytic material was controlled by the size and number of the liquid droplets.

(4)   Formation of the pellistor

The matrix with catalytic material was sintered again at 550 °C to form the Pt-Pb/Al$_2$O$_3$ layer, the platinum element being the pellistor and the sensing element (R_s).

3. Results and Discussion

3.1. Pattern of the Pellistor

The solvents including deionized water and alcohol volatilized during the additive process. It was necessary to control the droplet generating frequency and let solvents of ejected droplets have enough time to volatilize. While adding the Al$_2$O$_3$ suspension, the platinum coil was first placed with its axis in the same direction of the nozzle axis. After the internal space of the coil was filled with suspension (Figure 7a), the platinum coil was placed with its axis perpendicular to the nozzle axis (Figure 7b). If the Al_2O_3 suspension additive speed is too high for the solvents to volatilize, a hole defect may occur, as is shown in Figure 7c. Both the inside and outside of the platinum coil was coated with the Al_2O_3 matrix layer. If there is no matrix outside of the coil, the platinum exposed in an air environment will oxide rapidly since it is a micro-heater itself. Therefore, after the coil was filled with Al_2O_3 matrix, the suspension was ejected onto the lateral surface of the columniform coil. The platinum coil with a complete Al_2O_3 matrix layer is shown in Figure 7d. When the concentration of the suspension was relatively low, this lower viscosity suspension flowed out of the coil, as is shown in Figure 8. In this condition, it was not necessary to change the coil axis direction during the additive process. Satellite droplets [19] were inevitable during the droplet generating process, but did not affect the pattern of the pellistor.

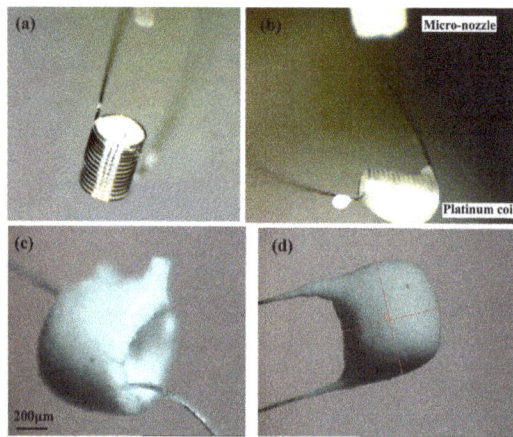

Figure 7. Pellistors: (**a**) filled with suspension, (**b**) with suspension on external surface, (**c**) with defects of shrinkage cavity, and (**d**) with perfect Carrier.

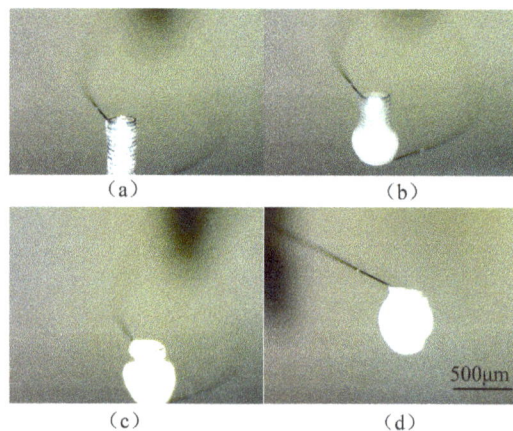

Figure 8. Process of coating Al_2O_3 suspension onto the platinum coil.

3.2. Performance of the Pellistor

3.2.1. Flameless Catalytic Combustion

Methane (CH_4), one of the most difficult-oxidative hydrocarbon combustible gases, was chosen as the testing object for the pellistor. The platinum coil heated the catalyst up to a sufficiently high temperature, at which methane gas molecules present produced flameless combustion and released combustion heat. The chemical equation for the flameless catalytic combustion is as follows, with the platinum palladium (Pd-Pt) solution as the catalyst:

$$CH_4 + 2O_2 \xrightarrow[\Delta]{Pt, Pd} CO_2 + 2H_2O + 795.5 \text{ KJ} \tag{2}$$

The reaction product was carbon dioxide (CO_2) and H_2O. The higher the concentration of methane gas molecules, the more heat was released. The resistance value of platinum coil increased with the raising of the ambient temperature. Therefore, a definite numerical relationship exists between methane concentration and platinum wire resistance value.

3.2.2. Activation of the Pellistor

When methane concentration was relatively high, typically above 80% LEL (low explosion limit), the $PdCl_2$ on the γ-Al_2O_3 layer decomposed to Pb and PbO, both of which having very high activities. An LEL of 80% LEL means that the volume fraction of CH_4 is 4%, as the LEL for CH_4 is 5%. It is a common method to enhance the output signal of sensor, the process of which is called sensor activation. The output signal of the pellistor was tested by a direct current bridge, as is shown in Figure 9a. The direct current bridge was composed of a pellistor (called the sensing element, R_s), an adjustable resistor (R), and two fixed resistors (R_1 and R_2) at room temperature. E, the applied voltage of the bridge, was set at 2.6 V. The output signal ($\Delta U = U_1 - U_2$) was set to zero by adjusting the resistance of R to that of R_s. When sample gas flowed to the sensor, gas molecules adsorbed onto the sensitive element, combusted, and induced a temperature increase in the presence of platinum palladium catalyst. The temperature increase was short-lived compared to the increase in resistance and the output signal (ΔU). Moreover, this circuit did not have any amplifiers, filters, or signal process circuits. Therefore, the output was the original signal.

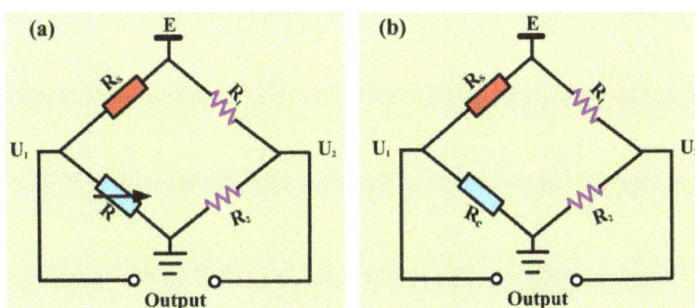

Figure 9. Direct current bridges: (**a**) for original signal testing, and (**b**) for steady signal testing with temperature compensation.

Therefore, the output signal ΔU can be calculated by Equation (3):

$$\Delta U = \frac{R_s}{R_s + R} E - \frac{R_2}{R_1 + R_2} E \tag{3}$$

Figure 10 shows the variation of original signal of the pellistor with porous matrix by size and the amount of Pd-Pt catalyst (sample number is 10). The porous matrix size was characterized by the diameter of the pellistor (*D*, μm), and the amount of Pd-Pt catalyst was characterized by its volume (*V*, nL). If the diameter of the pellistor was too small (*D* = 300 μm), the number of catalyst activated points was not enough for the pellistor to work a long time, and the original signal was also relatively low (below 100 mV); if the diameter of the pellistor was too large (*D* = 500 μm or 550 μm), the matrix and catalyst did not sinter completely and had poor heat transfer, which reduced the output voltage value of the sensor (below 110 mV). If the amount of the catalyst was too large, the matrix surface could not accommodate much catalyst, and consequently, the redundant catalyst caused a short-circuit while voltage was applied to the pellistor; if is the amount was too small, this small amount of catalyst (less than 7.0 nL) could not soak into whole matrix and remained only on surface. The temperature of the matrix surface layer was far below 750 °C while being sintered due to its poor heat transfer and could not form effective Pd-Pt/γ- Al$_2$O$_3$ layer. The maximum value of original signal was above 140 mV when the pellistor diameter and catalyst volume were 450 μm and 16.0 nL, respectively. The relative deviations of all testing data were below 5%, which means that the original signal of the pellistors had high reproducibility and stability.

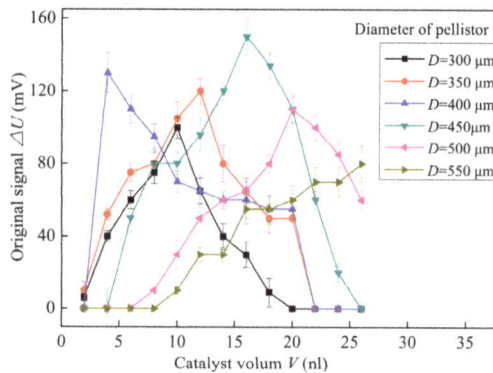

Figure 10. Variations in the original signals of the pellistor with porous matrix based on pellistor size and the amount of platinum palladium (Pd-Pt) catalyst.

3.2.3. Sensitivity of the Pellistor

Aging treatment was a key process to test the performance of sensor. In aging treatment experiments, the fabricated pellistor was placed in a testing chamber flushed by standard CH$_4$ gas with 0% LEL, 20% LEL, 35% LEL, 50% LEL, 75% LEL, and 100% LEL, in turn. CH$_4$ gas with 0% LEL was flushed through the chamber for 2 min, and then 35% LEL was flushed through until the output signal reached 90% the standard signal (original signal). The duration of this process was called the response time. Every pellistor was tested in different CH$_4$ concentrations. Each pellistor had its own chamber, and several pellistors were treated synchronously. All the pellistors were tested in the above standard CH$_4$ gas with different concentrations and tested in 50% LEL CH$_4$ gas for 6 h every day.

After aging treatment of 150 h in the environment of 50% LEL CH$_4$, the pellistor was connected in the direct current bridge with temperature compensation, as is shown in Figure 9b. The temperature compensation element (R_c) had the same diameter of matrix and catalytic volume as the sensing element. The operating temperature was 450 °C. The steady signal of the pellistor with an original signal of 140 mV was tested and remained at 80 mV. The output signal ΔU can be calculated by Equation (4):

$$\Delta U = \frac{R_s - R_c}{2(R_s + R_c)} E \qquad (4)$$

The output voltage value for per unit gas concentration was used to represent the sensitivity of the sensor. The unit of sensitivity was mV/% CH_4 for methane gas detection. The variation in output signal of the sensor versus variable methane concentrations is shown in Figure 11. All the data are the mean value of the above ten samples. While the methane concentration was in the range of 10% LEL CH_4 to 90% LEL CH_4, the output signal almost appeared to increase linearly. When the methane concentration was 50% LEL CH_4, the output signal was 80 mV, according to Figure 11. The low explosion limit (LEL) in air for CH_4 is 5%. This means that the output signal is 80 mV when CH_4 concentration in air is 2.5%. Therefore, the sensitivity of the sensor could be calculated as 32 mV/% CH_4. The larger the pellistor was, the greater the number of activated points in the catalytic material; the more activated points, the higher the sensitivity. In addition, the structure of the carrier after being sintered was filled without any defects in the shrinkage cavity via droplet generating technology. The sensitivity of 32 mV was relatively higher than that presented by Liu et al. [20].

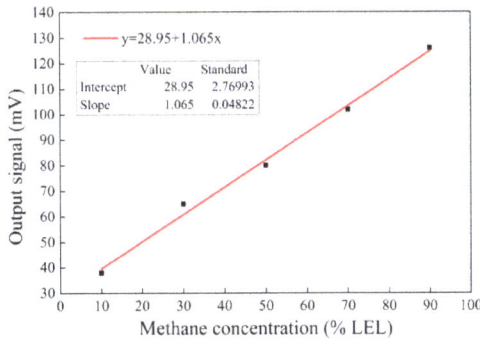

Figure 11. Variation in output signal of the sensor with variable methane concentrations.

3.2.4. Power Consumption

Power consumption is another important performance index, because thousands of sensors will be placed in a mine and all the sensors should work all the time for several years. When pellistor diameter and catalyst volume were 450 μm and 16.0 nL, respectively, the original signal obtained the maximum value of about 140 mV, which means that this pellistor had the best output performance among all the tested samples. After aging treatment, the output signal remained 80 mV, and the power consumption was 120 mW. The power consumption of pellistors only demonstrating the best output performances were measured. This can be calculated by Equation (5):

$$P = I^2 \times R_s \tag{5}$$

where I is the current applied to the sensing element, and R_s is the resistance value of the sensing element under working conditions. Experimental results have shown that the power consumption for the above sensor with the output signal of 80 mV was about 120 mW. The power consumption of the sensor was relatively high. The structure of the carrier after being sintered was filled without any defects in the shrinkage cavity via droplet generating technology. More energy was needed to heat the sensing element to the working temperature because the sensor in this paper had a slightly larger volume.

4. Conclusions

This paper presents a new method which uses a droplet generator based on pulse inertia force to introduce carrier and catalytic materials into a platinum coil and fabricate a micropellistor. The fabrication process of the pellistor includes four steps, which are adding Al_2O_3 suspension, forming the carrier, adding catalytic material, and forming the pellistor. The added amounts of both

Micromachines **2019**, *10*, 71

the carrier and catalytic materials can be controlled by the volume and rate of the ejected droplets. A bridge circuit is used to get the output signal of the gas sensors. Variation in the original signal of the pellistor with porous matrix size and amount of Pd-Pt catalyst was researched. The maximum value of original signal is above 140 mV when the pellistor diameter and catalyst volume are 450 μm and 16.0 nL, respectively. The steady output signal after aging treatment almost appeared to increase linearly with the increase of the methane concentration. The sensitivity and power consumption of the pellistor are 32 mV/% CH_4 and 120 mW, respectively.

Author Contributions: L.W. wrote the original draft; H.W. reviewed & edited the paper; T.Z. analyzed the experiment data; C.T. and L.Z. designed the experiment.

Funding: This work was funded by the Zhejiang Provincial Natural Science Foundation of China (grant number LQ17E050012) and the National Natural Science Foundation of China (grant numbers 51775154, 11841007).

Conflicts of Interest: The authors declare no conflict of interest.

References

1. Singh, M.; Yadav, B.C.; Ranjan, A.; Sonker, R.K.; Kaur, M. Detection of liquefied petroleum gas below lowest explosion limit (LEL) using nanostructured hexagonal strontium ferrite thin film. *Sens. Actuators B* **2017**, *249*, 96–104. [CrossRef]
2. Fonollosa, J.; Solorzano, A.; Marco, S. Chemical sensor systems and associated algorithms for fire detection: A review. *Sensors* **2018**, *18*, 553. [CrossRef] [PubMed]
3. Goldoni, A.; Alijani, V.; Sangaletti, L.; D'Arsiè, L. Advanced promising routes of carbon/metal oxides hybrids in sensors: A review. *Electrochim. Acta* **2018**, *266*, 139–150. [CrossRef]
4. Chen, Q.Y.; Dong, H.P.; Xia, S.H. A novel micro-pellistor based on nanoporous alumina beam support. *J. Electron.* **2012**, *29*, 469–472. [CrossRef]
5. Wang, Y.; Tong, M.M.; Zhang, D.; Gao, Z. Improving the Performance of Catalytic Combustion Type Methane Gas Sensors Using Nanostructure Elements Doped with Rare Earth Cocatalysts. *Sensors* **2011**, *11*, 19–31. [CrossRef] [PubMed]
6. Moseley, P.T. Solid state gas sensors. *Meas. Sci. Technol.* **1997**, *8*, 223–227. [CrossRef]
7. Famuyiro, S. Use of combustible gas detectors in Safety Instrumented Systems-A practical application case study. *J. Loss Prevent Proc.* **2018**, *54*, 333–339. [CrossRef]
8. Ma, H.; Qin, S.; Wang, L.; Wang, G.; Zhao, X.; Ding, E. The study on methane sensing with high-temperature low-power cmos compatible silicon microheater. *Sens. Actuators B* **2017**, *244*, 17–23. [CrossRef]
9. Weinberg, F.J. Combustion temperature: The future? *Nature* **1971**, *233*, 239–241. [CrossRef] [PubMed]
10. Xing, F.; Kumar, A.; Huang, Y.; Chan, S.; Fan, X. Flameless combustion with liquid fuel: A review focusing on fundamentals and gas turbine application. *Appl. Energy* **2017**, *193*, 28–51. [CrossRef]
11. Choi, Y.; Tajima, K.; Shin, W.; Izu, N.; Matsubara, I.; Murayama, N. Effect of pt/alumina catalyst preparation method on sensing performance of thermoelectric hydrogen sensor. *J. Mater.* **2006**, *41*, 2333–2338. [CrossRef]
12. Brauns, E.; Morsbach, E.; Kunz, S.; Baeumer, M.; Lang, W. Temperature modulation of a catalytic gas sensor. *Sensors* **2014**, *14*, 20372–20381. [CrossRef] [PubMed]
13. Brauns, E.; Morsbach, E.; Kunz, S.; Bäumer, M.; Lang, W. A fast and sensitive catalytic gas sensors for hydrogen detection based on stabilized nanoparticles as catalytic layer. *Sens. Actuators B* **2014**, *193*, 895–903. [CrossRef]
14. Bársony, I.; Ádám, M.; Fürjes, P.; Lucklum, R.; Hirschfelder, M.; Kulinyi, S. Efficient catalytic combustion in integrated micropellistors. *Meas. Sci. Technol.* **2009**, *20*, 124009. [CrossRef]
15. Sofija, V.M.; Frédéric, C.; Jürgen, V.; Jan, G.P.; David, N. Droplet generation and characterization using a piezoelectric droplet generator and high speed imaging techniques. *Crop Prot.* **2015**, *69*, 18–27.
16. Wang, T.; Lin, J.; Lei, Y.; Guo, X.; Fu, H.; Zhang, N. Dominant factors to produce single droplet per cycle using drop-on-demand technology driven by pulse electromagnetic force. *Vacuum* **2018**, *156*, 128–134. [CrossRef]
17. Ma, M.; Wei, X.; Shu, X.; Zhang, H. Producing solder droplets using piezoelectric membrane-piston-based jetting technology. *J. Mater. Process. Tech.* **2019**, *263*, 233–240. [CrossRef]
18. Wang, H.C.; Hou, L.Y.; Zhang, W.Y. A drop-on-demand droplet generator for coating catalytic materials on microhotplates of micropellistor. *Sens. Actuators B* **2013**, *183*, 342–349. [CrossRef]

19. Wang, T.; Lin, J.; Lei, Y.; Guo, X.; Fu, H. Droplets generator Formation and control of main and satellite droplets. *Colloids Surf. A* **2018**, *558*, 303–312. [CrossRef]

20. Liu, F.; Zhang, Y.; Yu, Y.; Xu, J.; Sun, J.; Lu, G. Enhanced sensing performance of catalytic combustion methane sensor by using Pd nanorod/γ-Al_2O_3. *Sens. Actuators B* **2011**, *160*, 1091–1097. [CrossRef]

micromachines

MDPI

Article

Implementation of a CMOS/MEMS Accelerometer with ASIC Processes

Yu-Sian Liu * and **Kuei-Ann Wen**

Institute of Electronic Engineering, National Chiao Tung University, Hsinchu 300, Taiwan;
stellawen@mail.nctu.edu.tw
* Correspondence: thomas.ee02g@nctu.edu.tw; Tel.: +886-3-573-1627

Received: 17 December 2018; Accepted: 8 January 2019; Published: 12 January 2019

check for updates

Abstract: This paper presents the design, simulation and mechanical characterization of a newly proposed complementary metal-oxide semiconductor (CMOS)/micro-electromechanical system (MEMS) accelerometer. The monolithic CMOS/MEMS accelerometer was fabricated using the 0.18 μm application-specific integrated circuit (ASIC)-compatible CMOS/MEMS process. An approximate analytical model for the spring design is presented. The experiments showed that the resonant frequency of the proposed tri-axis accelerometer was around 5.35 kHz for out-plane vibration. The tri-axis accelerometer had an area of 1096 μm × 1256 μm.

Keywords: accelerometer design; spring design; analytical model

1. Introduction

Micro-electromechanical system (MEMS) technology has enabled the substantial expansion of the inertial sensor market by decreasing power consumption, cost, and size. The complementary metal-oxide semiconductor (CMOS)/MEMS technology enables the integration of CMOS circuits with MEMS structures in a single chip [1]. CMOS/MEMS processes have the advantages of a mature foundry service for mass production, monolithic integration with CMOS circuitry to reduce the parasitic capacitance, and size reduction to decrease chip cost [2–7]. However, the composite thin-film structure of CMOS/MEMS technology suffers from residual stresses and limits the device's performance. The CMOS/MEMS structure consisted of multiple metal and dielectric stacking layers. After release from the substrate, the structure is deformed by the thin film residual stresses, which can significantly affect the device's performance [4,8]. The deformation of composite structures was predicted based on analytical models in [9].

A capacitive accelerometer can be implemented in the CMOS/MEMS process [10]. A capacitive CMOS/MEMS accelerometer typically consists of the proof-mass, springs, and sensing electrodes. The sensing electrodes are placed around the proof mass. The sensing technique involves using the gap-closing method [11,12]. For 3-axis integrated accelerometer design, the z-axis sensing element consists of an imbalanced proof mass, a torsional spring beam and comb fingers on both ends of the proof mass [13]. Using three individual sensing units to detect the tri-axis acceleration can reduce structural curling [14]. A single proof-mass tri-axis accelerometer can significantly reduce the chip size and improve the accelerometer sensitivity [15,16].

The sensitivity of accelerometers strongly depends on the spring constants of the suspension system [17]. Many types of springs can be utilized in the accelerometer design. Four commonly used flexures are: clamped-clamped flexure, crab-leg flexure, folded flexure and serpentine flexure. Among these four types, the serpentine flexure has the lowest stiffness and has reduced axial stress [18]. A serpentine spring is adopted in various MEMS sensors [15,19].

The design target refers to ADXL327. ADXL327 is a high precision, low power, tri-axis accelerometer with signal conditioned voltage outputs from Analog Devices Inc. (ADI, Norwood, MA, USA). ADXL327 measures acceleration with a full-scale range of ± 2 g, a range of 0.5 Hz to 1600 Hz for the x-and y-axis, and a range of 0.5 Hz to 550 Hz for the z-axis. In our design, the resonant frequency was targeted lower than 5 kHz and the sensing range was ± 1 g. ADXL327 can measure both the static acceleration of gravity in tilt-sensing applications and dynamic acceleration resulting from motion, shock, or vibration. The main application is for navigation and motion detection [20].

In this paper, the design, simulation and mechanical characterization of the proposed CMOS/MEMS accelerometer is presented. The 0.18 µm application-specific integrated circuit (ASIC)-compatible CMOS/MEMS process was adopted for sensor and circuit implementation. While the circuit as well as the electrical characterization was presented in our previous works [21,22]. Section 2 describes the process flow and structure design of the accelerometer, while the theory and simulation were also analyzed. The approximate analytical model for the spring design was also proposed. Section 3 describes the measurement results of the proposed CMOS/MEMS accelerometer. Section 4 presents the discussion of the proposed accelerometer, a comparison of the performance with a state-of-the-art alternative and presents the conclusions of this work.

2. Materials and Methods

2.1. Process Flow

The proposed CMOS/MEMS accelerometer was implemented in the 0.18 µm ASIC-compatible CMOS/MEMS process. The integrated circuit (IC) foundries were Taiwan Semiconductor Manufacturing Company (TSMC, Hsinchu, Taiwan) 0.18 µm mixed-signal/radio frequency (RF) CMOS process with an Asia Pacific Microsystems, Inc. (APM, Hsinchu, Taiwan) MEMS post-process and United Microelectronics Corporation (UMC, Hsinchu, Taiwan) 0.18 µm mixed-signal/RF CMOS process with a UMC MEMS post-process.

The ASIC compatible 1P6M process started with a 0.18 µm standard CMOS process. The CMOS process consisted of one poly-silicon layer and six metal layers that can be used for wiring and circuit integration.

The micromachining process was performed on the wafer of a standard CMOS process. The process flow is illustrated in Figure 1, where PO1 is the poly-silicon layer and metal1 (ME1) to metal6 (ME6) are the six metal layers. After the standard CMOS process, an additional patterned metal7 (ME7) layer and the passivation layers were deposited at the top of the structure and patterned as the etch-resistant mask (Figure 1a). A thick photoresist passivation layer was then deposited, which defined above the circuit and other regions for etch protection by using this mask, except the MEMS region (Figure 1b). The whole post-CMOS fabrication was performed using a dry etching processes. The area without photoresist protection was subjected to both anisotropic silicon oxide dry etching (Figure 1c) and isotropic silicon substrate dry etching (Figure 1d). The region without the photoresist passivation layer mask defined the MEMS etching region for the post-process. Metal 7 covered the whole microstructure to define the microstructures. The microstructures were released by isotropic silicon substrate dry etching. The passivation layer above the electronic circuits may have been slightly damaged after the post-micromachining process. The remaining photoresist layer was cleaned after the silicon etching.

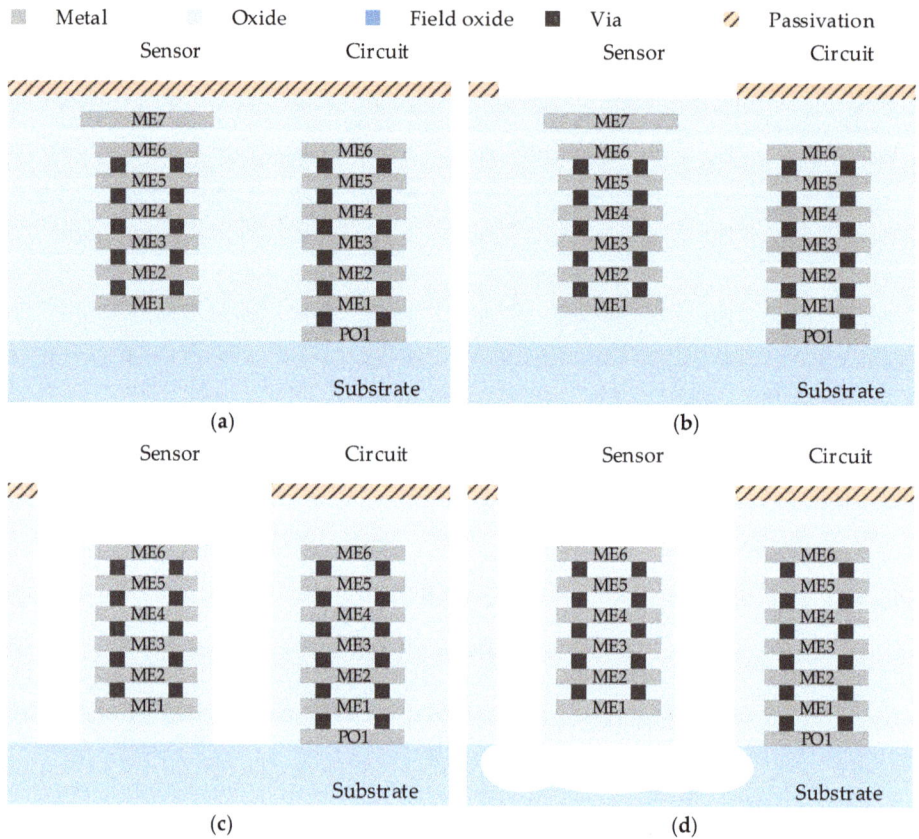

Figure 1. Cross-sectional view of the application-specific integrated circuit (ASIC)-compatible complementary metal-oxide semiconductor (CMOS)/micro-electromechanical system (MEMS) process flow: (**a**) The standard CMOS process with an additional patterned metal7 (ME7) layer; (**b**) The thick photoresist passivation layer is deposited for etch protection; (**c**) The anisotropic silicon oxide dry etching; (**d**) The isotropic silicon substrate dry etching.

2.2. Accelerometer Design

The single axis accelerometer was first implemented and the tri-axis accelerometer was later developed. The proposed CMOS/MEMS accelerometer consisted of a proof mass, sensing fingers, springs, and a curl matching frame.

2.2.1. Single Axis Accelerometer

Figure 2a is the top view of the proposed single axis accelerometer. The proof mass was suspended above the substrate by four sets of springs. The proof mass was a perforated structure that can be undercut etched to release the suspended structures. The density and size of the etching holes was limited by the etching condition of the undercut process. The design used a 6 μm × 6 μm etching hole and 6 μm spacing to form the proof mass based on the MEMS design rules from the manufacturers.

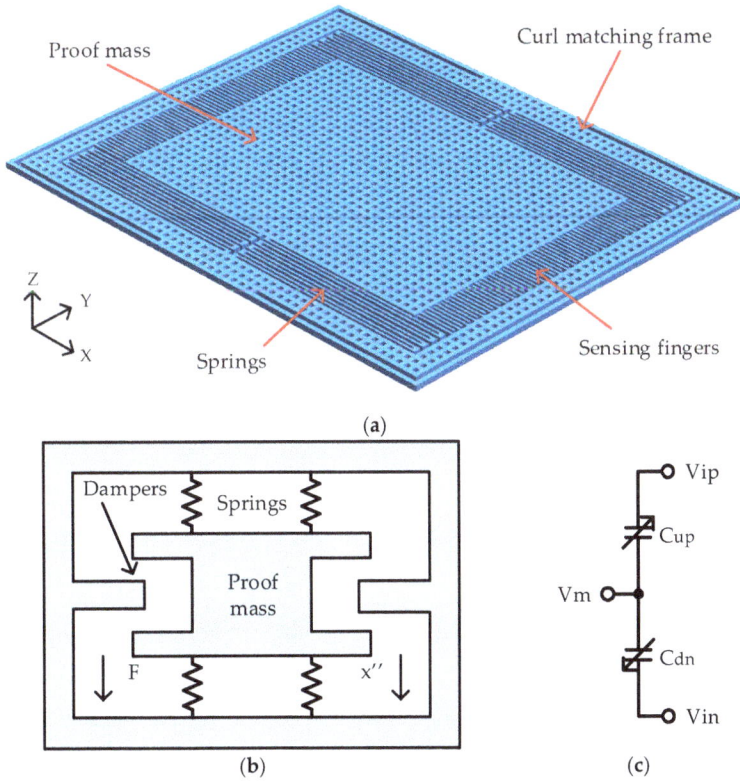

Figure 2. The proposed accelerometer: (**a**) The top view of the proposed accelerometer; (**b**) Mechanical model of the structure; (**c**) Circuit model of the structure.

The micro-accelerometer was equivalent to the mechanical model in Figure 2b. It is a second-order mass-spring-damper system modeled by the force balance equation, where F is applied force, m is the mass of suspended proof mass, x is the displacement, b is the damping coefficient and k is the spring constant:

$$F = mx'' + bx' + kx \tag{1}$$

The displacement (x) was transformed into capacitance (ΔC) by sensing fingers. The capacitance-to-voltage readout circuit transformed the capacitance to voltage. The circuit model in Figure 2c was simulated with a readout circuit. The circuit was simulated with Cadence Spectre simulator (Cadence Design Systems, Inc., San Jose, CA, USA).

The stiffness of the spring plays an important role in sensor design. Softer springs have less stiffness, and this means the device will have larger displacement, and hence larger capacitance (ΔC). The stiffness of the spring was decided by the width, length and turns of the springs.

Detailed models can be used to obtain more accurate results at the expense of speed of analysis. By developing a simplified analytical model, we gained insight regarding the mechanical behavior. Accurate results using elaborate models can be obtained using a finite element method (FEM) simulation. Simulations were carried out in CoventorWare 10 (Coventor, Inc., Cary, NC, USA).

MemMech is the FEM mechanical solver of CoventorWare, which is capable of computing displacement, reaction force and modal displacement. The material database provided contains characterized material properties for mechanical simulation. For a linear analysis, the displacement was calculated with the assumption that the stiffness is constant. For the nonlinear structural analysis in MemMech, the structure's stiffness changed as it deformed. The stiffness matrix of the structure was much more complicated to solve than a linear analysis. The nonlinearity caused by material nonlinearity, boundary nonlinearity and geometric nonlinearity were considered in the FEM simulations.

In this paper, serpentine springs are adopted for structure design, as in Figure 3a. By analyzing the structure, an approximate analytical model for the spring design is presented.

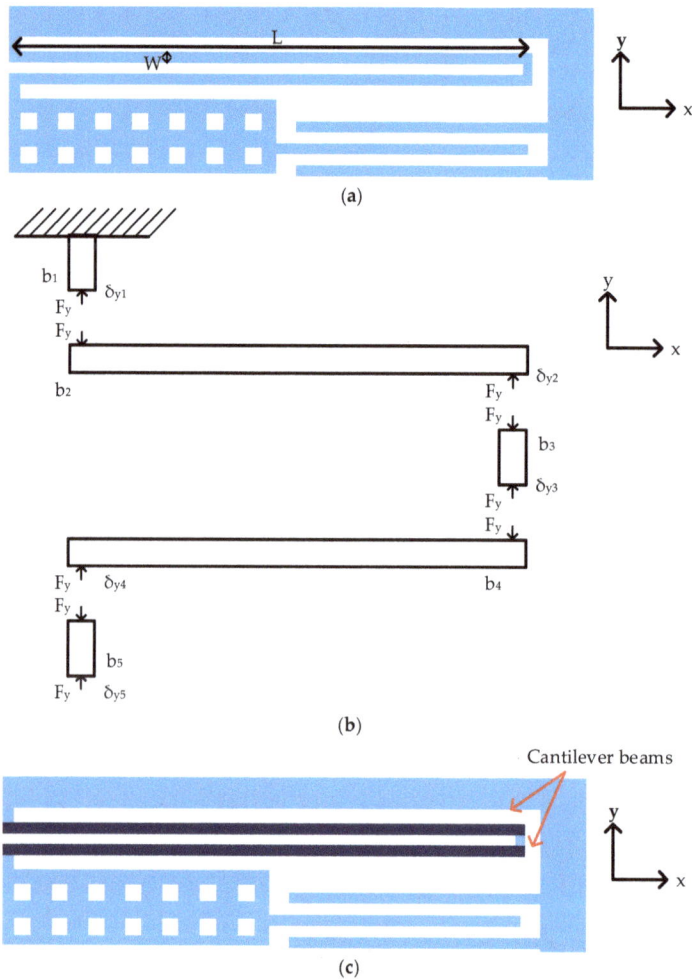

Figure 3. The schematic of serpentine spring: (**a**) Spring design parameters; (**b**) The free body diagram; (**c**) The proposed simplified model.

The schematic of proposed serpentine structure is shown in Figure 3a. Figure 3b shows the free body diagram of a serpentine spring. The beam segments were indexed from b_1 to b_5. The spring constant was found by applying a force balance to each beam segment. According to Hooke's law, the relation between applied force (F_y), spring constant along y-axis (k_y) and displacement along y-axis (δ_y) is formulated below:

$$F_y = k_y \cdot \delta_y \qquad (2)$$

As in Figure 3b, a lateral force along the y-axis (F_y) was applied at the end of the spring. The displacement along y-axis for each beam segment was given by:

$$\delta_{yi} = \frac{F_y}{k_{yi}} \tag{3}$$

where i is the index of beam segment from 1 to 5, δ_{yi} is the corresponding displacement along y-axis, and k_{yi} is spring constant of the segment along y-axis.

The spring constant was obtained by summing the displacement of each segment and then divided by the applied force F_y.

$$
\begin{aligned}
\delta_y &= \sum_{i=1}^{5} \delta_{yi} \\
&= \delta_{y1} + \delta_{y2} + \delta_{y3} + \delta_{y4} + \delta_{y5} \\
&= \frac{F_y}{k_{y1}} + \frac{F_y}{k_{y2}} + \frac{F_y}{k_{y3}} + \frac{F_y}{k_{y4}} + \frac{F_y}{k_{y5}}
\end{aligned}
\tag{4}
$$

Beam segment b_2 and b_4 were clamped-guided cantilever beams, hence spring constants k_{y2} and k_{y4} are listed below, where k_c is spring constant along the y-axis, E was Young's modulus of elasticity, t was the thickness of structure, W is the width of spring, n was the number of cantilever beam segments in series and L was the length of spring [23].

$$k_c = \frac{EtW^3}{L^3} \tag{5}$$

Beam segments b_1, b_3 and b_5 were rectangular beams hence spring constants k_{y1}, k_{y3} and k_{y5} are given by k_s [23]. The beam segments b_1, b_3 and b_5 were very stiff along the y-axis. There was almost no displacement along the y-axis. The width of the segment was deliberately selected two times larger than the cantilever beam to minimize the displacement of segments b_1, b_3 and b_5. The resulting spring constant was about 10^5 times larger than k_{y2} and k_{y4}.

$$k_s = \frac{EtW}{L} \tag{6}$$

By ignoring the displacement of beam segments b_1, b_3 and b_5, the serpentine spring only consisted of cantilever beams in series as in Figure 3c.

$$\delta_y \approx \frac{F_y}{k_{y2}} + \frac{F_y}{k_{y4}} = \frac{2F_y}{k_c} \tag{7}$$

$$k_y \approx \frac{F_y}{\delta_y} = \frac{F_y}{\frac{2F_y}{k_c}} = \frac{k_c}{2} \tag{8}$$

The spring constant of n cantilever beam segments in the y-axis was given by:

$$k_y = \frac{k_c}{n} = \frac{EtW^3}{nL^3} \tag{9}$$

The whole structure consisted of four sets of serpentine structures. Therefore, the spring constant of whole structure in the y-axis was four times that of a single serpentine structure.

$$k_y = \frac{4EtW^3}{nL^3} \tag{10}$$

Table 1 summarizes the spring design parameters of proposed single axis accelerometer.

Table 1. Spring design parameters.

Specifications	Design
Young's Modulus of Elasticity (E) (GPa)	70
Spring Width (W) (μm)	4
Spring Length (L) (μm)	370
Cantilever Beam Segments (n) (count)	8
Structure Thickness (t) (μm)	10.14

The displacement along the y-axis (δ_y) can be obtained by the following equation where m is mass of the proof mass and a_y was the acceleration along y-axis. The 1 g acceleration a_y was around 9.81 m/s^2. The dimension of proof mass was 606 μm $\times$ 462 μm $\times$ 10.14 μm and m was around 4.32 μg.

$$F_y = m \cdot a_y = k_y \cdot \delta_y \tag{11}$$

With mass and spring constant, the resonant frequency was given by:

$$f = \frac{1}{2\pi} \sqrt{\frac{k_y}{m}} \tag{12}$$

Table 2 compares the results predicted by FEM simulations and the proposed simplified analytical model. The predicted spring constant was slightly higher than the FEM results since the displacement was underestimated. From the simplified analytical model above, the y-axis spring constant was proportional to W^3, therefore the width of the spring (W), must be kept small to get higher sensitivity. The spring width (W), was limited to 4 μm by the CMOS/MEMS process. Increasing the spring length (L), or the number of cantilever beam segments in series n in a limited size can produce higher sensitivity. The proposed accelerometer had the displacement of 104.99 nm at 1 g. The FEM simulation results are listed in Table 3.

Table 2. Comparisons of design and finite element method (FEM) simulation results.

Specifications	Design	FEM	Error (%)
y-axis Spring Constant (k_y) (N/m)	0.45	0.41	8.18

Table 3. FEM simulation results.

Specifications	FEM
Displacement at 1 g (nm)	104.99
Initial Capacitance (C_0) (fF)	91.97
Capacitance (ΔC) (fF)	2.35
Resonant Frequency (f_0) (Hz)	1562.85
Mass (M) (μg)	4.32
Spring Constant (K) (N/m)	0.415

2.2.2. Tri-Axis Accelerometer

Figure 4 shows the proposed tri-axis single proof mass accelerometer. The tri-axis single proof mass accelerometer had an area of 1096 μm $\times$ 1256 μm. In order to suppress the structure curving effect, a curl matching frame was presented to achieve the same structure curling at the proof mass and the frame. The perforated structure and the layer combination were same for the proof mass and the frame to match the curling of the two parts. The layer combination ME1 and ME6 was chosen based on our previous work [9]. The z-axis sensor was embedded in the proof mass of the y-axis and the y-axis sensor was embedded in the proof mass of x-axis sensor. The springs of the x and y-axis were similar to a single axis design. Table 4 summarizes the in-plane (x-axis and y-axis) spring design parameters. Table 5 shows the results predicted by the FEM simulations and the proposed simplified analytical model.

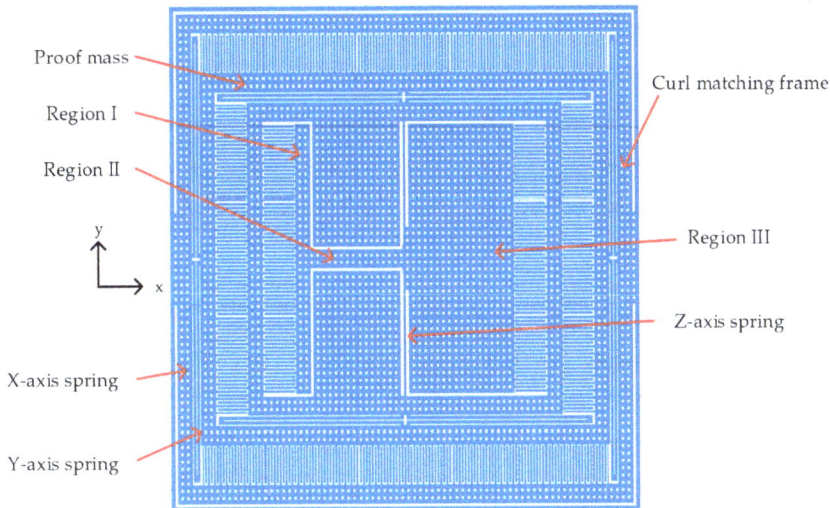

Figure 4. Top view of the proposed tri-axis accelerometer.

Table 4. In-plane spring design parameters.

Specifications	*x*-Axis	*y*-Axis
Young's Modulus of Elasticity (*E*) (GPa)	70	70
Spring Width (*W*) (μm)	5	5
Spring Length (*L*) (μm)	472	489
Cantilever Beam Segments (*n*) (count)	2	2
Structure Thickness (*t*) (μm)	10.14	10.14

Table 5. Comparisons of design and finite element method simulation results.

Specifications	Design	FEM	Error (%)
x-axis Spring Constant (k_x) (N/m)	1.69	1.63	3.45
y-axis Spring Constant (k_y) (N/m)	1.52	1.50	1.19

The torsion spring in Figure 5a was adopted for out-plane sensing. The imbalanced torsional *z*-axis sensing element was embedded in the in-plane proof mass. The design equation of the torsion spring was given by [23,24]:

$$k_\theta = \frac{GtW^3}{L}\left[\frac{1}{3} - 0.21\frac{W}{t}\left(1 - \frac{W^4}{12t^4}\right)\right] \tag{13}$$

where *G* is shear modulus, *W* is the width of the torsion beam, *L* is the length of the torsion beam, *t* is structure thickness. Table 6 summarizes the spring design parameters.

The whole structure consisted of two sets of torsional structures. Therefore, the spring constant of whole structure was two times that of a single torsional structure.

(a)

(b)

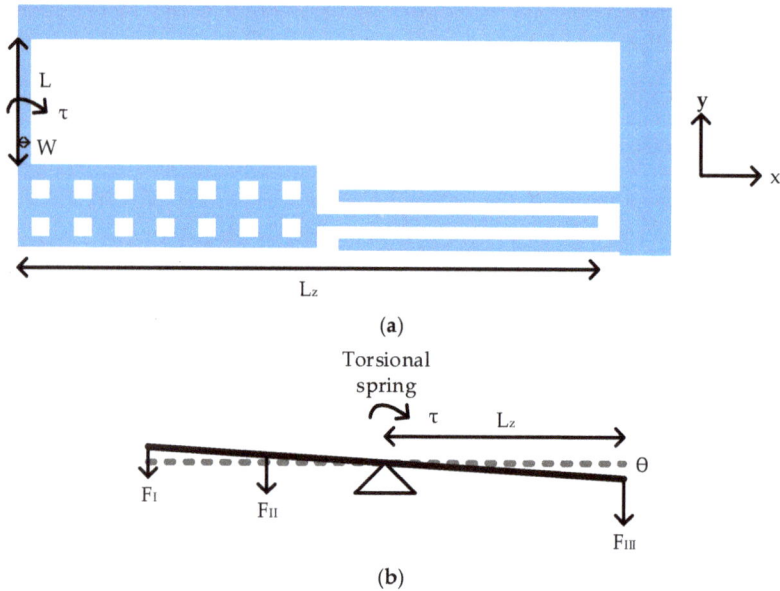

Figure 5. The schematic of torsion spring: (**a**) Spring design parameters; (**b**) The free body diagram.

Table 6. Torsion spring design parameters.

Specifications	Design
Shear Modulus (G) (GPa)	79
Spring Width (W) (µm)	4
Spring Length (L) (µm)	300
Structure Thickness (t) (µm)	10.14

The imbalanced sensing element consisted of three regions as in Figure 4. The design parameters are specified in Table 7.

Table 7. Proof mass design parameters.

Part	Area (µm × µm)	Moment Arm Length (µm)
Region I	700 × 40	280
Region II	260 × 40	130
Region III	700 × 30	150

Figure 5b shows the free body diagram of a torsion spring. F_I to F_{III} are force from these three parts. According to Hooke's law in angular form, the relation between applied torque (τ), torsion spring constant (k_θ) and rotation angle (θ) is formulated below:

$$\tau = k_\theta \theta \tag{14}$$

The displacement along z-axis (δ_z) was obtained by the following equation where L_z is the distance from the torsion spring to the sensing finger as in Figure 5. For 1 g acceleration τ was around 2.77×10^{-12} N·m, rotation angle was 3.23×10^{-5} rad and δ_z was around 16.24 nm. The displacement of FEM simulation was 20.08 nm. The FEM simulation results are listed in Table 8.

$$\delta_z = \theta \cdot L_z \tag{15}$$

Table 8. FEM simulation results.

Specifications	x-Axis	y-Axis	z-Axis
Displacement at 1 g (nm)	85.05	62.59	20.08
Resonant Frequency (f_0) (Hz)	1708.45	1991.43	2634.14
Mass (M) (μg)	14.15	9.58	3.88

3. Results

Figure 6 shows the chip photography for the sensors and readout circuitry. The single axis accelerometer had an area of 768 μm × 888 μm. The tri-axis accelerometer had an area of 1096 μm × 1256 μm.

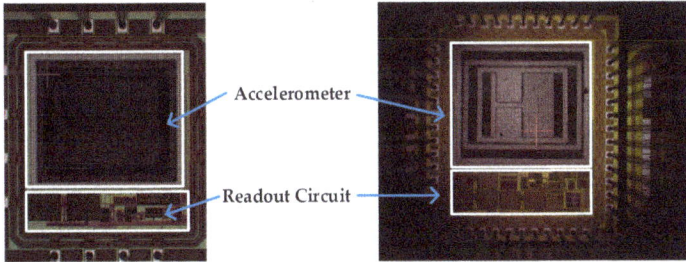

Figure 6. The chip die photo: (**a**) The single axis test chip; (**b**) The tri-axis test chip.

3.1. Surface Topography Measurement

The scanning electron microscope (SEM, TM3000, Hitachi, Tokyo, Japan) image of the whole structure is shown in Figure 7a. Figure 7b shows the curl matching frame. The serpentine spring in Figure 7c is used for x-axis sensing. Figure 7d shows the fabricated torsion spring for z-axis sensing.

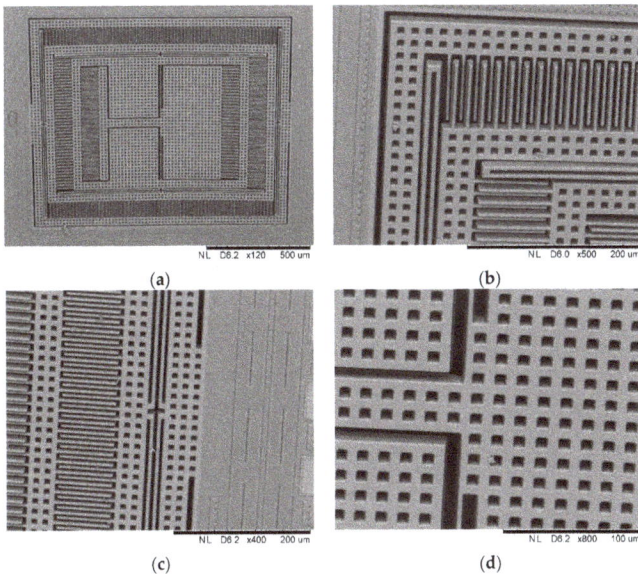

Figure 7. The SEM images of the proposed accelerometer: (**a**) The whole structure; (**b**) The curl matching frame; (**c**) The x-axis spring; (**d**) The torsion spring.

The white light interferometer (MSA-500, Polytec, Waldbronn, Germany) was used for surface topography measurement as in Figure 8. The white light interferometry measured the surface height and constructed three-dimensional surface profiles.

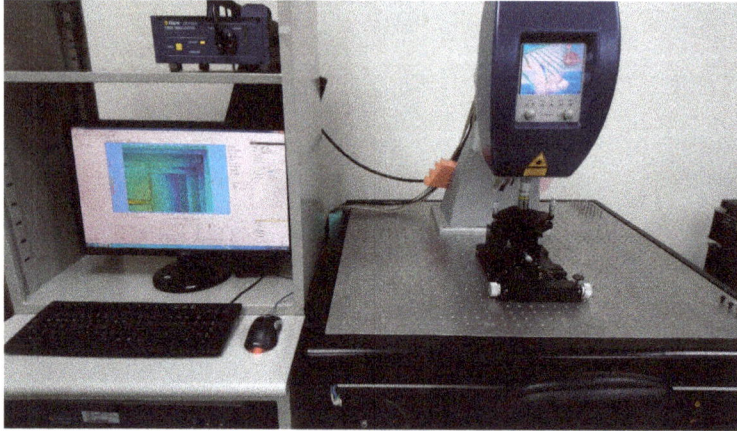

Figure 8. Measurement setup surface topography measurement.

Figure 9a shows the three-dimensional view of the structure. The structure curling is around 20 µm. Figure 9b shows the top view of the structure and the curvature. A and A′ are located at the curl matching frame. Figure 9c shows the curvature of AA′ cross section. The curl matching frame had the same curvature as the structure which compensated the curl effect as illustrated in Figure 9c.

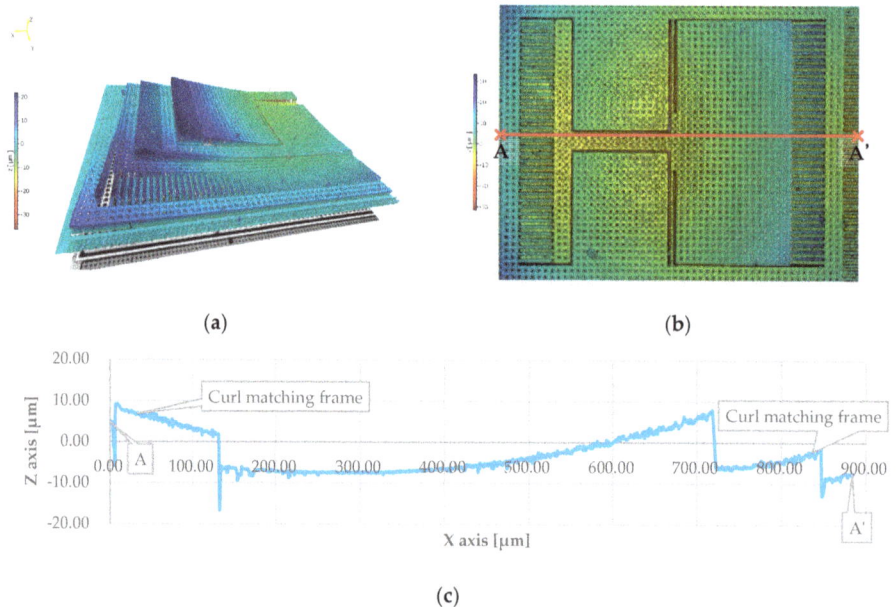

(a)

(b)

(c)

Figure 9. The surface profile of the proposed accelerometer: (**a**) The three-dimensional view; (**b**) The top view; (**c**) The cross-section view.

3.2. Mechanical Measurement

An in-plane vibration analyzer (MSA-500, Polytec, Waldbronn, Germany) was used to characterize the microstructure. The resonant frequency was detected optically at atmospheric pressure at room temperature with a laser Doppler vibrometer. Figure 10 shows the frequency response of the single axis accelerometer. The resonant was around 2 kHz. The in-plane resonant frequency of tri-axis accelerometer was around 2.5 kHz.

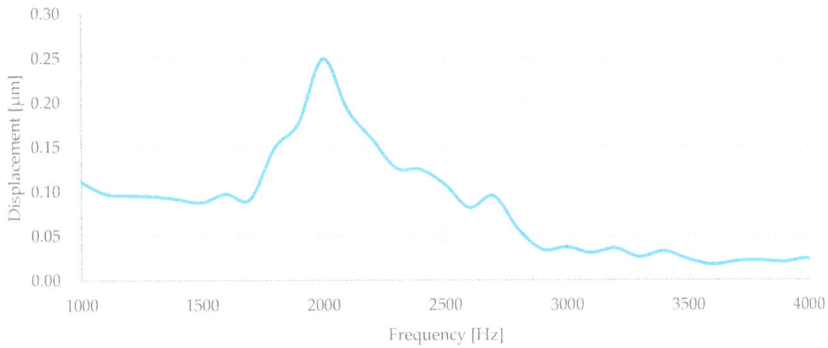

Figure 10. Frequency response of the proposed single axis accelerometer.

The out-plane (z-axis) motion was measured using a laser Doppler vibrometer (LV-1800, Ono Sokki, Yokohama, Japan). The output of the laser Doppler vibrometer was monitored using a network analyzer (Agilent 4395A, Keysight Technologies, Santa Rosa, CA, USA). Figure 11 shows the measurement setup for out-plane vibration characterization. The out-plane resonant frequency of tri-axis accelerometer was around 5.35 kHz, as in Figure 12.

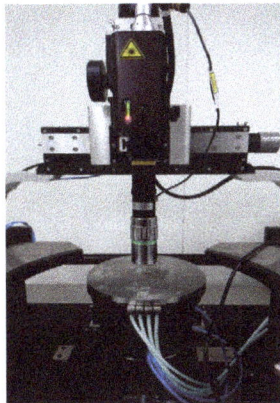

Figure 11. Measurement setup for out-plane vibration characterization.

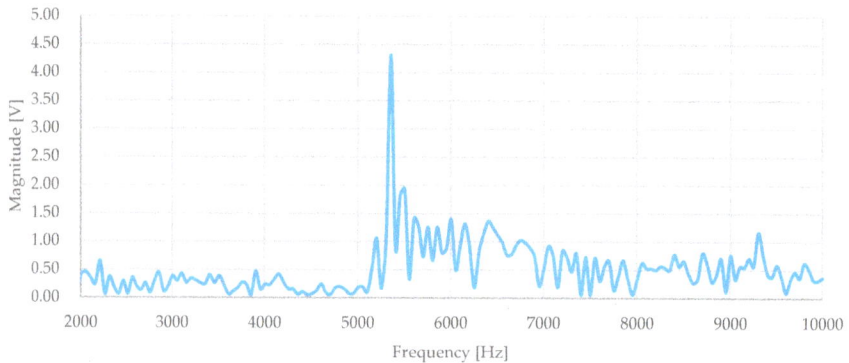

Figure 12. Out-plane frequency response of the proposed tri-axis accelerometer.

4. Discussion and Conclusions

The tri-axis single proof mass accelerometer had an area of 1096 μm $\times$ 1256 μm, while the single axis accelerometer had an area of 816 μm $\times$ 696 μm. Comparing the three individual sensing units, the tri-axis single proof mass accelerometer reduces 23.77% of the chip area. To solve the curving problem, the layer combination for residual stress reduction and the curl matching frame are presented. For in-plane sensing, the serpentine spring is used for both single and tri-axis design. For out-plane sensing, the torsion spring is adopted. Table 9 compares the performance of proposed model, FEM simulation and experimental results.

Table 9. Comparisons of proposed model, FEM simulation and experimental results.

Specifications	Proposed Model	FEM	Experimental Result
Single Axis: Resonant Frequency (f_0) (Hz)	1575.20	1562.85	2000.00
Tri-axis: In-plane Resonant Frequency (f_0) (Hz)	2036.05	1991.43	2500.00
Tri-axis: Out-plane Resonant Frequency (f_0) (Hz)	3910.12	2634.14	5354.65

For out-plane design, both of the models proposed and the FEM results show disagreement with the experimental results. The difference could be due to the dimensional variation of the process technology. The fabricated devices suffer from substantial parameter variations, from wafer to wafer and from lot to lot. Process variation of around 10% is usually considered in circuit design. The CMOS/MEMS process limits the minimum width of 4 μm. The 4 μm spring width was chosen to increase sensing capacitance. The design targets high sensitivity while process variation is anticipated. Spring width is the key design parameter to determine the spring constant and resonant frequency according to the proposed model. Geometries corresponding to 10% variation are depicted in Table 10. Increasing the spring width can lower the dimension variation at the cost of lowering sensitivity.

Table 10. Torsion spring constant and resonant frequency variation with spring width.

Specifications	Spring Width of 3.6 μm	Spring Width of 4 μm	Spring Width of 4.4 μm
Torsion Spring Constant (k_θ) (N·m/rad)	6.45×10^{-8}	8.57×10^{-8}	1.10×10^{-7}
Resonant Frequency (f_0) (Hz)	3392.79	3910.12	4436.83

This paper presents the design, simulation and mechanical characterization of a proposed CMOS/MEMS accelerometer. In this study, two accelerometer designs were evaluated, both theoretically and experimentally. The monolithic CMOS/MEMS accelerometer was fabricated using the 0.18 μm ASIC-compatible CMOS/MEMS process. An approximate analytical model for the spring design was presented. Surface topography measurement adopted both scanning electron microscope and white light interferometer for three-dimensional surface profiles observation. Mechanical measurement was performed for both in-plane and out-plane vibration analysis. The experiments show that the resonance frequency of the proposed tri-axis accelerometer was around 5.35 kHz for out-plane vibration.

Table 11 compares the performance of the proposed tri-axis accelerometer to the state-of-the-art CMOS/MEMS capacitive accelerometers. The proposed accelerometer was fabricated using a 0.18 μm CMOS/MEMS process. Compared with [10], which is a single axis accelerometer design, the proposed tri-axis accelerometer had an area of 1096 μm × 1256 μm.

Table 11. Comparison of the proposed accelerometer to the state-of-the-art.

	[10]	[15]	[16]	[13]	[11]	This Work
Sensing Range (g)	±6	0.8~6	-	-	0.25~6.75	±1
Resonant Frequency (Hz)	4.7	9.54	-	1.7	5.27	5.35
Sensor Area (μm × μm)	430 × 600	-	500 × 500	-	-	1096 × 1256
Process	UMC 0.18 μm CMOS/MEMS	TSMC 0.35 μm 2P4M process	TSMC 0.35 μm 2P4M process	TSMC 0.35 μm CMOS	0.35 μm CMOS/MEMS	TSMC/UMC 0.18 μm CMOS/MEMS

Author Contributions: Y.-S.L. conceived the idea, analyzed and performed the structure design, simulation and characterization. K.-A.W. conceived and supervised the project.

Funding: This research received no external funding.

Acknowledgments: The authors appreciate the National Chip Implementation Center (CIC), Taiwan, for supporting the CMOS/MEMS chip manufacturing. The authors also acknowledge Long-Sheng Fan for technical support with the CMOS/MEMS sensor.

Conflicts of Interest: The authors declare no conflicts of interest regarding the publication of this paper.

References

1. Qu, H. CMOS MEMS fabrication technologies and devices. *Micromachines* **2016**, *7*, 14. [CrossRef] [PubMed]
2. Tsai, M.; Liu, Y.; Sun, C.; Wang, C.; Cheng, C.; Fang, W. A 400 × 400 μm² 3-axis CMOS-MEMS accelerometer with vertically integrated fully-differential sensing electrodes. In Proceedings of the 2011 16th International Solid-State Sensors, Actuators and Microsystems Conference, Beijing, China, 5–9 June 2011; pp. 811–814. [CrossRef]
3. Yamane, D.; Konishi, T.; Takayasu, M.; Ito, H.; Dosho, S.; Ishihara, N.; Toshiyoshi, H.; Masu, K.; Machida, K. A sub-1g CMOS-MEMS accelerometer. In Proceedings of the IEEE Sensors, Busan, Korea, 1–4 November 2015; pp. 1–4. [CrossRef]
4. Chuang, W.; Hu, Y.; Chang, P. CMOS-MEMS test-key for extracting wafer-level mechanical properties. *Sensors* **2012**, *12*, 17094–17111. [CrossRef] [PubMed]
5. Wu, J.; Fedder, G.K.; Carley, L.R. A low-noise low-offset capacitive sensing amplifier for a 50-μg/√Hz monolithic CMOS MEMS accelerometer. *IEEE J. Solid-State Circuits* **2004**, *39*, 722–730. [CrossRef]
6. Boser, B.E.; Howe, R.T. Surface micromachined accelerometers. *IEEE J. Solid-State Circuits* **1996**, *31*, 366–375. [CrossRef]
7. Baltes, H.; Brand, O.; Hierlemann, A.; Lange, D.; Hagleitner, C. CMOS MEMS—Present and future. In Proceedings of the Technical Digest, MEMS 2002 IEEE International Conference, Fifteenth IEEE International Conference on Micro Electro Mechanical Systems (Cat. No. 02CH37266), Las Vegas, NV, USA, 24 January 2002; pp. 459–466. [CrossRef]
8. Yen, T.; Tsai, M.; Chang, C.; Liu, Y.; Li, S.; Chen, R.; Chiou, J.; Fang, W. Improvement of CMOS-MEMS accelerometer using the symmetric layers stacking design. In Proceedings of the IEEE Sensors, Limerick, Ireland, 28–31 October 2011; pp. 145–148. [CrossRef]
9. Kuo, F.Y.; Chang, C.S.; Liu, Y.S.; Wen, K.A.; Fan, L.S. Temperature-dependent yield effects on composite beams used in CMOS MEMS. *J. Micromech. Microeng.* **2013**, *23*. [CrossRef]

10. Tseng, S.; Lu, M.S.; Wu, P.; Teng, Y.; Tsai, H.; Juang, Y. Implementation of a monolithic capacitive accelerometer in a wafer-level 0.18 μm CMOS MEMS process. *J. Micromech. Microeng.* **2012**, *22*. [CrossRef]
11. Chiang, C. Design of a CMOS MEMS accelerometer used in IoT devices for seismic detection. *IEEE J. Emerg. Sel. Top. Circuits Syst.* **2018**, *8*, 566–577. [CrossRef]
12. Liu, Y.; Tsai, M.; Fang, W. Pure oxide structure for temperature stabilization and performance enhancement of CMOS-MEMS accelerometer. In Proceedings of the IEEE 25th International Conference on Micro Electro Mechanical Systems (MEMS), Paris, France, 29 January–2 February 2012; pp. 591–594. [CrossRef]
13. Qu, H.; Fang, D.; Xie, H. A single-crystal silicon 3-axis CMOS-MEMS accelerometer. In Proceedings of the IEEE Sensors, Vienna, Austria, 24–27 October 2004; Volume 2, pp. 661–664. [CrossRef]
14. Tseng, S.H.; Yeh, C.Y.; Chang, A.Y.; Wang, Y.J.; Chen, P.C.; Tsai, H.H.; Juang, Y.Z. A monolithic three-axis accelerometer with wafer-level package by CMOS MEMS process. *Proceedings* **2017**, *1*, 337. [CrossRef]
15. Sun, C.; Tsai, M.; Liu, Y.; Fang, W. Implementation of a monolithic single proof-mass tri-axis accelerometer using CMOS-MEMS technique. *IEEE Trans. Electron Devices* **2010**, *57*, 1670–1679. [CrossRef]
16. Tsai, M.; Liu, Y.; Sun, C.; Wang, C.; Fang, W. A CMOS-MEMS accelerometer with tri-axis sensing electrodes arrays. *Procedia Eng.* **2010**, *5*, 1083–1086. [CrossRef]
17. Jiang, K.; Chen, H.; Hsu, W.; Lee, Y.; Miao, Y.; Shieh, Y.; Hung, C. A novel suspension design for MEMS sensing device to eliminate planar spring constants mismatch. In Proceedings of the IEEE Sensors, Taipei, Taiwan, 28–31 October 2012; pp. 1–4. [CrossRef]
18. Karbari, S.R.; Kumari, U.; Pasha, R.C.; Gowda, V.K. Design and analysis of serpentine based MEMS accelerometer. *AIP Conf. Proc.* **2018**, *1966*, 020026. [CrossRef]
19. Fang, Y.; Mukherjee, T.; Fedder, G.K. SI-CMOS-MEMS dual mass resonator for extracting mass and spring variations. In Proceedings of the 2013 IEEE 26th International Conference on Micro Electro Mechanical Systems (MEMS), Taipei, Taiwan, 20–24 January 2013; pp. 657–660. [CrossRef]
20. ADXL327, Datasheet, Analog Devices, Inc. Available online: https://www.analog.com/ADXL327 (accessed on 17 December 2018).
21. Liu, Y.; Huang, C.; Kuo, F.; Wen, K.; Fan, L. A monolithic CMOS/MEMS accelerometer with zero-g calibration readout circuit. In Proceedings of the Eurocon 2013, Zagreb, Croatia, 1–4 July 2013; pp. 2106–2110. [CrossRef]
22. Liu, Y.; Wen, K. Monolithic Low Noise and Low Zero-g Offset CMOS/MEMS Accelerometer Readout Scheme. *Micromachines* **2018**, *9*, 637. [CrossRef] [PubMed]
23. Kaajakari, V. Beams as micromechanical springs. In *Practical MEMS: Analysis and Design of Microsystems, MEMS Sensors, Electronics, Actuators, RF MEMS, Optical MEMS, and Microfluidic Systems*; Small Gear Publishing: Las Vegas, NV, USA, 2009; pp. 56–65. ISBN 9780982299104.
24. Qu, H.; Fang, D.; Xie, H. A monolithic CMOS-MEMS 3-axis accelerometer with a low-noise, low-power dual-chopper amplifier. *IEEE Sens. J.* **2008**, *8*, 1511–1518. [CrossRef]

micromachines

MDPI

Article

Bonding Strength of a Glass Microfluidic Device Fabricated by Femtosecond Laser Micromachining and Direct Welding

Sungil Kim [1,2], Jeongtae Kim [1], Yeun-Ho Joung [1], Jiyeon Choi [2,*] and Chiwan Koo [1,*]

[1] Department of Electronics and Control Engineering, Hanbat National University, Daejeon 34158, Korea;
   sung1@hanbat.ac.kr (S.K.); Jeotae@daum.net (J.K.); Yeunho@gmail.com (Y.-H.J.)
[2] Department of Laser and Electron Beam Application, Korea Institute of Machinery and Materials,
   Daejeon 34103, Korea
*   Correspondence: jchoi@kimm.re.kr (J.C.); cwankoo@hanbat.ac.kr (C.K.); Tel.: +82-42-868-7536 (J.C.);
   +82-42-821-1168 (C.K.)

Received: 31 October 2018; Accepted: 30 November 2018; Published: 3 December 2018

check for
updates

Abstract: We present a rapid and highly reliable glass (fused silica) microfluidic device fabrication process using various laser processes, including maskless microchannel formation and packaging. Femtosecond laser assisted selective etching was adopted to pattern microfluidic channels on a glass substrate and direct welding was applied for local melting of the glass interface in the vicinity of the microchannels. To pattern channels, a pulse energy of 10 µJ was used with a scanning speed of 100 mm/s at a pulse repetition rate of 500 kHz. After 20–30 min of etching in hydrofluoric acid (HF), the glass was welded with a pulse energy of 2.7 µJ and a speed of 20 mm/s. The developed process was as simple as drawing, but powerful enough to reduce the entire production time to an hour. To investigate the welding strength of the fabricated glass device, we increased the hydraulic pressure inside the microchannel of the glass device integrated into a custom-built pressure measurement system and monitored the internal pressure. The glass device showed extremely reliable bonding by enduring internal pressure up to at least 1.4 MPa without any leakage or breakage. The measured pressure is 3.5-fold higher than the maximum internal pressure of the conventional polydimethylsiloxane (PDMS)–glass or PDMS–PDMS bonding. The demonstrated laser process can be applied to produce a new class of glass devices with reliability in a high pressure environment, which cannot be achieved by PDMS devices or ultraviolet (UV) glued glass devices.

Keywords: microfluidic; femtosecond laser; rapid fabrication; glass welding; bonding strength

---

1. Introduction

Microfluidic devices (or lab-on-a-chip) have been actively researched because they are able to provide rapid reaction and high-throughput screening of very small samples such as protein, DNA, cells, and tissues [1]. The material most commonly used for fabricating microfluidic devices is polydimethylsiloxane (PDMS) because it is transparent, biocompatible, and easy to fabricate. Glass-based microfluidic devices possess many advantages such as high transparency in the visible spectral range, high thermal resistance, and higher mechanical and chemical stability than PDMS [2]. Glass is biocompatible and does not absorb any organic compounds [3]. Glass micromachining, however, requires long fabrication time and efforts. To fabricate micro patterns on a glass sheet, there are several methods, such as wet/dry etching, mechanical fabrication, molding process, the use of photosensitive glass, and so on [4]. Those processes are complex and consist of multiple steps. Furthermore, they require special tools or a clean room facility. In addition, glass-to-glass bonding, such as thermal, fusion, anodic, and adhesive bonding, have disadvantages of long fabrication time and high cost [5]. Moreover, high temperature, high pressure, and high voltage conditions during these types of bonding do not provide an appropriate environment for devices with integrated electrical and mechanical parts [6]. Adhesive bonding using epoxy or ultraviolet (UV) cured glue is very simple, but most adhesives are vulnerable to solvent degradation of the bonding strength, and hence it is difficult to encapsulate devices [7]

Femtosecond laser processing of glass has attracted huge attention in recent years for applications utilizing glass such as photonics, displays, and optoelectronics [8,9]. It has revolutionized glass micromachining and glass-to-glass bonding to allow simple, rapid, and stable fabrication and packaging processes [10–19]. Femtosecond laser processing has also reduced the fabrication complexity as it can be utilized for both the formation of micro-patterns and the packaging of a glass chip. The fundamental mechanism of the femtosecond laser glass processing is to induce nonlinear absorption allowing direct photo-ionization in the glass, which is available only at very high laser intensity over a range of ~TW/cm^2. In this regime, transparent media can directly absorb photons, leading to local melting or structural modification at the exposed area. In particular, femtosecond laser direct welding of glass is one of the most innovative applications of femtosecond laser processing as there is no need to use sacrificial media or indirect heat transfer from intermediate absorbing layers. In most conventional transparent laser welding processes, it is crucial to have interfacial layers between glass substrates that are absorptive at the laser wavelength to absorb photons and deliver heat to the glass substrates [18]. Therefore, conventional transparent laser welding is not direct bonding of glass substrates or a single step process.

The first glass-to-glass welding by a femtosecond laser was reported by Tamaki et al. in 2005 [11]. They demonstrated the disappearance of Newton's ring at the femtosecond laser scanned area, which implies the removal of the gap between glass substrates. Thanks to intensive investigations, including those from Cvecek et al. and Okamoto et al., this method achieved higher throughput and process reliability by introducing efficient heat accumulation at MHz repetition, as well as optimized contact treatment methods [12–18,20,21].

Femtosecond laser glass welding provides several advantages for microfluidic device packaging. First, it has excellent chemical resistance due to the bonding of the base material, which is locally melted without any surface treatment or sacrificial layers [16]. Second, the width of the welding seam is typically a few tens of micrometers, keeping the welded lines as close as possible in a micrometer scale to the microfluidic channel, thus extending the useful space of the substrate [16]. Third, it has higher bonding strength than other bonding methods [16,18,20,21]. Although there is a lack of standardization and evaluation protocol of welding strength measurements thus far, a number of prior studies have already proven that femtosecond laser direct welding shows superior bonding strength compared with conventional bonding methods (e.g., shear stress measurement, three-point bending test, fracture strength, and simple leak check) [19,21–24].

To fabricate micro-patterns on a fused silica substrate, femtosecond laser assisted selective etching is presented [9,10,24–28]. Laser assisted selective etching is more effective for glass micromachining than a direct ablation process, because the former provides better surface roughness by minimizing debris and cracks. Laser-irradiated parts have a relatively high etch rate due to the laser induced material modification resulting in material density or phase changes, as well as the generation of nano-sized cracks increasing the surface area when immersed in etching fluids such as hydrofluoric acid (HF) and potassium hydroxide (KOH) [25,26]. The fabrication resolution is dependent on the configuration of the beam focusing optics and etching conditions. However, a few micron resolution is readily achievable for typical laser assisted selective etching.

In recent years, glass microfluidic device fabrication using a femtosecond laser has been actively researched. In particular, it has advantages in terms of fabrication compared with the conventional processing methods, including MEMS and photolithography. Although these conventional techniques have been well established, they are optimized for 2D planer surface microfabrication, while the laser process is able to fabricate a 3D microfluidic structure inside the glass without multiple complex process steps [27,28]. Furthermore, laser patterning is a direct-write scheme that can remove the need for a mask to transfer desired patterns onto the glass. It brings a huge benefit in terms of reducing manufacturing complexity and cost. However, long development time of laser exposed glass chips due to the low etch rate in KOH may be a drawback of the monolithic femtosecond laser fabrication of 3D internal microfluidic channels of glass. According to a prior study by J. Gottmann, the maximum anisotropic etch rate of femtosecond laser exposed glass is typically about 300 µm/h [26]. This implies that the appropriate use of the combination of 2D planar structures with laser direct welding may be reasonable for manufacturing embedded 3D structures in practical applications.

In this notion, we present a rapid and reliable glass microfluidic device fabrication method for a simple 2D planer channel structure using femtosecond laser assisted selective etching with HF and direct welding by optimizing laser parameters. After fabricating a simple microfluidic device, the endurable pressure of the device was measured to characterize the glass-to-glass welding efficiency. To measure the pressure in a microfluidic channel, a customized hydraulic pressure measurement system was developed and the measured endurable pressure data was compared with that of other bonding methods. In addition, as a demonstration of the application of our proposed method, a droplet generator was fabricated using our method in this work.

2. Materials and Methods

2.1. Fabrication Procedure

We propose a rapid and highly reliable fabrication method for glass microfluidic devices that consists of three steps, as shown in Figure 1. In the first step, a microfluidic channel pattern was directly laser written onto a fused silica glass substrate (JMC glass, Ansan, Korea) with a size of 25 mm × 25 mm × 1 mm and a surface roughness (Ra) of 25 nm. A Yb-doped femtosecond laser amplifier (SatsumaHP2, Amplitude systèmes, Pessac, France) was coupled with a Galvano scanner (IntelliSCAN III, ScanLab, Puchheim, Germany), which was then focused by a f-theta lens (focal length 100 mm, Sill Optics, Wendelstein, Germany). The center wavelength of the laser was 1030 nm and the laser beam profile was Gaussian with a beam quality M^2 of 1.2. The pulse duration was adjustable from 300 fs to 10 ps. In this step, the pulse duration was kept at the shortest (~300 fs) period to maximize the laser intensity at the used pulse energy of 10 µJ. The pulse energy was sufficient to modify the structure of the glass. The laser beam scanning speed was 100 mm/s at a pulse repetition rate of 500 kHz. The focused laser beam size was about 20 µm and the laser writing was repeated 30 times with 10 µm pitch in the lateral direction to make the width of the microfluidic channel equal to 300 µm. Two holes for inlet and outlet of the microfluidic channel were then laser drilled with a scan speed of 2 m/s. The diameter of the holes was 1 mm.

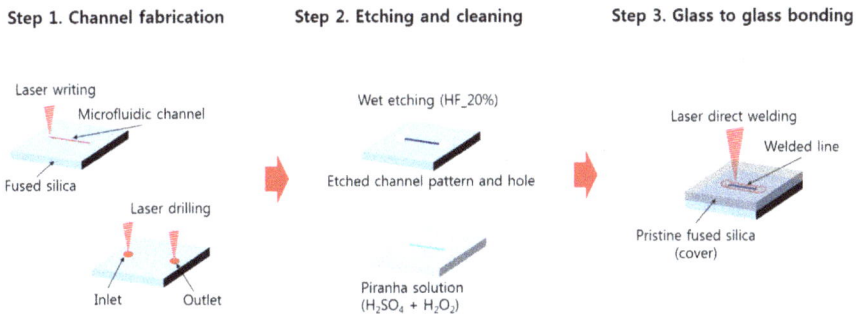

Figure 1. Illustration of the rapid and mask-less glass microfluidic channel device fabrication procedure by using a femtosecond laser. HF—hydrofluoric acid.

In the second step, the laser exposed fused silica substrate was etched in 20% HF acid for 30 min to obtain the microchannel pattern and clear inlet/outlet holes. After the etch process, the substrate was rinsed with DI water and was immersed in a 3:1 mixture of concentrated sulfuric acid (H_2SO_4) with hydrogen peroxide (H_2O_2), known as a piranha solution. This cleaning procedure is required for the next step to create highly hydrophilic surfaces and to provide optical contact [29] for better laser welding quality. Optical contact is not a mandatory condition for laser welding [30], however, we opted for optical contact for more rigid bonding. In the third step, another glass substrate was placed on top of the etched glass with patterns and they were bonded by femtosecond laser direct welding.

Figure 2 shows the femtosecond laser direct welding setup. Direct welding was performed by inducing heat accumulation at a high repetition rate of 2 MHz in the focused volume using an objective lens 20× (378-867-5, Mitutoyo, Kawasaki, Japan) with a numerical aperture of 0.40. The advantage of high repetition rate (MHz) ultrafast laser irradiation is that the molten layer is uniformly formed and thus bonding stability and strength increase [12]. The focal position is placed slightly below the interface of the two glass substrates. We used a 3D motorized machining stage to translate the glass substrates mounted on a custom tilting jig that compensates for the flatness within 5 μm. The laser pulse energy and welding speed were 2.7 μJ and 20 mm/s, respectively.

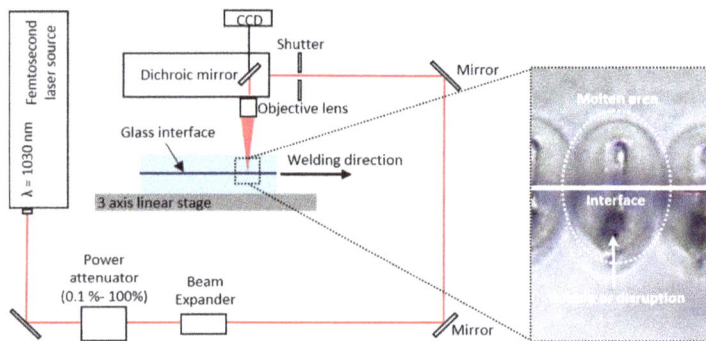

Figure 2. Schematic of laser direct welding experiment setup. The focal position is placed slightly below the interface of two glass substrates and bubbles or disruptions are formed at the position (as shown in the dotted box).

2.2. Characterization of the Fabricated Glass Microfluidic Device

To investigate the reliability of the proposed rapid glass microfluidic device packaging method, we performed internal pressure measurement and a leakage test. In addition, a simple droplet generator was fabricated with glass sheets and tested as a demonstration of our fabrication method on microfluidic applications.

2.2.1. Measurement of Internal Pressure and Leakage Test

To measure the endurable pressure of the fabricated glass microfluidic device packaging, we configured an internal pressure measurement system (Figure 3A). While injecting water into the inlet of the microfluidic channel and measuring the pressure at the outlet of the channel, the internal pressure built up until it reached the maximum pressure at which the welding of the glass device was cracked. The internal pressure was continuously measured as a voltage signal by an analogue voltage output pneumatic/fluidic pressure sensor (PX309-200G5V, OMEGA™, Sunbury, OH, USA) and the signal was converted from analogue to digital (ADC) and transferred to a computer. A syringe pump (Fusion 200, Chemyx Inc., Stafford, TX, USA) and a plastic syringe filled with water were used to inject water into the microfluidic channel. The connections between the syringe and the microfluidic channel and between the microfluidic channel and the pressure sensor were made with tubing (Figure 3B). To interface the glass microfluidic device with tubing, PDMS (Sylgard 184, Dow Corning, Midland, MI, USA) blocks were fabricated and bonded to the surface of the glass microfluidic device using oxygen plasma treatment [31]. The epoxy was applied to hold tubing tightly and prevent leakage between it and the PDMS blocks (Figure 3C). In addition, we fabricated glass microfluidic devices bonded using an adhesive glue to compare the bonding strength of our packaging method with that of gluing. While other glass-to-glass bonding methods use entire surface bonding or large bonding area [32], our packaging uses only the small welding area around the patterned microfluidic channel (Figure 4A).

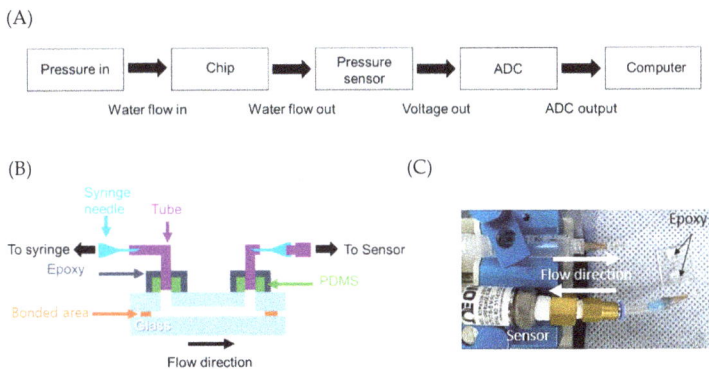

Figure 3. (**A**) Block diagram of the internal pressure measurement test procedure. (**B**) Configuration of interface between chip and peripheral devices and (**C**) the internal pressure measurement system. polydimethylsiloxane (PDMS) blocks were used as an interfacing material between tubing and the glass microfluidic device. Epoxy was applied around the PDMS block and tubing connection area. ADC—analogue to digital.

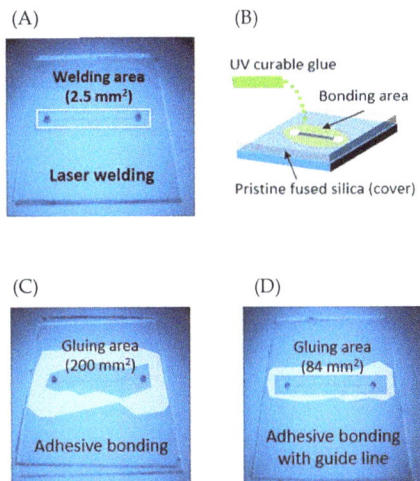

Figure 4. (**A**) Glass microfluidic device welded using the laser. The welding line (white line) was found around the microchannel. (**B**) Schematic of the ultraviolet (UV) curable glue application to bond two glass sheets. (**C,D**) Glass microfluidic devices bonded using UV curable glue. The left one had larger gluing area (200 mm^2) than the right (84 mm^2).

Similarly, UV curable glue is able to be applied on a small bonding area and takes short amount of time to bond two materials, so the glue (Loctite3321, Loctite, Dusseldorf, Germany) was selected for comparison with our packaging.

The glue was manually applied around the microchannel and a clean glass substrate was placed on it. We then waited until the glue stopped spreading out at the glass interface and exposed it to UV light (Figure 4B). Figure 4C shows the glass microfluidic device bonded with UV glue. Applying the glue on a glass substrate with microchannels was not simple and it was necessary to apply a moderate amount of glue at the proper place. Otherwise, the direction of the glue's spreading at the glass interface was not controlled and the bonding area was not uniform. From time to time, the glue went into microchannels and blocked the channel. When we used a small amount of glue to prevent the blockage, the bonding strength was weak and the two glass substrates easily came apart. Therefore, another glass microfluidic device was fabricated with a glue guide trench around the microfluidic channel (Figure 4D). The UV glue was applied in the guide line and it was guided well into the trench, not into the microfluidic channel. After fabricating the glass microfluidic devices bonded using the UV curable glue, their bonding strength was tested with an internal pressure measurement system.

2.2.2. Droplet Generator Experiment

To demonstrate our proposed rapid glass microfluidic device fabrication, we chose a droplet generator with a cross junction (focused flow), because it is widely used in many fields for generating highly reproducible micro- or nano-droplets of water, oil, and other materials [33]. A cross junction for generating droplets and a large chamber for collecting droplets were designed and fabricated on a glass substrate. For making the surface of the microfluidic channel hydrophobic, a water repellent agent (47100, Aquapel, Pittsburgh, PA, USA) was coated. To make water droplets, water mixed with blue dye was injected through the mid inlet, and oil mixed with a surfactant at a 2 wt.% ratio was injected through two side inlets.

3. Results and Discussion

3.1. Fabrication of the Glass Microfluidic Device

A simple microfluidic channel was patterned on a fused silica glass substrate and successfully bonded with another fused silica glass substrate. Figure 5A shows the welding seams around the microfluidic channel. The glass substrates were welded at about 500 μm intervals and the width of the welding seam line was 150 μm. The dark circular spots in Figure 5A are bubbles or disruptions of fused silica substrates [34]. A microfluidic channel 15 mm long and 300 μm wide was fabricated using laser assisted selective etching and scanning electron microscope (S-4800, Hitachi, Tokyo, Japan) images were observed, as shown Figure 5B. The surface of the channel was smooth because of the HF wet etch and the cross-sectional shape of the channel was a rounded channel, which is difficult to achieve by normal microfluidic channel fabrication methods such as soft lithography. The inlet/outlet area showed a smooth surface without glass particles. The roughness of the microfluidic channel surface and non-patterned glass surface depended on the time of HF wet etch. If the etch time was less than 10 min, the microfluidic channel surface was still rough to be used as a microfluidic channel, and thus the minimum etch time should be longer than 20 min. When it was etched over 60 min, the glass etch rate dropped dramatically, and hence a long etch time was not necessary. Thus, the etch time was optimized; 30 min was selected and we obtained 360 nm (Ra) roughness for the patterned microfluidic channel. During the wet etch, the non-patterned glass surface is also exposed to the etchant and the surface roughness may affect the welding quality and efficiency. Therefore, we compared the surface roughness using an atomic force microscope (XE-100, Park systems, Gyeonggi-do, Korea) before and after the HF wet etch. Figure 6 shows the measurement results of the surface roughness near the welding area by changing the etching time up to 40 min. According to published papers, optical contacted fused silica welding with ultrafast laser can bridge a 1 to 3 μm gap [20,35]. In conclusion, our etching time of

30 min, generating a surface roughness of 10 nm, which is still less than a hundredth of the wavelength of the welding laser, will not affect the welding efficiency [30].

(A)   (B)

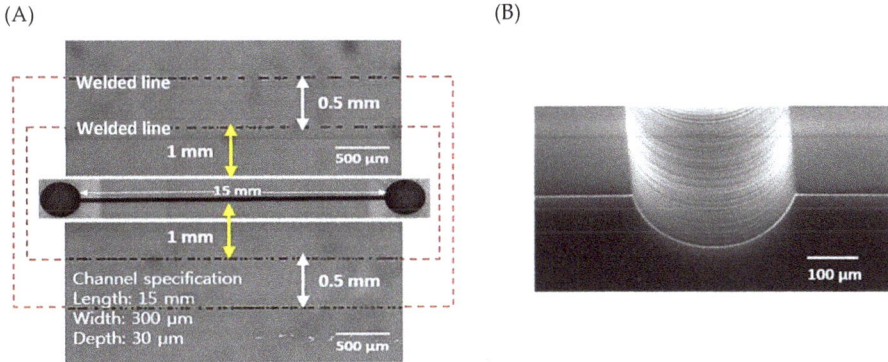

Figure 5. (**A**) Top view of glass microfluidic device using laser assisted selective etching (LASE) and welding. Channel specifications: length 15 mm, width 300 μm, and depth 30 μm. Red dash line welding area. (**B**) Scanning electron microscope (SEM) image showing the formed microfluidic channel with a surface roughness of 360 nm (Ra).

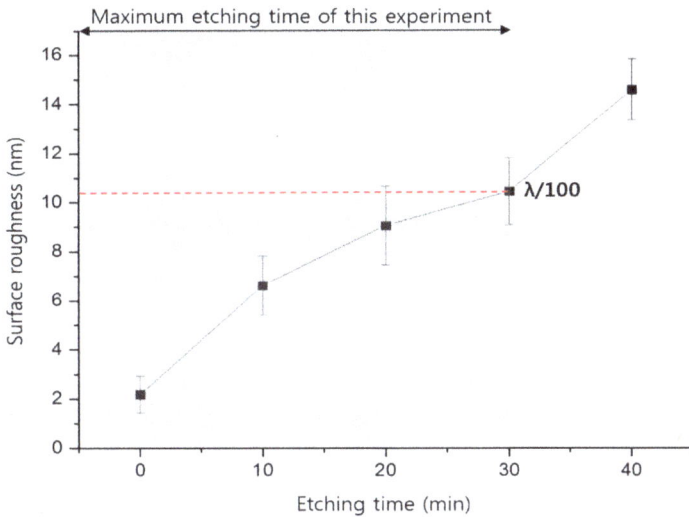

Figure 6. Effect of etching time on the surface roughness of the glass substrate (fused silica) (5 sections, the size of the examined area is 30×30 μm^2).

3.2. Internal Pressure Measurement

To confirm the reliability of the laser welding on a glass microfluidic device, internal pressure measurement was conducted. Upon slowly injecting water into the microfluidic channel in the glass device, the internal pressure was built up at a rate of 100 kPa/min and suddenly dropped after a point of time. The highest pressure was 1.4 MPa. However, the glass-to-glass welding still remained, but the epoxy sealing on the tubing connection between the glass device and the tubing was broken. Therefore, we carried out a tensile test (Instron 5848 Instron, Norwood, MA, USA) and the measured value was 7.5 MPa. The glass microfluidic device with an 84 mm^2 gluing area could prevent blockage of the microchannel by the UV glue with help of a UV glue guide line around the microchannel, but the bonding was broken at 1.1 MPa internal pressure. The glass device with a larger gluing area (200 mm^2) was able to withstand internal pressure of at least 1.4 MPa. Because of failure of the tubing connection, higher pressures could not be tested. The UV glue bonding method is able to provide a strong bond of two glass substrates, but it has disadvantages of requiring a large area of UV glue at the interface of the two glass substrates and an additional process for fabricating the UV glue guide line to prevent the applied UV glue from entering the microchannel. In addition, it requires preparation time of about 20 to 30 min for the UV glue to spread as well as UV curing time. Table 1 lists the reported bonding strength of the conventional bonding methods and our bonding method together. The PDMS-to-glass and PDMS-to-PDMS plasma bonding, which are widely used for fabricating microfluidic devices, endured 510 kPa and 551 kPa, respectively, in the liquid injection test [31,36]. The bonding strength of our method was at least two-fold higher that that provided by other methods. The glass-to-glass bonding method using a microwave kiln, removable ceramic paper, and a microwave oven showed about 1 MPa bonding strength in the liquid injection test [37]. Higher pressures could not be applied because of the failure of the tubing connection and it was assumed that their bonding strength is extremely high, corresponding with the glass strength. However, the method requires specific ceramic papers and the microchannel shape was deformed, while the entire bonding area melted and was bonded. The fabrication time for microchannels and bonding in a microwave oven was about 2 to 4 min, but it required a long cooling time, exceeding 45 min. In the cases of anodic bonding at room temperature and glass-to-glass rapid bonding using Pyrex glass, the bonding strength was measured as 29.7 MPa and 2.5 MPa, respectively, with a tensile test [38,39]. When comparing the strength of our method and anodic bonding and the rapid Pyrex glass bonding, the strength of our bonding was lower than that of the anodic bonding, but higher than that of the rapid Pyrex bonding. Except for the anodic bonding method, our laser welding method could endure higher force than other methods. Thus, we believe that the laser welding method provides reliable bonding. In addition, in terms of time consumption, our method was at least 10 times faster than other bonding methods.

Table 1. Comparison of glass bonding method. PDMS—polydimethylsiloxane.

Materials	Bonding Method	Bonding Time (h)	Maximum Bonding Strength (MPa)	Test Method
PDMS–Glass [31]	Plasma	0.5~2	0.51	Pressure injection
PDMS–PDMS [36]	Plasma	0.5~2	0.55	Pressure injection
Glass–Glass [37]	Microwave oven	1	>1 (assume: 1 to 30)	Pressure injection
Glass–Glass [38]	Anodic	24	29.7	Tensile test
Glass–Glass [39]	Pyrex	1.3	2.5	Tensile test
Glass–Glass	UV adhesive	0.5	1.1	Pressure injection
Glass–Glass	Laser welding (this work)	<0.08 (5 min)	>1.4	Pressure injection
			7.5	Tensile test

3.3. Liquid Leakage and Droplet Generator

We fabricated a droplet generator to investigate the possibility of applying our fabrication method to microfluidic devices in different fields. Before the droplet generation test, three dyes were injected into the microfluidic channels to generate a laminar flow and to observe the area of the laser welding seam to check if there was leakage. No leak was observed and a laminar flow was formed successfully (Figure 7A). Therefore, the microchannel surface roughness was smooth, allowing generation of a laminar flow, and the bonding tightness was enough to test for a normal microfluidic test. For the droplet generator test, the water with blue dye was injected through the mid inlet and the oil mixed with the surfactant at 2 wt.% ratio was injected through two side inlets. Micro-droplets were generated sequentially and successfully collected in a reservoir (Figure 7B–D). The developed fabrication method provides not only reliable packaging, but also the capability for application to glass microfluidic devices.

(A)

(B)

(C)

(D)

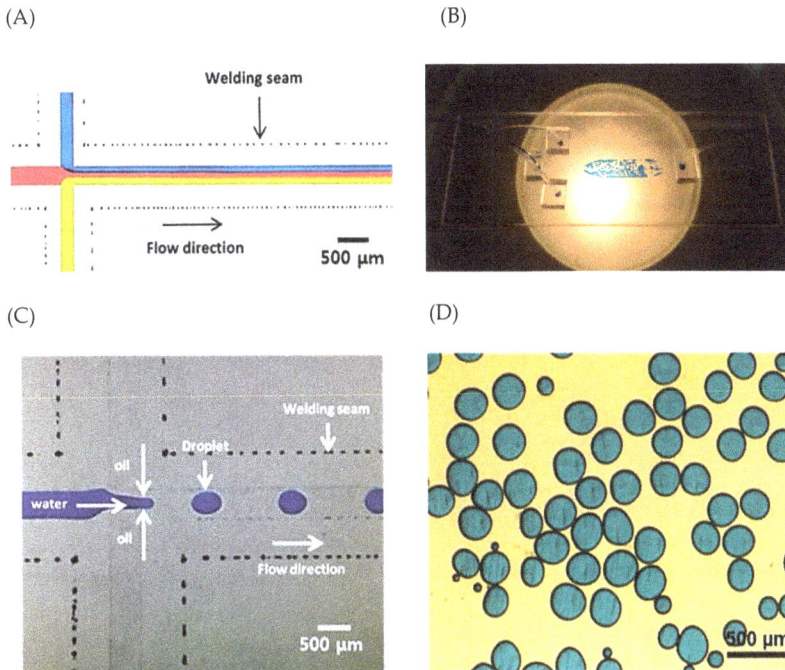

Figure 7. (**A**) Picture of the laminar flow generated at the microfluidic channel fabricated by laser test. (**B**) Picture of the fabricated droplet generator. (**C**) Picture of droplet generating at the cross-section. Blue dye was mixed with water to show the droplets clearly. (**D**) Generated droplets collected at a reservoir.

4. Conclusions

All glass microfluidic devices have advantages over polymer based microfluidic devices, but they are difficult to fabricate and can only be used for specific experiments. However, in this paper, we propose a rapid and reliable glass microfluidic device fabrication strategy based on laser assisted selective etching and direct welding of simple 2D planar channels, similar to most PDMS devices using optimized laser parameters. It even provides high bonding strength compared with conventional bonding methods. We believe that the fabrication method introduced here has potential to be used for glass microfluidic devices in harsh environments that PDMS microfluidic devices cannot endure. In future work, we will develop a glass microfluidic device using solvents or stains (e.g., Nile red and Rhodamine B) that cannot be applied to PDMS devices and a glass lab-on-a-chip enduring high pressure conditions.

Author Contributions: S.K. design of experiments, fabrication, and writing this paper; J.K. bonding strength test, analyzing the data, and writing this paper; J.C. developed the fabrication method and reviewed laser micromachining; Y.-H.J. and C.K. supervised and reviewed microfluidic pressure; S.K. and J.K. contributed equally to this work.

Acknowledgments: This work supported by the Ministry of Trade Industry and Energy (MOTIE) as Industrial Technology Innovation Program (Project No. 10052668), Global Excellent Technology Innovation (Project No. 10053877), Global collaborative R&D program (Project No. N0002095), and the Center for Advanced Meta-Materials (CAMM) funded by the Ministry of Science and ICT (MSIT) as Global Frontier (Project No. CAMM2014M3A6B3063707).

Conflicts of Interest: The authors declare no conflict of interest.

References

1. Chin, C.; Linder, V.; Sia, S. Lab-on-a-chip devices for global health: Past studies and future opportunities. *Lab Chip* **2007**, *7*, 41–57. [CrossRef] [PubMed]
2. Yalikun, Y.; Hosokawa, Y.; Iino, T.; Tanaka, Y. An all-glass 12 μm ultra-thin and flexible microfluidic chip fabricated by femtosecond laser processing. *Lab Chip* **2016**, *16*, 2427–2433. [CrossRef] [PubMed]
3. Jagannadh, V.; Mackenzie, M.; Pal, P.; Kar, A.K.; Gorthi, S.S. Imaging Flow Cytometry with Femtosecond Laser-Micromachined Glass Microfluidic Channels. *IEEE J. Sel. Top. Quantum Electron.* **2015**, *21*, 370–375. [CrossRef]
4. Wang, T.; Chen, J.; Zhou, T.; Song, L. Fabrication microstructures on glass for microfluidic chips by glass molding process. *Micromachines* **2018**, *9*, 269. [CrossRef] [PubMed]
5. Hulsenberg, D.; Harnisch, A.; Bismarck, A. *Micro Structuring of Glass*; Springer: Berlin, Germany, 2008; pp. 263–276. ISBN 978-3-540-49888-9.
6. Wang, C.; Wang, Y.; Tian, Y.; Wang, C.; Suga, T. Room-temperature direct bonding of silicon and quartz glass wafers. *Appl. Phys. Lett.* **2017**, *110*, 221602. [CrossRef]
7. Gong, Y.; Park, J.M.; Lim, J. An Interference-Assisted Thermal Bonding Method for the Fabrication of Thermoplastic Microfluidic Devices. *Micromachines* **2016**, *7*, 211. [CrossRef] [PubMed]
8. Zimmermann, F.; Richter, S.; Döring, S.; Tünnermann, A.; Nolte, S. Ultrastable bonding of glass with femtosecond laser bursts. *Appl. Opt.* **2013**, *52*, 1149–1154. [CrossRef]
9. Gattass, R.R.; Mazur, E. Femtosecond laser micromachining in transparent materials. *Nat. Photonics* **2008**, *2*, 219–2253. [CrossRef]
10. Sugioka, K.; Cheng, Y. Ultrafast lasers—Reliable tools for advanced materials processing. *Light Sci. Appl.* **2014**, *3*, e149. [CrossRef]
11. Tamaki, T.; Watanabe, W.; Nishii, J.; Itoh, K. Welding of transparent materials using femtosecond laser pulses. *JJAP* **2005**, *44*, 20–23. [CrossRef]
12. Tamaki, T.; Watanabe, W.; Itoh, K. Laser micro-welding of transparent materials by a localized heat accumulation effect using a femtosecond fiber laser at 1558 nm. *Opt. Express* **2006**, *14*, 10460–10468. [CrossRef] [PubMed]
13. Miyamoto, I.; Horn, A.; Gottmann, J.; Wortmann, D.; Yoshino, F.J. Novel fusion welding technology of glass using ultrashort pulse lasers. *J. Laser Micro/Nanoeng.* **2007**, *2*, 483–493. [CrossRef]
14. Horn, A.; Mingareev, I.; Werth, A.; Kachel, M.; Brenk, U. Investigations on ultrafast welding of glass–glass and glass–silicon. *Appl. Phys. A* **2008**, *93*, 171–175. [CrossRef]
15. Cvecek, K.; Miyamoto, I.; Strauss, J.; Wolf, M.; Frick, T.; Schmidt, M. Sample preparation method for glass welding by ultrashort laser pulses yields higher seam strength. *Appl. Opt.* **2011**, *50*, 1941–1944. [CrossRef] [PubMed]
16. Okamoto, Y.; Miyamoto, I.; Cvecek, K.; Okada, A.; Takahashi, K.; Schmidt, M. Evaluation of molten zone in micro-welding of glass by picosecond pulsed laser. *J. Laser Micro/Nanoeng.* **2013**, *8*, 65–69. [CrossRef]
17. Tan, H.; Duan, J.A. Welding of glasses in optical and partial-optical contact via focal position adjustment of femtosecond-laser pulses at moderately high repetition rate. *Appl. Phys. B* **2017**, *123*, 1–9. [CrossRef]
18. Huang, H.; Yang, L.M.; Liu, J. Direct welding of fused silica with femtosecond fiber laser. *Proc. SPIE* **2012**, *8244*, 824403.
19. Carvalho, R.R.; Reuvekamp, S.; Zuilhof, H.; Blom, M.T.; Vrouwe, E.X. Laser welding of pre-functionalized glass substrates: A fabrication and chemical stability study. *J. Micromech. Microeng.* **2018**, *28*, 015002. [CrossRef]
20. Cvecek, K.; Odato, R.; Dehmel, S.; Miyamoto, I.; Schmidt, M. Gap bridging in joining of glass using ultra short laser pulses. *Opt. Express* **2015**, *23*, 5681–5693. [CrossRef]
21. Schaffer, C.B.; García, J.F.; Mazur, E. Bulk heating of transparent materials using a high-repetition-rate femtosecond laser. *Appl. Phys. A* **2003**, *76*, 351–354. [CrossRef]
22. Hélie, D.; Gouin, S.; Vallée, R. Assembling an endcap to optical fibers by femtosecond laser welding and milling. *Opt. Mater. Express* **2013**, *3*, 1742–1754. [CrossRef]
23. Gstalter, M.; Chabrol, G.; Bahouka, A.; Serreau, L.; Heitz, J.-L.; Taupier, G.; Dorkenoo, K.-D.; Rehspringer, J.-L.; Lecler, S. Stress-induced birefringence control in femtosecond laser glass welding. *Appl. Phys. A* **2017**, *123*, 714. [CrossRef]

24. Kim, S.I.; Kim, J.; Koo, C.; Joung, Y.; Choi, J. Rapid prototyping of 2D glass microfluidic devices based on femtosecond laser assisted selective etching process. *Proc. SPIE* **2018**, *10522*, 105221V.

25. Yang, Q.; Tong, S.; Chen, F.; Deng, Z.; Bian, H.; Du, G.; Yong, J.; Hou, X. Lens-on-lens microstructures. *Opt. Lett.* **2015**, *40*, 5359–5362. [CrossRef] [PubMed]

26. Gottmann, J.; Hermans, M.; Repiev, N.; Ortmann, J. Selective Laser-Induced Etching of 3D Precision Quartz Glass Components for Microfluidic Applications—Up-Scaling of Complexity and Speed. *Micromachines* **2017**, *8*, 10. [CrossRef]

27. Sugioka, K.; Cheng, Y.; Midorikawa, K. Three-dimensional micromachining of glass using femtosecond laser for lab-on-a-chip device manufacture. *Appl. Phys. A* **2005**, *81*, 1–10. [CrossRef]

28. Cheng, Y. Internal Laser Writing of High-Aspect-Ratio Microfluidic Structures in Silicate Glasses for Lab-on-a-chip Applications. *Micromachines* **2017**, *8*, 59. [CrossRef]

29. Lord Rayleigh, F.R.S. A study of glass surfaces in optical contact. *Proc. Soc. Lond. A* **1936**, *156*, 326–349. [CrossRef]

30. Richter, S.; Zimmermann, F.; Eberhardt, R.; Tünnermann, A.; Nolte, S. Toward laser welding of glasses without optical contacting. *Appl. Phys. A* **2015**, *121*, 1–9. [CrossRef]

31. Bhattacharya, S.; Datta, A.; Berg, J.M.; Gangopadhyay, S. Studies on surface wettability of poly(dimethyl) siloxane (PDMS) and glass under oxygen-plasma treatment and correlation with bond strength. *J. MEMS* **2005**, *14*, 590–597. [CrossRef]

32. Lin, C.H.; Lee, G.B.; Lin, Y.H.; Chang, G. A fast prototyping process for fabrication of microfluidic systems on soda-lime glass. *J. Micromech. Microeng.* **2001**, *11*, 726–732. [CrossRef]

33. Weibe, D.B.; Whitesides, G.M. Applications of microfluidics in chemical biology. *Curr. Opin. Chem. Biol.* **2006**, *10*, 584–591. [CrossRef] [PubMed]

34. Cvecek, K.; Miyamoto, I.; Schmidt, M. Gas bubble formation in fused silica generated by ultra-short laser pulses. *Opt. Express* **2014**, *22*, 5877–15893. [CrossRef] [PubMed]

35. Chen, J.; Carter, R.M.; Thomson, R.R.; Hand, D.P. Avoiding the requirement for pre-existing optical contact during picosecond laser glass-to-glass welding. *Opt. Express* **2015**, *23*, 18645–18657. [CrossRef] [PubMed]

36. Eddings, M.A.; Johnson, M.A.; Gale, B.K. Determining the optimal PDMS–PDMS bonding technique for microfluidic devices. *J. Micromech. Microeng.* **2008**, *18*, 067001. [CrossRef]

37. Kopparthy, V.L.; Crews, N.D. Microfab in a Microwave Oven: Simultaneous Patterning and Bonding of Glass Microfluidic Devices. *J. MEMS* **2018**, *27*, 434–439. [CrossRef]

38. Howlader, M.M.R.; Suehara, S.; Suga, T. Room temperature wafer level glass/glass bonding. *Sens. Actuators A Phys.* **2006**, *127*, 31–36. [CrossRef]

39. Akiyama, Y.; Morishima, K.; Kogi, A.; Kikutani, Y.; Tokeshi, M.; Kitamori, T. Rapid bonding of Pyrex glass microchips. *Electrophoresis* **2007**, *28*, 994–1001. [CrossRef]

micromachines

MDPI

Article

Monolithic Low Noise and Low Zero-g Offset CMOS/MEMS Accelerometer Readout Scheme

Yu-Sian Liu *[iD] and Kuei-Ann Wen

Institute of Electronic Engineering, National Chiao Tung University, Hsinchu 300, Taiwan; stellawen@mail.nctu.edu.tw
* Correspondence: thomas.ee02g@nctu.edu.tw; Tel.: +886-3-573-1627

Received: 17 November 2018; Accepted: 28 November 2018; Published: 30 November 2018

check for updates

Abstract: A monolithic low noise and low zero-g offset CMOS/MEMS accelerometer and readout scheme in standard 0.18 μm CMOS mixed signal UMC process is presented. The low noise chopper architecture and telescopic topology is developed to achieve low noise. The experiments show noise floor is 421.70 μg/$\sqrt{Hz}$. The whole system has 470 mV/g sensitivity. The power consumption is about 1.67 mW. The zero-g trimming circuit reduces the offset from 1242.63 mg to 2.30 mg.

Keywords: Accelerometer readout; low noise; low zero-g offset

1. Introduction

Micro-electromechanical system (MEMS) products are widely used in our daily life. One of them is the MEMS accelerometer. The accelerometers have many applications in automobiles, navigation, vibration monitoring, and even portable electronics [1]. In most cases, measuring the accelerations and the additional signal processing are necessary. The CMOS (Complementary Metal-Oxide-Semiconductor)/MEMS process has the advantage of integration. The process can integrate MEMS devices as well as CMOS circuitry.

The main noise sources of readout circuit are thermal noise and flicker noise. The thermal noise comes from the random motion of electrons due to thermal effects. The 1/f noise or flicker noise is a low-frequency noise. The power spectral density (PSD) of flicker noise is inversely proportional to frequency. The accelerometer operates at low frequency. Hence, flicker noise is dominant.

Due to the process limitation, the sensing capacitor of the accelerometers at 1 g for the CMOS/MEMS process is in the order of few femto farads. The sensing signal may be damaged by electronic noise. Therefore low noise circuit is needed.

A low-noise feedforward noise reduction scheme is presented in Reference [2], which is a simple two-phase correlated double sampling (CDS) scheme to suppress the offset voltage and flicker noise. Both chopper stabilization (CS) and correlated double sampling are adopted in Reference [3]. The chopper stabilization modulate the sensing signal to high frequency. After amplification, the output chopper demodulates back to low frequency. The modulation and demodulation is simply implemented by CMOS switches driven by clock signals.

A pseudo-random chopping scheme is presented in Reference [4], which spreads the interference over a wide bandwidth, reducing its in-band portion to the level of the noise floor. Other circuit architectures such as dual-chopper amplifier (DCA), which employs two fundamental chopping clocks, have been reported in References [1,5–7].

Sensor readout circuits for capacitive accelerometers suffer from a signal offset due to production mismatch [8]. The offset from process variation can appear at the sensor output. It reduces dynamic range and causes the DC output level to vary from die to die [9]. The two-part correction is demonstrated in Reference [8], which consists of a capacitor array and a current digital to analog converter (DAC). A capacitor array is used to apply a signal correction by placing digitally controlled capacitors in parallel to the sensor capacitors. The second part consists of a current DAC placed within a differential amplifier to balance out asymmetric currents caused by the signal offset.

Standard CMOS process is suitable for implementing digital offset trimming. Hence, the offset trimming mechanism is presented to overcome the offset from sensor and interface circuit.

The design target refers to ADXL103. ADXL103 is a high precision, low power single-axis accelerometer with a signal conditioned voltage outputs from Analog Devices. ADXL103 measures acceleration with a full-scale range of ±1.7 g, sensitivity of 1000 mV/g, noise floor of 110 $\mu g/\sqrt{Hz}$, and power of 3.5 mW. ADXL103 can measure both dynamic acceleration and static acceleration. The main application is for navigation and motion detection [10].

The design target of our readout circuit is ±1 g sensing range, noise floor of 10 $\mu g/\sqrt{Hz}$, and power of milli-watt scale, which is suitable for navigation and motion detection. Based on our previous work [11], the UMC 0.18 μm CMOS/MEMS process is adopted for sensor and circuit implementation. This paper presents a low noise and low zero-g offset CMOS/MEMS accelerometer and readout scheme. Section 2 describes the CMOS/MEMS accelerometer and the circuit design of the low noise and low zero-g offset readout. In Section 3, describes the measurement results of the proposed readout scheme. Section 4 presents the discussion of the proposed readout scheme by comparison of performance with the state-of-the-art and presents the conclusions of this work.

2. Materials and Methods

2.1. CMOS/MEMS Accelerometer

In this work, the application-specific integrated circuit (ASIC) compatible 1P6M process of UMC 0.18 μm mixed-signal/RF CMOS process is adopted. The micromachining process is performed on the wafer of standard CMOS process. Figure 1a shows the top view of the proposed accelerometer. The proposed CMOS/MEMS accelerometer consists of proof mass, sensing fingers, single-folded springs, and a curl-matching frame. It is equivalent to a second-order mass-spring-damper mechanical model, as in Figure 1b. The displacement is transformed into capacitance ΔC by sensing fingers. The circuit model in Figure 1c is simulated with readout circuit. The circuit is simulated in Cadence design environment by Spectre simulator. The side view of the CMOS process with micromachining post process is shown in Figure 1d.

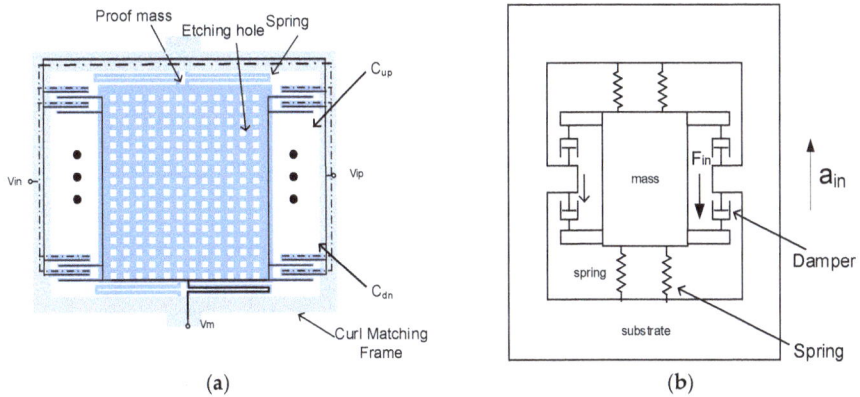

Figure 1. *Cont.*

(c)

(d)

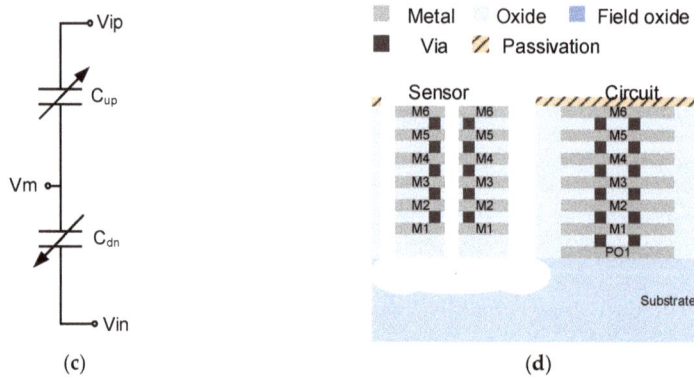

Figure 1. The proposed accelerometer: (**a**) The top view of the proposed accelerometer; (**b**) mechanical model of the structure; (**c**) circuit model of the structure; and (**d**) the side view of the structure.

2.2. Readout Circuit Design

The readout circuit has two main parts. The first part is a low noise unit and contains the main amplifier and pre-amplifier. The low noise chopper architecture and telescopic topology is developed to achieve low noise. The second part is a sensor-trimming unit that is an 8-bit trimming capacitor.

The architecture is shown in Figure 2. The sensing signal is modulated to 333 kHz and passes through amplification stages, track-and-hold amplifier (THA), output stage, and band limiting RC filter. The overall performance summery is listed in Table 1.

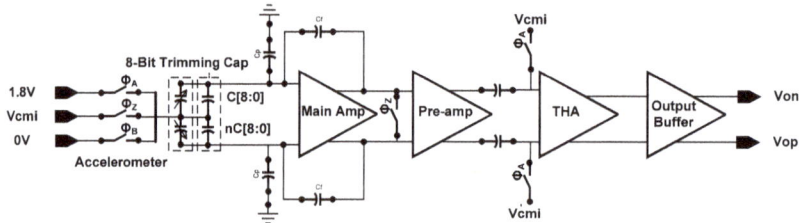

Figure 2. The system architecture.

Table 1. System specifications.

Specifications	Post-Sim	Measurement
System frequency (MHz)	1	1
Chopper frequency (kHz)	333.33	333.33
Overall sensitivity (mV/g) (at 1 g 100 Hz)	434.93	470
Noise ($\mu g/\sqrt{Hz}$) (at 1 g 100 Hz)	10	421.70
Resonance Frequency (kHz)	4.16	1.25
Brownian noise ($\mu g/\sqrt{Hz}$)	7.9	-
Sensing Range (g)	±1	±1
Power (mW)	2	1.67

Figure 3 shows the working principle of the sensor readout with the simplified modulation signal, which can be found in Figure 3b. The modulation frequency is 333.33 kHz. The sensing signal in Figure 3a is modulated by modulation clock signal and passes through amplifier stages in Figure 3c,d. The demodulation is achieved by track-and-hold stage, which is equivalent to multiply the demodulation signal in Figure 3e. The demodulated signal in Figure 3f passes through the zero-order hold and the signal in Figure 3g is obtained.

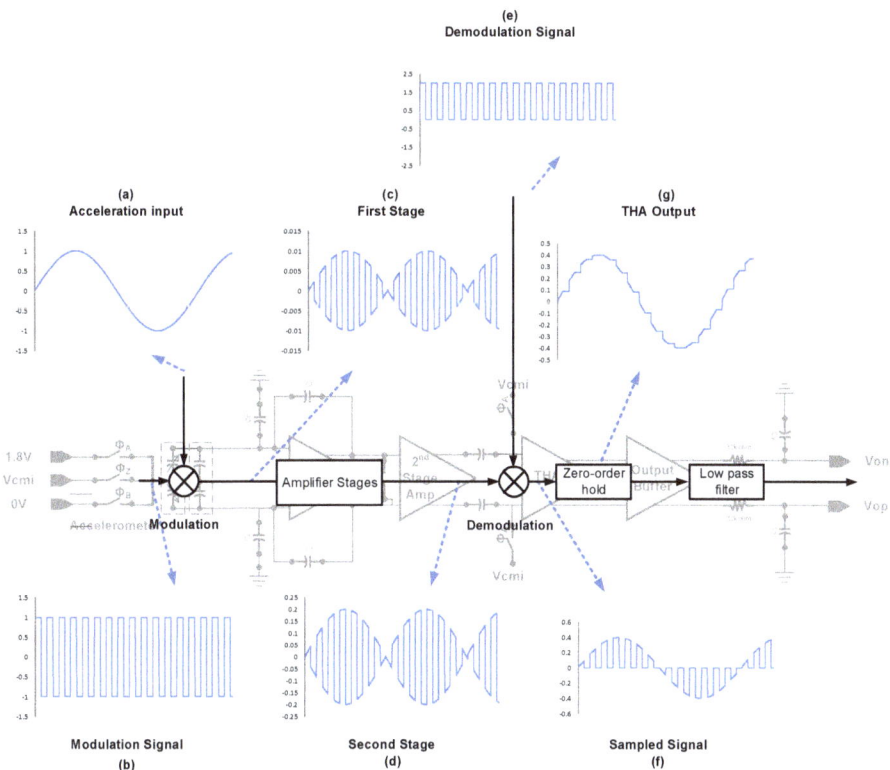

Figure 3. System function blocks with simplified modulation signal of the proposed readout circuit.

2.2.1. Low Noise Chopper Architecture

Since flicker noise is inversely proportional to frequency, the operation frequency determines the noise performance. The chopper architecture modulates the signal to chopping frequency to suppress flicker noise. The quantitative analysis is carried out at the transistor level to verify the effectiveness of the proposed architecture.

The sensing signal is modulated to 333.33 kHz by the switches (ϕ_A, ϕ_Z and ϕ_B), which is known as signal chopping. In this work, both the main amplifier and pre-amplifier are working at high frequency (at 333.33 kHz chopping frequency). For the conventional design in Reference [1], the sensing signal is demodulated using ϕ_H (at 1 MHz chopping frequency) in Figure 4. After demodulation by ϕ_H, the signal is further boosted by an amplifier.

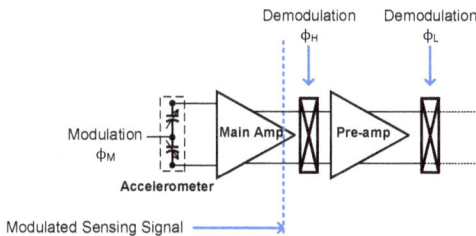

Figure 4. The simplified architecture in Reference [1].

The proposed low noise interface circuit is presented in Figure 5. The modulated sensing signal is first amplified by main amplifier and further boosted by the pre-amplifier. The amplified signal is demodulated by ϕ_A.

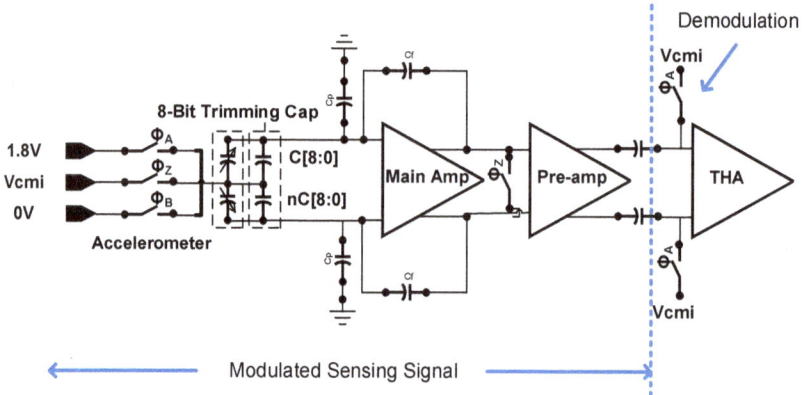

Figure 5. The proposed architecture for noise reduction.

The noise figure F of the network is defined as the ratio of the available signal-to-noise ratio at the signal-generator terminals to the available signal-to-noise ratio at its output terminals as the following equation [12].

$$F = \frac{SNR_{in}}{SNR_{out}} \tag{1}$$

Total noise figure of the whole system can be expressed by Friis' Formula:

$$F_{sys} = F_1 + \frac{F_2 - 1}{G_1} + \frac{F_3 - 1}{G_1 G_2} + \frac{F_4 - 1}{G_1 G_2 G_3} + \ldots + \frac{F_n - 1}{G_1 G_2 G_3 \ldots G_{n-1}} \tag{2}$$

where F_n is the noise figure for the n-th device and G_n is the power gain (linear, not in dB) of the n-th device. The design target is lower than the whole system F_{sys}. For the two amplifier stages, the noise of main amplifier is F_1, the power gain of main amplifier is G_1, and the noise of pre-amplifier is F_2. The noise figure of the third stage F_3 will be divided by the gain of the first two stages (the $G_1 G_2$ term). Thus, the noise figures of the first two stages must be considered [12]. The simplified noise figure is given by:

$$F_{sys} \approx F_1 + \frac{F_2 - 1}{G_1} \tag{3}$$

Two strategies are applied to lower the noise figure F_{sys}. The proposed circuit architecture minimize the F_1 and F_2 terms. First, modified the circuit architecture operates the second stage amplifier at 333.33 kHz to lower the F_2 term, which reduces noise contribution from the second stage amplifier. Second, the noise factor of the first amplifier F_1 is significant for the readout circuit since the F_1 term is directly added to F_{sys}. The gain G_1 is determined by the overall sensitivity. G_1 is around 7.88 V/V.

The telescopic amplifier is shown in Figure 6a, Q_1 and Q_2 form the input differential pair, and Q_3–Q_6 are the cascode transistors. Cascading transistors increase the voltage gain at the cost of output voltage headroom. Since the output swing requirements are very small at the first stage, on the order of several millivolts, a telescope may be used. For telescopic topology, the Q_1, Q_2, Q_7, and Q_8 are the primary noise sources. The folded-cascode topology is a popular amplifier architecture as in Figure 6b. Q_1 and Q_2 form the input differential pair, and Q_5 and Q_6 are the cascode transistors, which are folded, as compared to telescopic topology. For folded-cascode topology, the Q_1, Q_2, Q_7,

Q_8, Q_9, and Q_{10} are the primary noise sources. Assuming the transistors exhibit similar noise levels, folded-cascode topology suffers from greater noise than its telescopic counterpart. The telescopic topology is desirable since it has fewer noise-contributing transistors, and hence F_1 is reduced.

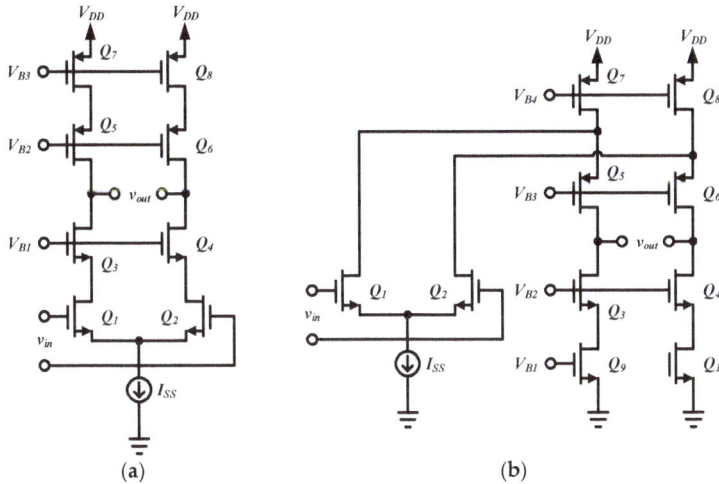

Figure 6. Schematic of telescopic amplifier.

Spectre PNoise simulation is used for noise characterization. The PNoise simulation gives the noise response of main amplifier and pre-amplifier. Table 2 shows the comparison of noise and power at each stage. For main amplifier, Reference [1] modulates the sensing signal to 1 MHz, while the proposed architecture modulates to 333.33 kHz. For pre-amplifier, Reference [1] demodulates the sensing signal to 20 kHz, while the proposed architecture is still working at 333.33 kHz. Comparing the two frequency arrangements, the proposed architecture has 8% less noise than Reference [1] at the cost of 10% more power consumption. The simulation results verify the effectiveness of the proposed reduction architecture.

Table 2. Analysis and comparison of noise and power at the two stages.

Stage	This Work			Reference [1]		
	Signal Frequency (Hz)	Power (µW)	Noise (µg/$\sqrt{\text{Hz}}$)	Signal Frequency (Hz)	Power (µW)	Noise (µg/$\sqrt{\text{Hz}}$)
Main Amp.	333.33 kHz	37.18	40.45	1 MHz	29.93	30.18
Pre-Amp	333.33 kHz	40.17	59.84	20 kHz	39.40	171.80
Total		77.34	48.04		69.33	51.97

2.2.2. Low Zero-g Offset Design

The sensing capacitive mismatch needs to be compensated. Small capacitor in sub femto farad scale is placed in parallel with the sensing capacitors to cancel the sensor offsets. A segmented split capacitor structure is proposed to realize small capacitor, as in Figure 7. Figure 7b shows the 7-bit trimming capacitance. The most significant bit (MSB) C[7] controls the switch in Figure 7a, which determines adding trimming capacitance to the upper plane or lower plane of the accelerometer. The C_{tm1} and C_{tm2} are the 7-bit trimming capacitance in Figure 7b. Trimming capacitance is estimated by using the equation below:

$$C \approx \frac{b_0 C_{b0} + b_1 C_{b1} + b_2 C_{b2} + b_3 C_{b3} + b_4 C_{b4} + b_5 C_{b5} + b_6 C_{b6}}{(C_{b0} + C_{b1} + C_{b2} + C_{b3} + C_{b4} + C_{b6}) + C_{t1}} \left(\frac{C_{t2}}{C_{t2} + C_{t3}} \right) C_{t4} \tag{4}$$

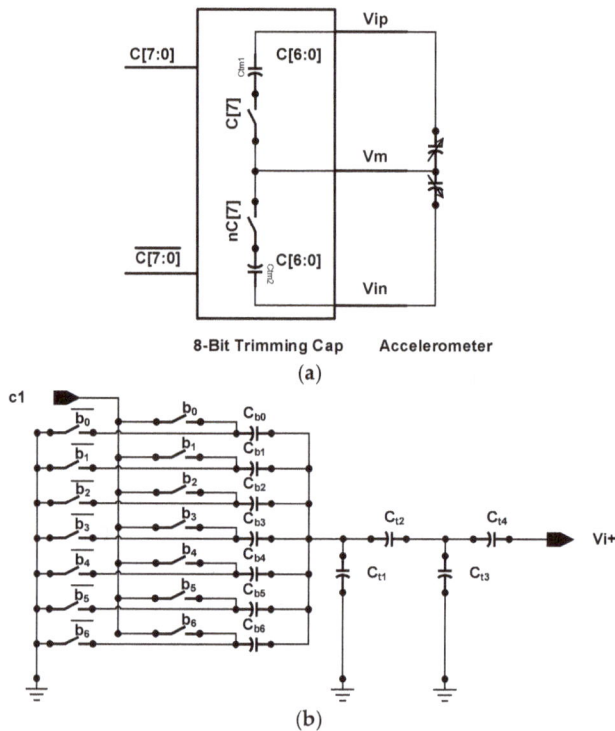

Figure 7. Schematic of 8-bit trimming capacitance: (**a**) The 8-bit trimming capacitor with the CMOS/MEMS accelerometer circuit model; and (**b**) the 7-bit segmented split capacitor structure.

The ratio of capacitance C_{t2} and C_{t3} make the overall capacitance C smaller to get sub femto farad scale capacitance.

3. Results

The circuit is implemented in UMC 0.18 μm process. In this work, low noise readout scheme is presented. The trimming capacitor is added for zero-g offset compensation. The die photo and chip layout is shown in Figure 8.

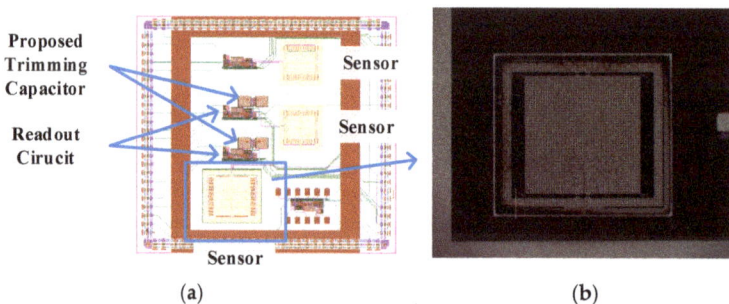

Figure 8. The chip layout and die photo.

3.1. CMOS/MEMS Accelerometer

The dry-etch-based post-process are used after standard CMOS process for microstructure fabrication. Figure 9 shows the cross section of the CMOS/MEMS accelerometer. The curl matching frame and the proof mass of accelerometer have the same curling.

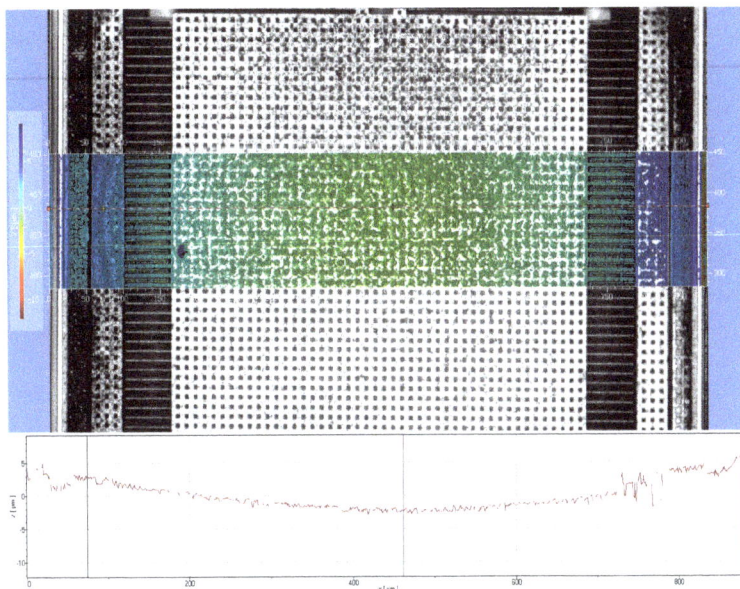

Figure 9. Cross section of the accelerometer.

3.2. Low Noise Design

Figure 10 shows the evaluation board schematic for acceleration readout measurement. The fabricated chip directly mounts on the printed circuit boards. On board oscillator generates 1 MHz clock for acceleration readout. The 1.8 V supply is generated by regulator for digital power (V_{ddD}) and analog power (V_{ddA}). The calibration readout is controlled by on-board switches. The fabricated chip directly mounts on the printed circuit boards, as shown in Figure 11.

Noise Considerations in Board Design

A digital circuit can produce noise at 1 MHz. Circuit noise decoupling capacitors are added at power line for digital noise reduction (1 MHz) (power line filter). Since the power line is 60 Hz, which is near 100 Hz of the sensing signal and cannot be easily filter by conventional filter. Power line noise is isolated by using battery power. The battery power passes though voltage regulator into readout circuit. The voltage regulator LM1117 is adopted, which reported RMS output noise is 0.003% of V_{OUT} at frequency 10 Hz $\leq$ f $\leq$ 10 kHz, where V_{OUT} is 1.8 V.

The evaluation board is placed on the LDS V408 shaker, as shown in Figure 12 for noise and sensitivity measurement. The shaker generates 1 g signal 1 kHz acceleration input. Figure 13 shows the spectrum of output voltage at the excitation. The noise floor is 421.70 µg/$\sqrt{Hz}$. The signal-to-noise ratio (SNR) is around 67.5 dB.

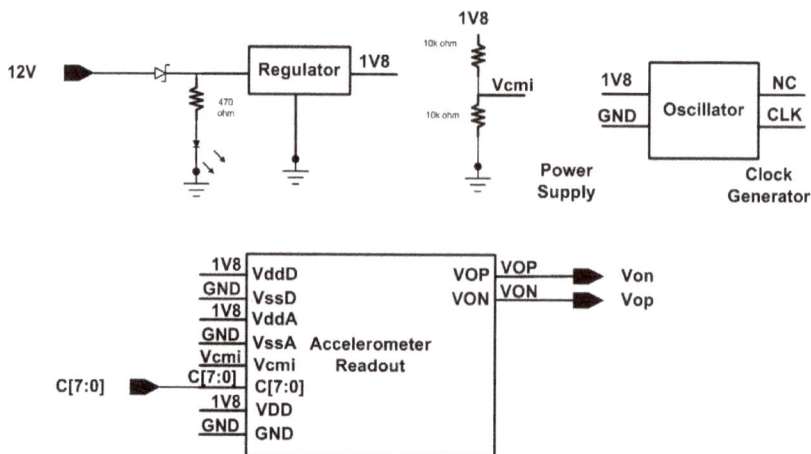

Figure 10. Simplified evaluation board schematics for accelerometer and readout circuit characterization.

Figure 11. Evaluation board photo.

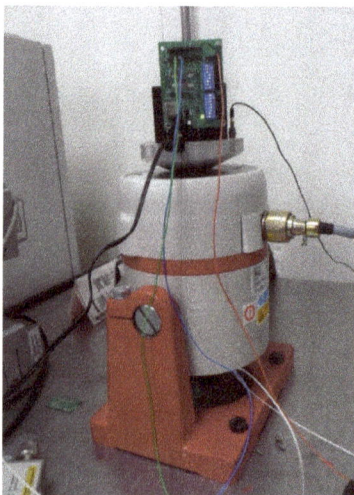

Figure 12. Measurement setup for accelerometer and readout circuit characterization.

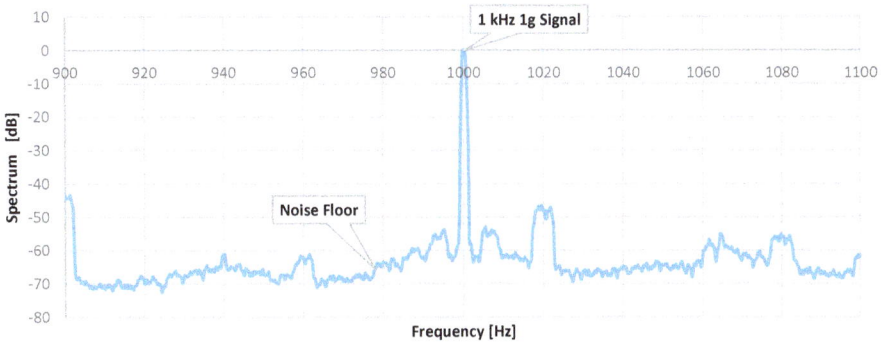

Figure 13. The output noise spectrum.

The sensitivity of the system is characterized for the two aspects, linearity and frequency response. The shaker generates 0.25 g signal to characterize the frequency response of the system as in Figure 14. The frequency range from 10 Hz to 1333.33 Hz is limited by the shaker. For the frequency around 1 kHz, the sensitivity increases due to the resonance of accelerometer.

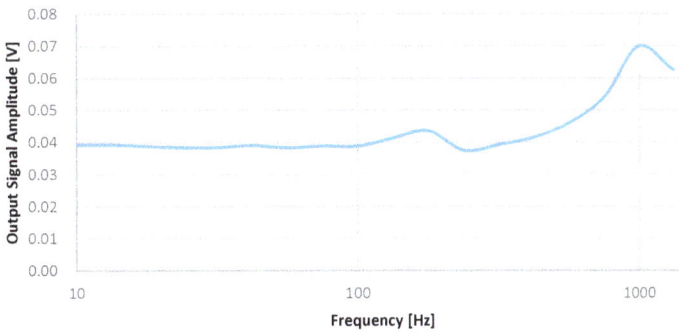

Figure 14. The frequency response of the system.

The sensing range of readout circuit is designed for ±1 g. The readout circuit is characterized using 1 kHz signal from zero to 1.5 g as in Figure 15. The linear regression is performed for zero to 1 g input signal. For a signal larger than 1 g, the output saturates and deviates from linear operation.

3.3. Low Zero-g Offset Design

The trimming capacitor is controlled by the digital value from the evaluation board to eliminate the zero g offset. The zero-g offset of the system is characterized for the two aspects, static and dynamic operation.

For static operation, the system output measured without external excitation that is the zero g output. For the ideal case, the zero-g output should be zero. The difference of positive output (VOP) and negative output (VON) represents the accelerometer readout. The differential output of the sensing signal VOP and VON should be the same. Figure 16 shows the output voltage with different configurations of the trimming capacitor. For the 8′b0000_0000 configuration, the 0 fF trimming capacitor is in parallel to the sensor capacitors, which stands for zero g offset value without trimming. The circuit output is saturated. The zero-g offset is 745.06 mV, as in Figure 16a. For the 8′b1111_1111 configuration, the maximum trimming capacitance is in parallel to the sensor capacitors. The offset is 77.08 mV, as in Figure 16b. For the 8′b1001_0000 configuration, the trimming capacitance is in parallel to eliminate the zero g offset. The offset is reduced from 745.06 mV to 1.38 mV. That is, the zero g offset is reduced from 1242.63 mg to 2.30 mg, as in Figure 16c.

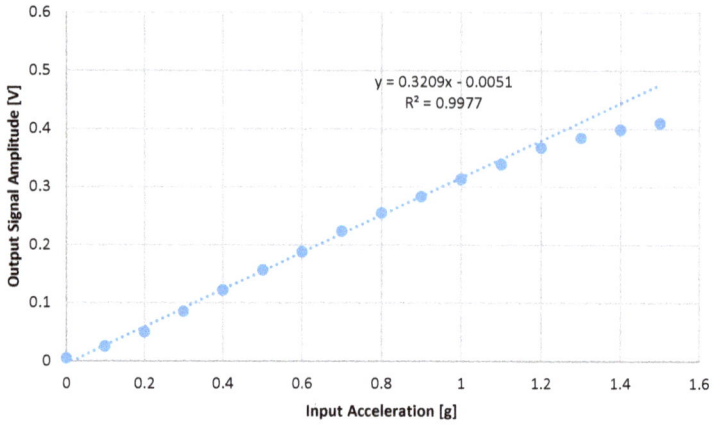

Figure 15. The output signal amplitude versus acceleration.

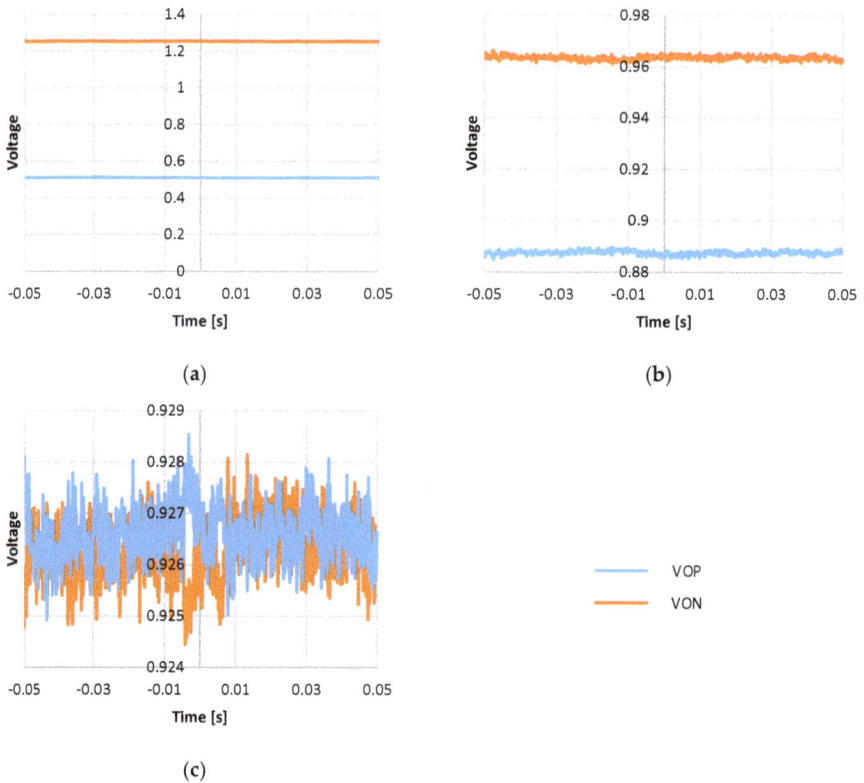

Figure 16. Output measurement without external excitation for various offset trimming configuration: (**a**) Trimming capacitor with 8′b0000_0000 configuration; (**b**) trimming capacitor with 8′b1111_1111 configuration; and (**c**) trimming capacitor with 8′b1001_0000 configuration.

For dynamic operation, the excitation of 1 g 1 kHz is applied with different configurations of the trimming capacitor. For the 8′b0000_0000 configuration, the circuit output is saturated. Sensitivity is degraded to 1.61 mV/g, as in Figure 17a. For the 8′b1111_1111 configuration, the sensitivity is around 706.32 mV/g. The output exhibits nonlinear distortion, which is obviously undesirable, as in Figure 17b. For the 8′b1001_0000 configuration, the trimming capacitance is in parallel to eliminate the zero g offset. The measurement shows sensitivity around 599.58 mV/g, as in Figure 17c.

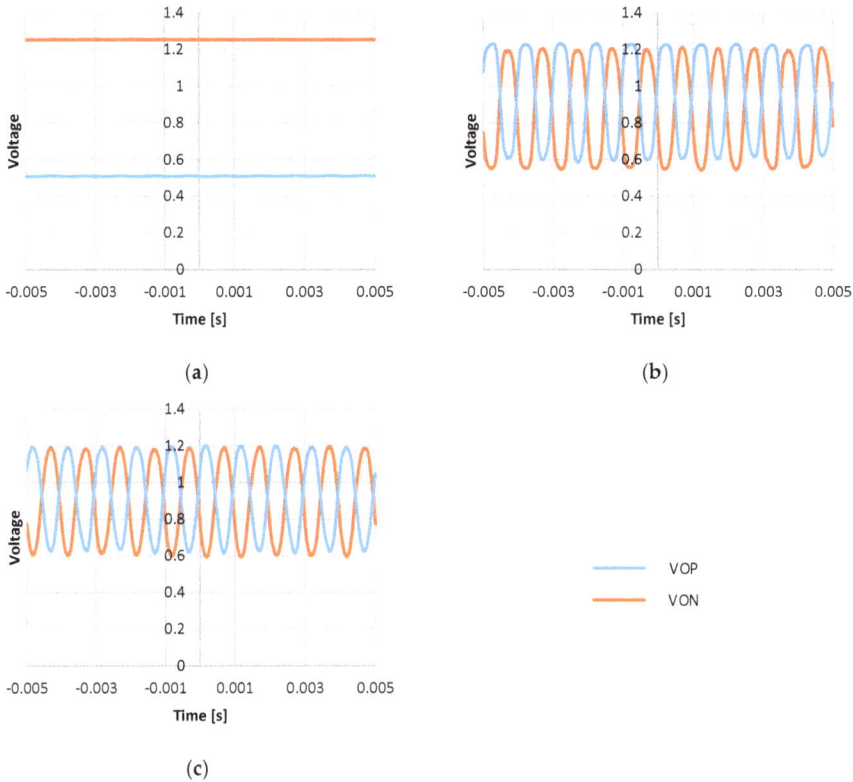

(a)

(b)

(c)

Figure 17. Output measurement with 1 g 1 kHz excitation for various offset trimming configuration: (a) Trimming capacitor with 8′b0000_0000 configuration; (b) trimming capacitor with 8′b1111_1111 configuration; and (c) trimming capacitor with 8′b1001_0000 configuration.

4. Discussion and Conclusions

A monolithic low noise and low zero-g offset CMOS/MEMS accelerometer and readout scheme in standard 0.18 μm CMOS mixed signal UMC process is presented. For 1 g 100 Hz acceleration input, the whole system has 470 mV/g sensitivity. The power consumption is about 1.67 mW. Table 3 compares the performance of the work proposed here to the state-of-the-art. Comparing with Reference [2], using the same 0.18 μm process node, the noise floor and zero g offset is reduced, while the overall power consumption is increased.

Table 3. Comparison of the proposed readout scheme to the state-of-the-art.

Parameters	[3]	[4]	[13]	[2]	This Work
Sensing Range (g)	±4	±49	±30	±8	±1
Noise Floor ($\mu g/\sqrt{Hz}$)	930	380	1	970	421.70
Zero-g Offset (mg)	N/A	N/A	N/A	±33	2.30
Supply Power (V)	1.2	1.9/3.3 [1]	1.5	1	1.8
Power (W)	25.44 μ	1.4 m	2.7 m	181 n	1.67 m
Chip Area (Readout Circuit) (mm^2)	1.73	1.1	10.9	1.14	1.23
Process	0.25 μm CMOS process	0.18 μm CMOS process	0.35 μm CMOS process	0.18 μm CMOS process	UMC 0.18 μm CMOS/MEMS

[1] A 1.9-V supply is regulated from an external 3.3-V supply.

The low noise chopper architecture and telescopic topology is developed to achieve low noise. The experiments show that noise floor is 421.70 $\mu g/\sqrt{Hz}$. The trimming capacitors are used for offset calibration. The zero g trimming circuit reduces the offset from 1242.63 mg to 2.30 mg.

Author Contributions: Y.-S.L. conceived the idea, designed and implemented sensor and readout circuit. K.-A.W. conceived and supervised the project.

Funding: This research received no external funding.

Conflicts of Interest: The authors declare no conflicts of interest regarding the publication of this paper.

References

1. Qu, H.; Fang, D.; Xie, H. A monolithic CMOS-MEMS 3-axis accelerometer with a low-noise, low-power dual-chopper amplifier. *IEEE Sens. J.* **2008**, *8*, 1511–1518. [CrossRef]
2. Akita, I.; Okazawa, T.; Kurui, Y.; Fujimoto, A.; Asano, T. A 181 nW 970 $\mu g/\sqrt{Hz}$ accelerometer analog front-end employing feedforward noise reduction technique. In Proceedings of the 2018 IEEE Symposium on VLSI Circuits, Honolulu, HI, USA, 18–22 June 2018; pp. 161–162. [CrossRef]
3. Paavola, M. A Micropower $\Delta\Sigma$-Based Interface ASIC for a Capacitive 3-Axis Micro-Accelerometer. *IEEE J. Solid-State Circuits* **2009**, *44*, 3193–3210. [CrossRef]
4. Petkov, V.P.; Balachandran, G.K.; Beintner, J. A fully differential charge-balanced accelerometer for electronic stability control. *IEEE J. Solid-State Circuits* **2014**, *49*, 262–270. [CrossRef]
5. Fang, D.; Qu, H.; Xie, H. A 1mW dual-chopper amplifier for a 50-$\mu g/\sqrt{Hz}$ monolithic CMOS-MEMS capacitive accelerometer. In Proceedings of the 2006 IEEE Symposium on VLSI Circuits, Honolulu, HI, USA, 15–17 June 2016. [CrossRef]
6. Sun, H.; Fang, D.; Jia, K.; Maarouf, F.; Qu, H.; Xie, H. A low-power low-noise dual-chopper amplifier for capacitive CMOS-MEMS accelerometers. *IEEE Sens. J.* **2011**, *11*, 925–933. [CrossRef]
7. Sun, H.; Maarouf, F.; Fang, D.; Jia, K.; Xie, H. An improved low-power low-noise dual-chopper amplifier for capacitive CMOS-MEMS accelerometers. In Proceedings of the 2008 3rd IEEE International Conference on Nano/Micro Engineered and Molecular Systems, Sanya, China, 6–9 January 2008.
8. Bergert, J.; Strache, S.; Wunderlich, R.; Droste, D.; Heinen, S. Offset correction in dual-chopper read-out circuits for half-bridge capacitive accelerometers. In Proceedings of the 2012 8th Conference on Ph.D. Research in Microelectronics & Electronics (PRIME), Aachen, Germany, 12–15 June 2012.
9. Lemkin, M.; Boser, B.E. A three-axis micromachined accelerometer with a CMOS position-sense interface and digital offset-trim electronics. *IEEE J. Solid-State Circuits* **1999**, *34*, 456–468. [CrossRef]
10. ADXL103, Datasheet, Analog Devices, Inc. Available online: https://www.analog.com/ADXL103 (accessed on 26 November 2018).
11. Liu, Y.; Huang, C.; Kuo, F.; Wen, K.; Fan, L. A monolithic CMOS/MEMS accelerometer with zero-g calibration readout circuit. In Proceedings of the Eurocon 2013, Zagreb, Croatia, 1–4 July 2013; pp. 2106–2110. [CrossRef]

12.   Friis, H.T. Noise figures of radio receivers. *Proc. IRE* **1944**, *32*, 419–422. [CrossRef]

13.   Zhao, J. A 0.23-µg bias instability and 1-µg/√Hz acceleration noise density silicon oscillating accelerometer with embedded frequency-to-digital converter in PLL. *IEEE J. Solid-State Circuits* **2017**, *52*, 1053–1065. [CrossRef]

micromachines

MDPI

Article

A 5 *g* Inertial Micro-Switch with Enhanced Threshold Accuracy Using Squeeze-Film Damping

Yingchun Peng *, Guoguo Wu, Chunpeng Pan, Cheng Lv and Tianhong Luo *

College of Mechanical and Electrical Engineering, Chongqing University of Arts and Sciences, Chongqing 402160, China; 20140014@cqwu.edu.cn (G.W.); 20170118@cqwu.edu.cn (C.P.); 20170017@cqwu.edu.cn (C.L.)
* Correspondence: 20170073@cqwu.edu.cn (Y.P.); 20180012@cqwu.edu.cn (T.L.); Tel.: +86-023-6116-2716 (Y.P.)

Received: 16 August 2018; Accepted: 21 October 2018; Published: 23 October 2018

Abstract: Our previous report based on a 10 *g* (gravity) silicon-based inertial micro-switch showed that the contact effect between the two electrodes can be improved by squeeze-film damping. As an extended study toward its potential applications, the switch with a large proof mass suspended by four flexible serpentine springs was redesigned to achieve 5 *g* threshold value and enhanced threshold accuracy. The impact of the squeeze-film damping on the threshold value was theoretically studied. The theoretical results show that the threshold variation from the designed value due to fabrication errors can be reduced by optimizing the device thickness (the thickness of the proof mass and springs) and then establishing a tradeoff between the damping and elastic forces, thus improving the threshold accuracy. The design strategy was verified by FEM (finite-element-method) simulation and an experimental test. The simulation results show that the maximum threshold deviation was only 0.15 *g*, when the device thickness variation range was 16–24 μm, which is an adequately wide latitude for the current bulk silicon micromachining technology. The measured threshold values were 4.9–5.8 *g* and the device thicknesses were 18.2–22.5 μm, agreeing well with the simulation results. The measured contact time was 50 μs which is also in good agreement with our previous work.

Keywords: MEMS (micro-electro-mechanical system); inertial switch; acceleration switch; threshold accuracy; squeeze-film damping

1. Introduction

Inertial micro-switches based on MEMS (micro-electro-mechanical system) technology have been widely used for acceleration sensing applications [1,2] due to their small size, high integration level, and low or even no power consumption [3,4]. The inertial micro-switches are typically designed with a proof mass that is anchored to a substrate through flexible springs. The proof mass serves as a moveable electrode, and it is separated by a certain distance from a fixed electrode on the substrate. At a pre-selected threshold acceleration, the moveable electrode moves toward the substrate and it comes into contact with the fixed electrode, turning on the switch and triggering the external circuit. Thus, the inertial micro-switches require a reliable contact effect of the two electrodes, such that the turn-on signal can be recognized by the external circuit. From the perspective of application convenience, since most of the switches are mass produced in the industry sector, a high degree of device-to-device threshold uniformity of the same production batch is needed. As such, a high threshold accuracy is also essential for the inertial micro-switches.

Since the first inertial micro-switch was reported in 1972 [5], a great number of inertial micro-switches based on various working mechanisms and manufacturing methods have been developed. However, most of the switches reported in the past have been mainly designed to improve the contact effect and the threshold accuracy is rarely considered. This is because the inherent issues of the methods employed to improve the contact effect usually lead to a low threshold accuracy.

The main methods to improve the contact effect of the inertial micro-switches are designing the switches with a keep-close function, or flexible electrodes. The switches with a keep-close function can keep it closed after the acceleration event is over, thus improving the contact reliability. However, the keep-close function requires special constructions, such as the hook-shaped electrodes of the latching switches [6–8], the V-shaped beams of the bi-stable switches [9,10], and the valve-channel of the micro-fluidic switches [11,12], complicating the structure topology or working mechanism or fabrication method, thus reducing the threshold accuracy, as shown in Table 1. The latching switch with a 50.59 g designed threshold value in Ref. [7], for instance, was first switched on when the applied acceleration was between 28 g and 43.7 g, and it was completely closed when a higher acceleration was applied. This is due to the collision and friction contact process of the two hook-shaped electrodes.

The contact effect of the inertial micro-switches with flexible electrodes can be improved by the deformation of the flexible electrodes during the contact process. In this case, the contact time of the two electrodes is generally longer than 50 µs [13,14]. This kind of switch is usually fabricated by a multi-layer nickel-electroplating process, based on the surface micromachining technology, since the conventional bulk silicon micromachining technology mainly results in rigid structures. However, the often-repeated electroplating processes might cause unexpected fabrication errors such as dimension variations, an inhomogeneous Young's modulus, and structural deformation induced by residual stresses between each electroplating layer, leading to a threshold deviation from the designed value [4,13–17], as shown in Table 1. In Ref. [15], the measured threshold of the switch with 240 g designed threshold was 288 g, and in Ref. [16], the actual thresholds increased from 32 g to 38 g, while the intended target was 38 g. The multi-layer electroplating process is not applicable to the case of the switches with a designed threshold value below 10 g (also named as low-g switch in this paper) because the threshold deviation may be more seriously caused by the fabrication errors. According to the static equilibrium equation $a_{th} = kx_0/m$ (where a_{th} is the threshold acceleration, k is the spring constant, x_0 is the distance between the two electrodes, and m is the mass of the proof mass), a low-g switch requires flexible springs and large proof mass, because the minimum size of x_0 is usually limited by the fabrication process. The large proof mass should be fabricated by a great number of electroplating processes, resulting in serious fabrication errors.

In recent years, several researchers have paid great attention to improving the threshold accuracy of the inertial micro-switches, as shown in Table 1. McNamara and Gianchandani [18] presented an array redundancy design of the inertial micro-switch to broaden the sensing range of acceleration (10–150 g in 10 g increments) and allow fault latitude, wherein multi switches at each threshold level were employed. By weighting the measured results on the majority status of these redundant switches, the measured thresholds were 80–90% of the target values. Jr and Epp [19] proposed a stochastic dynamics model to modify the device dimensions based on the experimental results, reducing the threshold deviation caused by the fabrication errors. Currano et al. [4] demonstrated an inertial micro-switch that could detect identical accelerations in the *x*, *y*, and *z* axes using a single mass/spring assembly. To reduce the threshold deviation due to the fabrication errors, they modified the 2 µm width spring (the original designed value) to 5 µm, and changed the spring lengths to tune the in-plane (*x*/*y*) thresholds to the target acceleration levels. Then, the in-plane threshold values were generally close to the designed values, but the thresholds in the *z*-axis were much lower than the target levels (~10–40 g, as opposed to ~90–230 g). Du et al. [20] modified the device thickness of the switch, based on the sizes of the pre-fabricated structure components. The designed and measured thresholds were 38 g and 35–40 g, respectively. Zhang et al. [21] fabricated a 5.5 g inertial micro-switch on a SOI (silicon-on-insulator) wafer to accurately define the device thickness, thus improving the threshold accuracy. The measured threshold values were 4.77–5.97 g.

Table 1. Comparisons of the main research results reported in the past.

Switch Category	Enhancement Method	Fabrication Method	Designed Threshold	Measured Threshold	Contact Effect
Contact effect enhanced switch	Latching [7]	Bulk silicon	50.59 g	28–43.7 g	Keep closed
	Bi-stable [9]	Nickel electroplating	35 g	32.38 g	Keep closed
	Micro-fluidic [11]	Bulk silicon	9 g	8.525 g	Keep closed
	Flexible electrodes [14]	Nickel electroplating	500 g	466 g	390 μs
	Flexible electrodes [15]	Nickel electroplating	240 g	288 g	150 μs
	Flexible electrodes [16]	Nickel electroplating	38 g	32–38 g	230 μs
Threshold accuracy enhanced switch	Redundancy design [18]	Nickel electroplating	10–150 g	80–90% of the target	$-$[1]
	Dimension modification [4]	Nickel electroplating	90–230 g	10–40 g	$-$[1]
	Dimension compensation [20]	Nickel electroplating	38 g	35–40 g	102 μs
	SOI wafer [21]	Bulk silicon	5.5 g	4.77–5.97 g	$-$[1]

[1] The data was not presented in the paper.

In our previous report [22], an inertial micro-switch with a threshold value of 10 g and a high damping ratio of 2 was presented based on a typical silicon-on-glass process. The contact effect (40 μs contact time) was significantly improved using the squeeze-film damping effect compared with the typical switches with rigid electrodes (the contact time usually less than 20 μs). In this paper, the impact of the squeeze-film damping on the threshold acceleration was studied by theoretical analysis, FEM (finite-element-method) simulation and experimental test. The study was implemented based on our previous device structure but with a lower threshold value of 5 g. The experimental results show that squeeze-film damping can not only prolong the contact time, but also improve the threshold accuracy. The study is significant for the applications of the inertial micro-switches, where low-g-sensing, long contact time, and high threshold accuracy are required.

2. Theory and FEM Simulation

2.1. Device Structure

The switch consists of a proof mass that is suspended by four flexible serpentine springs, which serves as the sensing element moving toward the substrate to sense out-of-plane acceleration. The vertical direction sensitivity enables the switch to employ the squeeze-film damping effect, since the slid-film damping effect involved by laterally driven switches is so weak that it is usually neglected. A small size of protrusion positioned at the bottom center of the proof mass is defined as the movable contact electrode, reducing the contact area. Two separated metal strips on the substrate serve as the double-contact-configuration fixed electrode. When an environmental acceleration exceeding the preset threshold is applied to the switch in the sensitive direction, the proof mass moves toward the substrate, traveling the electrode gap and making the movable electrode contact with the fixed electrode, thus turning the switch on. The movement of the proof mass in the horizontal insensitive directions is limited by four fixed pillars. Figure 1 shows a sketch of the designed switch, wherein the proof mass and springs are set as transparent structures to display the two electrodes under them.

According to the static equilibrium equation mentioned in the introduce section, the electrode gap height (h_e) was set to 1 μm, which is near our process limit to reduce the required volume of the proof mass for a low threshold value. In this case, the width and length of the proof mass were both set to 2300 μm. The four flexible serpentine springs with a 30 μm width have a much lower equivalent spring constant than the typical cantilever beam. This structure feature enables the switch to easily respond to the target threshold of 5 g. The distance between the proof mass and the substrate ($h_a = 35$ μm) was designed to achieve the required damping ratio (ca. 2.1) by changing the height of the protrusion (h_p)

when the other structure parameters were determined. Due to the small size of the protrusion, the impact of changing h_p on the equivalent mass of the proof mass can be ignored.

The thickness of the proof mass is designed to be identical to that of the springs (t_b) for facilitating fabrication. More importantly, in order to reduce the threshold deviation due to fabrication errors, t_b was defined as a crucial dimension, and then theoretically studied to establish a tradeoff between the damping and elastic forces that the switch is subjected to, while operating under an over-damping condition (as explained later).

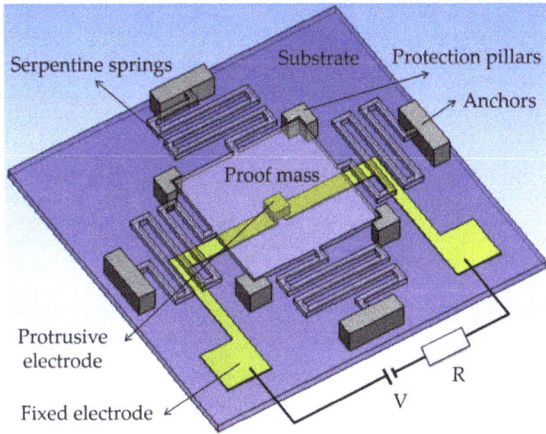

Figure 1. Scheme of the designed inertial micro-switch.

2.2. Theoretical Analysis of the Threshold Accuracy

As a typical inertial sensor, the inertial micro-switch can be modeled by a mass-spring-damping system to represent the mechanical behavior of the device. Considering that the acceleration signal applied to the switch in practical work is a half-sine wave, the governing mechanical equation is:

$$m\ddot{x} + c\dot{x} + kx = ma\sin\omega t, \tag{1}$$

where x is the relative displacement of the proof mass with respect to the substrate, m is the proof mass, c is the squeeze-film damping viscous coefficient, k is the equivalent elastic stiffness of the four serpentine springs, ω is the angular frequency of the acceleration, a is the amplitude of the acceleration, and its minimum value that can close the switch is regarded as the threshold acceleration (a_{th}).

The solution to Equation (1) given in [23] consists of two parts: the transient item, which will decrease exponentially with time determined by the damping condition, and the steady-state item, wherein the oscillation of the proof mass is the same frequency as that of the applied acceleration. Considering the over-damping-condition design in this paper, the steady-state item is primary, and the transient item is second of the solution. As such, the solution to Equation (1) given in [23] can be simplified as:

$$x(t) = \frac{a\sin(\omega t - \phi)}{\sqrt{(\omega_n{}^2 - \omega^2)^2 + (c\omega/m)^2}}, \tag{2}$$

where $\omega_n = \sqrt{k/m}$ is the natural angular frequency of the switch, and ϕ is the displacement phase of the proof mass with respect to the substrate. Then, the threshold acceleration can be represented in Equation (3), according to its own definition as mentioned above:

$$a_{th} = \sqrt{(\omega_n{}^2 - \omega^2)^2 + (c\omega/m)^2} \cdot h_e. \tag{3}$$

In fact, ω_n is usually higher than ω to minimize the threshold discrepancy of the switch, due to the duration of the input acceleration [24]; thus, Equation (3) can be rewritten as:

$$\tilde{a}_{th} \propto \sqrt{(k/m)^2 + (c\omega/m)^2}, \tag{4}$$

where $\tilde{a}_{th}$ is the value of a_{th} normalized by h_e, i.e., $\tilde{a}_{th} = a_{th}/h_e$. Substituting the expressions of k (which is derived by equivalently taking the four serpentine springs as a cantilever beam) in Equation (5) and c in Equations (6) [25] into (4), the expression of the normalized threshold is rewritten in Equation (7):

$$k = \frac{Ew_b t_b{}^3}{4l_b{}^3}, \tag{5}$$

$$c = \frac{\mu l_m w_m{}^3}{h_a{}^3}\gamma, \tag{6}$$

$$\gamma = \left\{ 1 - \frac{192}{\pi^5}\frac{w_m}{l_m} \sum_{n=1,3,5,}^{\infty} \frac{1}{n^5} \tanh\left(\frac{n\pi l_m}{2w_m} \right) \right\},$$

$$\tilde{a}_{th} \propto \sqrt{\left(\frac{Ew_b t_b{}^3}{4l_b{}^3 \rho l_m w_m t_m} \right)^2 + \left(\frac{\mu w_m{}^2}{h_a{}^3 \rho t_m}\gamma\omega \right)^2}, \tag{7}$$

where E and ρ are the Young's modulus and the density of silicon, respectively; l_b, w_b, and t_b are the length, width, and thickness of the equivalent cantilever beam, respectively; l_m, w_m, and t_m are the length, width, and thickness of the proof mass, respectively; μ is the viscosity coefficient of air, γ is a correction factor determined by w_m/l_m, and is equal to 0.42 when $w_m = l_m$.

From the perspective of fabrication, the spring thickness (20 μm in this paper) is much thinner than the silicon wafer (ca. 500 μm thick), and such that it is defined by a several hours of deep back-etch fabrication process (a KOH etching process for low cost and high etch rate). Due to the long duration process, the thickness non-uniformity within the wafer may become obvious induced by several factors, such as the hydrogen generation, and the diffusion of the etchant and reaction products (e.g., the maximal height difference over the back-etched surface of a 18 μm thick sieve is 4.6 μm [26]). Therefore, the variation of the spring thickness due to the fabrication errors is usually larger than that of the spring width in the similar size. Moreover, according to Equation (5), the changing of the spring thickness has greater influence on the spring stiffness than that of the spring width, thus leading to a larger threshold deviation of the switch. Therefore, we define the spring thickness t_b as the crucial dimension that may cause the primary threshold deviation to the switch, due to the fabrication errors. For facilitating design and manufacturing, the thickness of the proof mass t_m is set to be equal to t_b, and Equation (7) can then be further rewritten as:

$$\tilde{a}_{th} \propto \sqrt{\eta_1^2 \cdot t_b^4 + \eta_2^2/t_b^2}, \tag{8}$$

where $\eta_1 = Ew_b/4l_b{}^3\rho l_m w_m$ and $\eta_2 = \mu w_m{}^2 \gamma\omega/h_a{}^3\rho$ can be regarded as constants when the structure parameters of the switch have been determined, except for t_b. Equation (8) indicates that $\eta_1^2 \cdot t_b^4$ and η_2^2/t_b^2 change in opposite directions with t_b, meaning that there should be a specific interval value of t_b in which the sum of $\eta_1^2 \cdot t_b^4$ and η_2^2/t_b^2 keeps relatively stable, and then $\tilde{a}_{th}$ as well. In this case, the threshold deviation can be significantly reduced, despite the variations of t_b caused by fabrication errors.

This design strategy can be explained as follows: (i) according to Equation (1), the forces that the switch is subjected to while operating contain the external force $ma\sin\omega t$, the inertial force $m\ddot{x}$, the squeeze-film damping force $c\dot{x}$, and the elastic force kx; (ii) normalizing the four forces by m, then $a_{th}(\omega)$ (its value is influenced by the angular frequency of the acceleration ω), c/m and k/m represent the threshold acceleration, the coefficients of the damping force and elastic force, respectively, and the coefficient of the inertial force is a constant that has no effect on the threshold deviation;

(iii) subsequently, the threshold is mainly proportional to the sum of the damping and elastic forces; (iv) according to Equations (5) and (6) and the design of $t_b = t_m$, k/m and c/m are proportional to t_b^2 and t_b^{-1}, respectively. Then, k/m increases and c/m decreases with increasing t_b, and the two items will be equal for a specific value of t_b (labeled as t_b'); (v) at $t_b < t_b'$, along with the increase of t_b, the increment of k/m is smaller than the decrement of c/m, thus decreasing the threshold value, and it is the opposite at $t_b > t_b'$. (vi) When the value of t_b changes around that of t_b', the absolute variations of k/m and c/m are roughly equal; i.e., establishing a tradeoff between the damping and elastic forces, thus keeping the threshold value relatively maintained. It should be noted that this design strategy is only valid when (1) the squeeze-film damping condition is over-dampened such that the damping force is large enough to be comparable with the elastic force, and (2) the value of t_b is optimized.

2.3. FEM Simulation

FEM transient analysis (ANSYS Workbench, 15.0, ANSYS Inc., Pittsburgh, PA, US) was performed to obtain the proper value of t_b. The FEM model was meshed by the method of Hex Dominant. The end sections of the four suspended springs were constrained to be zero in all degrees of freedom. In the analysis settings, the damping effect was represented by the so called Rayleigh damping coefficients—the alpha damping coefficient $\alpha = 2\zeta\omega_{n1}\omega_{n2}/(\omega_{n1} + \omega_{n2})$ and the beta damping coefficient $\beta = 2\zeta/(\omega_{n1} + \omega_{n2})$, where $\zeta = c/2m\omega_n$ is the squeeze-film damping ratio, ω_{n1} and ω_{n2} are the first and second order natural angular frequencies of the switch, respectively [27]. The frequency of the applied half-sine wave is 500 Hz, i.e., 1 ms duration acceleration, which is lower than that of the designed switch (808 Hz) such that it is enough for explaining the design strategy. The total step number of the acceleration load is 30, and the numbers of the substep (including initial substep, minimum substep, and maximum substep) are 4, 2, and 6, respectively, which are enough for guaranteeing computational accuracy. During the transient analysis, by presetting the electrode-gap height h_e, the required acceleration amplitude (i.e., the threshold acceleration) applied to the switch for different values of t_b can be obtained. The main geometric parameters and the material properties of the switch for the FEM study are shown in Tables 2 and 3, respectively.

Table 2. Main geometrical parameters of the designed switch.

Component	Geometric Parameter	Value (μm)
Proof mass	Length l_m	2300
	Width w_m	2300
	Thickness t_m	$=t_b$
Protrusion	Length l_p	50
	Width w_p	50
	Height h_p	34
Serpentine spring	Span beam length l_s	1600
	Connector beam length l_c	150
	Width w_b	30
	Thickness t_b	Variable
Electrode gap	Height h_e	Variable

Table 3. Main material properties of the device structure.

Material	Density	Young's Modulus	Poisson's Ratio
Silicon	2330 kg/m^3	169 GPa	0.28
Glass	2200 kg/m^3	70 GPa	0.17

Figure 2 shows the changes of the threshold acceleration with the spring thickness t_b, for the cases when the electrode-gap height $h_e = 0.5$ μm, 1 μm, and 1.5 μm, respectively. It can be seen that the threshold tends to level off when the value of t_b changes around 20 μm, being in good agreement with the theoretical study. In the case of $h_e = 1$ μm, the maximum threshold deviation is only 0.15 g (4.91–5.15 g), when t_b increases from 16 to 24 μm. Since the 8 μm (16–24 μm) fabrication latitude is wide enough for the current bulk silicon micromachining technology, the threshold deviation of the switch caused by fabrication errors can be significantly reduced. As such, the values of t_b (t_m) and h_e were set to 20 μm and 1 μm, respectively.

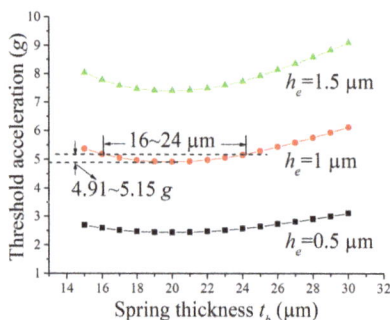

Figure 2. Threshold acceleration changes with the spring thickness in cases of different electrode-gap heights.

3. Experiments and Discussion

The switch was fabricated using a typical silicon-on-glass process. The silicon (<100> *n*-type) and glass wafers used in this work are both 4 inch in diameter and 500 μm in thickness. The main fabrication processes shown in Figure 3 have been presented in our previous study [22], except for a pre-release recess process (Figure 3e). Due to the anodic bonding process (Figure 3d), there is a pressure difference between the bonded chamber (10^{-2} mbar) and the outside air (ca. 1 bar), which may probably cause the protrusion to stick to the glass substrate, or even fracture the springs upon post-etch release of the structure (Figure 3f). Our experimental results show that this issue becomes evident as we reduce the electrode gap height from 2 μm to 1 μm. As such, several recesses (ca. 10 μm deep) on the edge of the chamber were constructed by an inductive plasma (ICP) etch (Figure 3e), which were firstly penetrated during the final structure release process (Figure 3f), thus eliminating the pressure difference. Figure 4 shows the fabricated and packaged micro-switches.

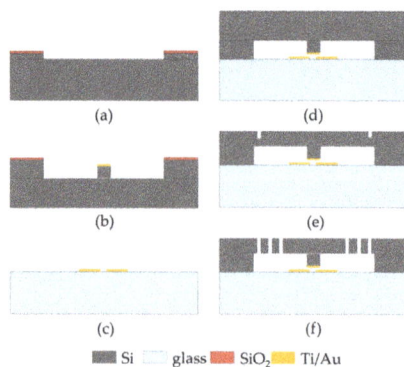

Figure 3. Process sequence for the fabrication of the micro-switch. (**a**) electrode gap; (**b**) protrusive electrode; (**c**) double-contact-configuration fixed electrode; (**d**) anodic bonding; (**e**) thinning and pre-release recess; (**f**) structure releasing.

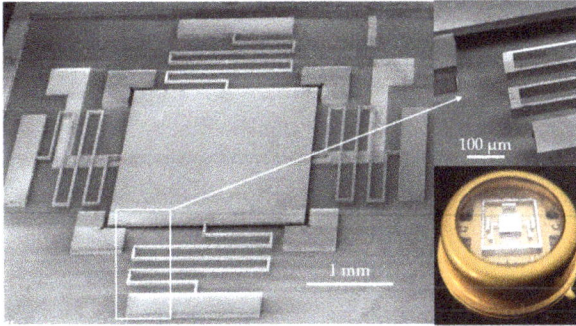

Figure 4. SEM (scanning electron microscope) and optical photographs of the fabricated and packaged micro-switches.

The fabricated switch was tested by a vibration measurement system, as shown in Figure 5. Figure 6 shows the schematic of the test circuit. The packaged switch in a device holder was mounted on the vibration table, thus establishing a connection to a 5 V DC power supply, and a 1000 Ω resistor. A signal generator was used to offer a half-sine wave to simulate the acceleration load, which was amplified by a power amplifier and then fed into the vibration system to produce the required vibration. The vibration was detected by a standard accelerometer (100 mV/g) fixed on the vibration table beside the switch. The output of the accelerometer was amplified by a charge amplifier for facilitating the signal collection. The applied acceleration with various amplitudes (0.1~30 g with 0.1 g accuracy) and durations (>0.1 ms) was controlled by a computer. When the applied acceleration reaches the threshold value, the moveable electrode shorts the fixed electrode and thus turns on the test circuit, outputting voltage signal of ca. 5 V. The outputs of the switch and the accelerometer were recorded by a multichannel oscilloscope, and each experimental datum was repeated at least 10 times.

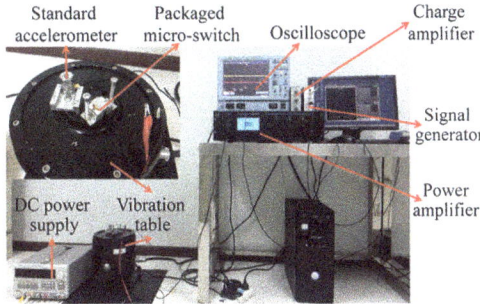

Figure 5. Experiment setup for testing the fabricated micro-switch.

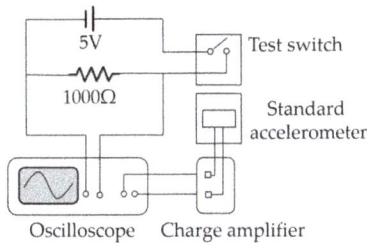

Figure 6. Schematic diagram of the test circuit.

Figure 7a shows the typical test results of the fabricated switches with a 4.9 g threshold under 1 ms acceleration duration. The threshold values of 10 randomly selected switches were tested as 4.9, 5.1, 5.2, 5.8, 5.3, 5.1, 5.5, 5.2, 5.5, and 5.6 g (hereafter, the switch is named after its threshold value under 1 ms acceleration), respectively. The SEM (scanning electron microscope, Carl Zeiss AG, Stuttgart, Baden-Württemberg, Germany) measurement results showed that the spring thicknesses between each switch range from 18.2 μm to 22.5 μm. The measured results were in accordance with the simulation results, and they are comparable to those of the switch based on a SOI wafer, wherein the device thickness can be accurately controlled [21]. Considering that the SOI wafer requires additional cost, compared with the common wafer used in this paper, the results show that the threshold accuracy enhancement presented in this paper is effective and low-cost.

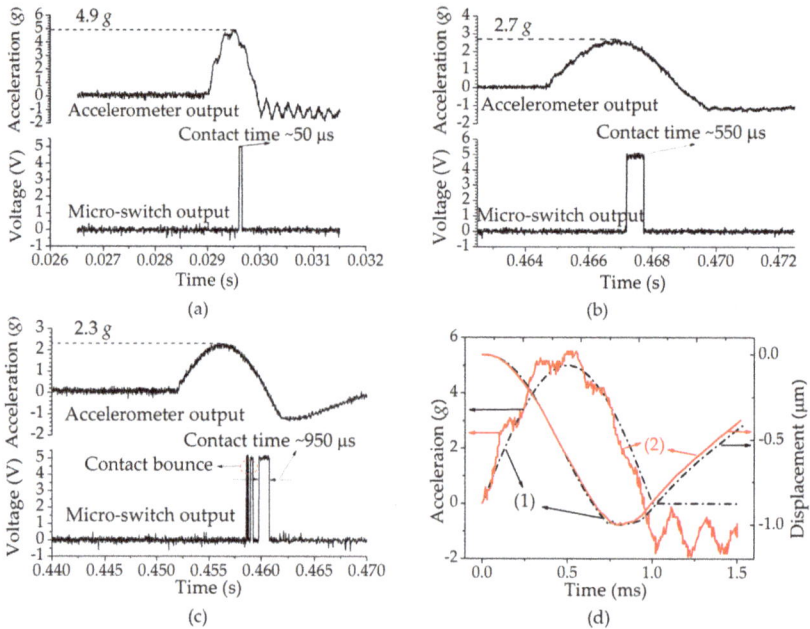

Figure 7. (**a–c**) The typical test results of the fabricated switches under (**a**) 1 ms, (**b**) 5 ms, and (**c**) 10 ms accelerations. (**d**) Simulation results of the displacement response of the switch under the cases of (1) a 30.9 μm spring width and a 5.5 g practical acceleration and (2) a 30 μm spring width and a 5 g standard half-sine wave.

The measured thresholds of the 4.9 g switch under 5 ms and 10 ms accelerations were 2.7 g and 2.3 g, respectively, as shown in Figure 7b,c. The test results of all 10 selected switches show that the thresholds were 2.2–3.8 g and 1.9–3.7 g for the cases of 5 ms and 10 ms accelerations, respectively, while their corresponding simulation results were 2.8 g and 2.6 g, respectively. It can be seen that with prolonging the acceleration duration, the measured threshold values were generally decreased, and the threshold deviation from the designed value became larger. Considering that the damping force that the switch is subjected to while operating is in proportion to the acceleration frequency as seen in Equation (8), the squeeze-film damping effect was obviously decreased, as the acceleration was prolonged from 1 ms to 5 ms, or 10 ms. Therefore, the lower threshold value and larger threshold deviation were both due to the weak squeeze-film damping effect, spotlighting the vital role of the squeeze-film damping on the improvement of threshold accuracy.

It should be noted that a higher threshold value measured under 1 ms acceleration was not necessarily relatively higher in the case of 5 ms or 10 ms acceleration. For example, under 10 ms acceleration, the measured threshold value of the 4.9 g switch was 2.7 g, which was lower than that of the 5.2 g switch (3.4 g), but was higher than that of the 5.1 g switch (2.3 g). This may be due to the randomness of the device thickness, leading to different squeeze-film damping effects, or it might be because unexpected fabrication errors.

Knowing that the variations of the measured spring thicknesses (18.2–22.5 µm) were all within the allowable latitude of 16–24 µm as obtained by the simulation, the threshold discrepancy in the measurement results between each device could contribute to the dimension errors of the spring width and the applied un-standard acceleration. In fact, the spring width of the fabricated switch was defined by the final structure release process by using an ICP etch. Since the silicon-on-glass bonded structure of the switch is similar to a SOI wafer, the etch profile of the springs is sensitive to the process parameters of the ICP etch, such as the RF power, the sample stage temperature, and the O_2 gas flow rate [28]. SEM measurements showed that the spring widths of the fabricated switches were in the range of 29.6–30.9 µm. In order to further investigate the impacts of the dimension errors of the spring width and the applied non-standard acceleration on the threshold value, the response of the switch with the measured dimension of the spring width (30.9 µm) and under the actual acceleration (extracted from the oscilloscope) was simulated by the ANSYS Workbench. Figure 7d compares the obtained results (the solid line) with that of the switch with the designed spring width of 30 µm and under the standard half-sine wave (the dotted line). As seen in the figure, the practical acceleration was much rougher than the standard wave, and the maximum amplitude of the practical acceleration (5.5 g) which was regarded as the threshold value was higher than that of the standard wave (5 g). The results indicate that the dimensions errors of the spring width and the applied un-standard acceleration may led to the increment of the threshold value by 0.5 g. In addition, some other factors such as the residual stresses in the springs might also influence the threshold value.

The measured contact time of the 4.9 switch under 1 ms, 5 ms, and 10 ms accelerations was 50 µs, 550 µs and 950 µs, respectively, as seen in Figure 7a–c. The results were in good agreement with our previous study [22], and they were significantly longer than the typical switches with rigid electrodes, based on bulk silicon micromachining technology (usually less than 20 µs) [29,30]. The contact bounce occurring in Figure 7c was due to the rigid contact process of the two electrodes when the squeeze-film damping effect was weak [31].

4. Conclusions

In this work, the impact of squeeze-film damping on the threshold of a 5 g inertial micro-switch was studied by theoretical analysis, FEM simulation, and experimental test, based on our previous research. The theoretical analysis results indicate that the threshold variation due to the fabrication errors can be reduced by establishing a tradeoff between the damping and elastic forces. The design strategy was achieved by optimizing the device thickness (the thickness of the proof mass and springs) and verified by a FEM simulation. The simulation results show that the maximum threshold deviation was only 0.15 g when the variation range of the device thickness was 16–24 µm, which is an adequate wide latitude for the current bulk silicon micromachining technology. The switch was fabricated by a typical silicon-on-glass process and tested by a vibration measurement system. The test results indicate that the threshold values under 1 ms acceleration were between 4.9–5.8 g, and the device thicknesses were 18.2–22.5 µm. The enhanced threshold accuracy is comparable to that of the switch fabricated on a SOI wafer, wherein the device thickness can be accurately controlled. The threshold accuracy was generally decreased when the acceleration duration was prolonged from 1 ms to 5 ms, and then to 10 ms, wherein the squeeze-film damping effect is obviously decreased, spotlighting the vital role of the squeeze-film damping on the improvement of threshold accuracy. The contact time was also significantly prolonged (50 µs), compared with the typical silicon-based switches (usually less than 20 µs), showing a good agreement with our previous study. The study is beneficial in various inertial micro-switches where low-g-sensing, long contact time, and high threshold accuracy are required.

Author Contributions: Conceptualization, Y.P. and T.L.; Methodology, Y.P., C.L., and G.W.; Software, C.L.; Validation, C.P.; Investigation, Y.P. and C.P.; Resources, T.L. and G.W.; Writing—Original Draft Preparation, Y.P.; Writing—Review & Editing, C.P. and G.W.; Supervision, T.L.; Project Administration, Y.P. and C.L.; Funding Acquisition, Y.P. and C.L.

Funding: This research was funded by the Foundation for High-level Talents of Chongqing University of Arts and Sciences (Grant No. 2017RJD15 and 2017RJD14) and the Scientific and Technological Research Program of Chongqing Municipal Education Commission (KJQN201801329).

Conflicts of Interest: The authors declare no conflict of interest.

References

1. Whitley, M.R.; Kranz, M.S.; Kesmodel, R.; Burgett, S.J. Latching shock sensors for health monitoring and quality control. In Proceedings of the MEMS/MOEMS Components and Their Applications II, San Jose, CA, USA, 22–27 January 2005.
2. Ongkodjojo, A.; Tay, F.E.H. Optimized design of a micromachined G-switch based on contactless configuration for health care applications. *J. Phys. Conf. Ser.* **2006**, *34*, 1044–1052. [CrossRef]
3. Ma, W.; Li, G.; Zohar, Y.; Wong, M. Fabrication and packaging of inertia micro-switch using low-temperature photo-resist molded metal-electroplating technology. *Sens. Actuators A Phys.* **2004**, *111*, 63–70. [CrossRef]
4. Currano, L.J.; Becker, C.R.; Lunking, D.; Smith, G.L.; Thomas, L. Triaxial inertial switch with multiple thresholds and resistive ladder readout. *Sens. Actuators A Phys.* **2013**, *195*, 191–197. [CrossRef]
5. Frobenius, W.D.; Zeitman, S.A.; White, M.H.; O'Sullivan, D.D. Microminiature ganged threshold accelerometers compatible with integrated circuit technology. *IEEE Trans. Electron Device* **1972**, *19*, 37–40. [CrossRef]
6. Zhou, Z.J.; Nie, W.R.; Xi, Z.W.; Wang, X.F. Electrical contact performance of MEMS acceleration switch fabricated by UV-LIGA technology. *Microsyst. Technol.* **2015**, *21*, 2271–2278. [CrossRef]
7. Lee, Y.; Sim, S.M.; Kim, H.; Kim, Y.K.; Kim, J.M. Silicon mems acceleration switch with high reliability using hooked latch. *Microelectron. Eng.* **2016**, *152*, 10–19. [CrossRef]
8. Dellaert, D.; Doutreloigne, J. A thermally-actuated latching mems switch matrix and driver chip for an automated distribution frame. *Mechatronics* **2016**, *40*, 287–292. [CrossRef]
9. Zhao, J.; Liu, P.B.; Tang, Z.A.; Fan, K.F.; Ma, X.S.; Gao, R.J.; Bao, J.D. A Wireless MEMS Inertial Switch for Measuring Both Threshold Triggering Acceleration and Response Time. *IEEE Trans. Instrum. Meas.* **2014**, *63*, 3152–3161. [CrossRef]
10. Gao, R.; Li, M.; Wang, Q.; Zhao, J.; Liu, S. A novel design method of bistable structures with required snap-through properties. *Sens. Actuators A Phys.* **2018**, *272*, 295–300. [CrossRef]
11. Liu, T.; Wei, S.; Tao, Y.; Yuan, X. Vibration interference analysis and verification of micro-fluidic inertial switch. *AIP Adv.* **2014**, *4*, 32–33. [CrossRef]
12. Liu, T.T.; Su, W.; Wang, C.; Yang, T. Threshold Model of Micro-Fluidic Inertial Switch Based on Orthogonal Regression Design. *Key Eng. Mater.* **2015**, *645–646*, 455–461. [CrossRef]
13. Zhang, Q.; Yang, Z.; Xu, Q.; Wang, Y.; Ding, G.; Zhao, X. Design and fabrication of a laterally-driven inertial micro-switch with multi-directional constraint structures for lowering off-axis sensitivity. *J. Micromech. Microeng.* **2016**, *26*, 055008. [CrossRef]
14. Xu, Q.; Sun, B.; Li, Y.; Xiang, X.; Lai, L.; Li, J.; Ding, G.; Zhao, X.; Yang, Z. Design and characterization of an inertial microswitch with synchronous follow-up flexible compliant electrodes capable of extending contact duration. *Sens. Actuators A Phys.* **2018**, *270*, 34–45. [CrossRef]
15. Xu, Q.; Yang, Z.; Fu, B.; Li, J.; Wu, H.; Zhang, Q.; Sun, Y.; Ding, G.; Zhao, X. A surface-micromachining-based inertial micro-switch with compliant cantilever beam as movable electrode for enduring high shock and prolonging contact time. *Appl. Surf. Sci.* **2016**, *387*, 569–580. [CrossRef]
16. Wang, Y.; Feng, Q.; Wang, Y.; Chen, W.; Wang, Z.; Ding, G.; Zhao, X. The design, simulation and fabrication of a novel horizontal sensitive inertial micro-switch with low g value based on mems micromaching technology. *J. Micromech. Microeng.* **2013**, *23*, 105013. [CrossRef]
17. Chen, W.; Wang, Y.; Wang, Y.; Zhu, B.; Ding, G.; Wang, H.; Zhao, X.; Yang, Z. A laterally-driven micromachined inertial switch with a compliant cantilever beam as the stationary electrode for prolonging contact time. *J. Micromech. Microeng.* **2014**, *24*, 065020. [CrossRef]

18. McNamara, S.; Gianchandani, Y.B. LIGA fabricated 19-element threshold accelerometer array. *Sens. Actuators A Phys.* **2004**, *112*, 175–183. [CrossRef]

19. Field, R.V., Jr.; Epp, D.S. Development and calibration of a stochastic dynamics model for the design of a mems inertial switch. *Sens. Actuators A Phys.* **2007**, *134*, 109–118. [CrossRef]

20. Du, L.; Li, Y.; Zhao, J.; Wang, W.; Zhao, W.; Zhao, W.; Zhu, H. A low-g mems inertial switch with a novel radial electrode for uniform omnidirectional sensitivity. *Sens. Actuators A Phys.* **2018**, *270*, 214–222. [CrossRef]

21. Zhang, F.; Yuan, M.; Jin, W.; Xiong, Z. Fabrication of a silicon based vertical sensitive low-g inertial micro-switch for linear acceleration sensing. *Microsyst. Technol.* **2017**, *23*, 2467–2473. [CrossRef]

22. Peng, Y.; Wen, Z.; Li, D.; Shang, Z. A low-g silicon inertial micro-switch with enhanced contact effect using squeeze-film damping. *Sensors* **2017**, *17*, 387. [CrossRef] [PubMed]

23. Thomson, W.T. *Theory of Vibration with Applications*, 2nd ed.; George Allen & Unwin: London, UK, 1983; pp. 48–52.

24. Cai, H.; Yang, Z.; Ding, G.; Zhao, X. Fabrication of a MEMS inertia switch on quartz substrate and evaluation of its threshold acceleration. *Microelectron. J.* **2008**, *39*, 1112–1119. [CrossRef]

25. Bao, M.H. *Analysis and Design Principles of MEMS Devices*, 1st ed.; Elsevier: Amsterdam, The Netherlands, 2005; pp. 135–136.

26. Schurink, B.; Berenschot, J.W.; Tiggelaar, R.M.; Luttge, R. Highly uniform sieving structure by corner lithography and silicon wet etching. *Microelectron. Eng.* **2015**, *144*, 12–18. [CrossRef]

27. Liu, M.; Gorman, D.G. Formulation of Rayleigh damping and its extensions. *Comput. Struct.* **1995**, *57*, 277–285. [CrossRef]

28. Jiang, F.; Keating, A.; Martyniuk, M.; Prasad, K.; Faraone, L.; Dell, J.M. Characterization of low-temperature bulk micromachining of silicon using an SF_6/O_2 inductively coupled plasma. *J. Micromech. Microeng.* **2012**, *22*, 2–5. [CrossRef]

29. Gerson, Y.; Schreiber, D.; Grau, H.; Krylov, S. Meso scale MEMS inertial switch fabricated using an electroplated metal-on-insulator process. *J. Micromech. Microeng.* **2014**, *24*, 405–412. [CrossRef]

30. Xi, Z.; Ping, Z.; Nie, W.; Du, L.; Yun, C. A novel MEMS omnidirectional inertial switch with flexible electrodes. *Sens. Actuators A Phys.* **2014**, *212*, 93–101. [CrossRef]

31. Yang, Z.; Cai, H.; Ding, G.; Wang, H.; Zhao, X. Dynamic simulation of a contact-enhanced MEMS inertial switch in Simulink. *Microsyst. Technol.* **2011**, *17*, 1329–1342. [CrossRef]

micromachines

MDPI

Article

Thermal Performance of Micro Hotplates with Novel Shapes Based on Single-Layer SiO$_2$ Suspended Film

Qi Liu, Guifu Ding *, Yipin Wang and Jinyuan Yao

National Key Laboratory of Science and Technology on Micro/Nano Fabrication, School of Electronic Information and Electrical Engineering, Shanghai Jiao Tong University, Shanghai 200240, China; Liuqi94@sjtu.edu.cn (Q.L.); yipinwang@sjtu.edu.cn (Y.W.); jyyao@sjtu.edu.cn (J.Y.)
* Correspondence: gfding@sjtu.edu.cn; Tel.: +86-21-3420-6686

Received: 15 September 2018; Accepted: 1 October 2018; Published: 11 October 2018

check for updates

Abstract: In this paper, two kinds of suspended micro hotplate with novel shapes of multibeam structure and reticular structure are designed. These designs have a reliable mechanical strength, so they can be designed and fabricated on single-layer SiO$_2$ suspended film through a simplified process. Single-layer suspended film helps to reduce power consumption. Based on the new film shapes, different resistance heaters with various widths and thicknesses are designed. Then, the temperature uniformity and power consumption of different micro hotplates are compared to study the effect of these variables and obtain the one with the optimal thermal performance. We report the simulations of temperature uniformity and give the corresponding infrared images in measurement. The experimental temperature differences are larger than those of the simulation. Experimental results show that the lowest power consumption and the minimum temperature difference are 43 mW and 50 °C, respectively, when the highest temperature on the suspended platform (240×240 μm^2) is 450 °C. Compared to the traditional four-beam micro hotplate, temperature non-uniformity is reduced by about 30–50%.

Keywords: suspended micro hotplate; single-layer SiO$_2$; temperature uniformity; power consumption; infrared image

---

1. Introduction

A micro hotplate (MHP), a heater isolated on a membrane, can be used in gas sensors, micro heat meters, gas flow meters, infrared light sources, and so on [1–3]. It presents a range of advantages, such as a miniaturized size, low power consumption, fast thermal response, high sensitivity, compatibility with CMOS technology, ease of integration with other microelectronic devices, and so on [4].

There are two types of micro hotplate structures: The closed-membrane type, and the suspended-membrane type [5,6]. The closed-membrane type has a faster heat dissipation, and more power loss under the same temperature [7]. The closed-membrane type structure generally requires multilayer stacking to balance the excessive internal stress attained after backside etching. We designed a closed-membrane type micro hotplate with only a layer of 2 μm SiO$_2$; the simulation results are shown in Figure 1. The maximum stress on 2 μm SiO$_2$ film obtained by backside etching was 1.83 GPa (compressive stress), which was far beyond the stable residual stress limit of 0.1 GPa [8]. Therefore, a closed-membrane type structure with single-layer SiO$_2$ can easily be damaged during fabrication, as shown in Figure 1b. To balance stress, many of the previously designed micro hotplates use multilayer composite films, because SiO$_2$ has compressive stress, while Si$_3$N$_4$ has tensile stress. S.Z. Ali designed a micro hotplate including passivation, a metal heat spreading plate, a tungsten heater and silicon heat spreading plate in silicon dioxide, buried silicon dioxide, and substrate from top to bottom [9]. G. Saxena came up with a micro hotplate containing insulation nitride, a heater, an Si_3N_4/SiO_2 composite membrane, and substrate [10]. I. Simon presented a design with passivation, metal heater and electrodes, and a dielectric layer of SiO_2 and Si_3N_4 [11]. For low stress Si_3N_4, the closed-membrane type micro hotplate can be successfully fabricated; however, due to the large size of the micro hotplate, power consumption is as high as 250 mW [12].

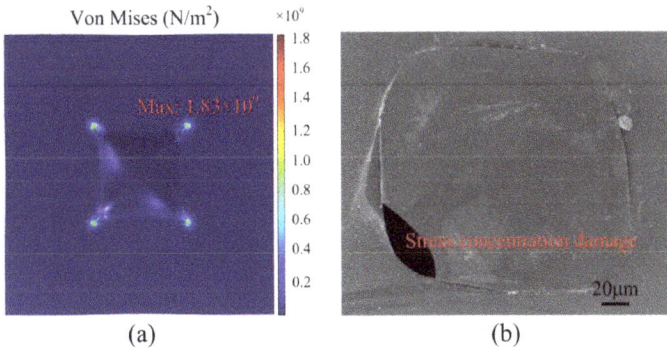

Figure 1. Closed-membrane type structure micro hotplate: (**a**) Stress distribution; (**b**) SEM image of SiO_2 film.

The suspended-membrane type micro hotplate has also been widely investigated, but the shape of suspended film is monotonous and there are still problems with the strength and temperature uniformity. F. Samaeifer performed complicated processes of five photolithography with four masks to fabricate the suspended-membrane micro hotplate with Si island [13]. M. Kaur developed a double spiral hotplate with non-uniform temperature distribution on the suspended platform and stress concentration at the connection between the beam and the platform [14]. M. Prasad designed a micro hotplate with 120×120 μm^2 SiO_2 film; the etching depth was only 56 μm, but the temperature difference was more than 100 °C [15]. Suspended films with different numbers of beams were designed in [16], which had the drawback of fragility; the stress on the beams and the vertical displacement of the plate were quite large. We performed simulation calculations of the traditional four-beam structure in Figure 2, and the deformation under residual stress was up to 14.5 μm, while the temperature difference on the platform exceeded 250 °C. A comparison of various micro hotplates is summarized in Table 1. For the purpose of solving stress concentration and non-uniform temperature distribution, novel shapes of suspended SiO_2 film were designed. Novel shapes of suspended films have been demonstrated in previous paper for their better mechanical properties [17].

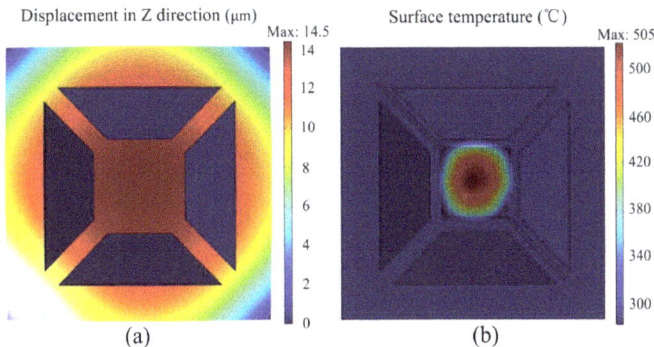

Figure 2. Four-beam suspended structure micro hotplate: (**a**) Deformation; (**b**) Temperature distribution.

Table 1. Comparison of various micro hotplates.

Structure	Membrane size/Heater size (μm²)	Maximum Temperature/Temperature Difference (°C)	Power Consumption (mW)	Type	Ref.
SiN/Cr/CrN/Pt/CrN/Cr/ SiO/SiN/Glass/SiN	2500 × 2500 1000 × 910	498 22	2350	closed membrane	[1]
Pt/PI film/PI sheet	200 × 200 100 × 100	300 22	15.5	closed membrane	[2]
SiO/Poly-Si/SiN/SiO/ Si/SiO/SiN	1400 × 1600 1100 × 1100	460 90	250	closed membrane	[3]
Poly-Si/SiN/SiO/Si/SiO	500 × 500 340 × 340	645 –	62.8	closed membrane	[7]
Insulation nitride/heater/ SiN/SiO/Si	120 × 120 100 × 100	421 22	30	Closed membrane	[10]
Pt/SiO/TaN/SiO/SiN/SiO/ Si/ SiO/SiN/SiO	300 × 300 100 × 100	450 30	100	Closed membrane	[18]
Pt/SiO/Si/SiO	50 × 50 –	400 –	11.8	suspended membrane	[6]
Pt/SiO/Si/SiO (with Si island)	1000 × 1000 440 × 440	500 30(with Si island) >80(without Si island)	50	suspended membrane	[13]
Pt/SiO/Si/SiO	500 × 500 –	600 >100	–	Suspended membrane	[14]
Pt/SiO/Si/SiO	120 × 120 –	500 >100 °C	20	suspended membrane	[15]

In this paper, micro hotplates with novel shapes of single-layer SiO_2 suspended film (multibeam structure and reticular structure) were designed and fabricated using extremely simplified processing steps. The goal of new designs was to reduce power consumption and improve temperature uniformity, based on a single layer of a suspended supporting membrane with a reliable mechanical performance. The novel structures can work stably with only a layer of SiO_2. Finite element simulation guides the design of membrane structure and resistance heaters. The thermal characteristics of these new designs were tested and clear infrared temperature distribution images are given to obtain a design with the optimal comprehensive performances.

2. Materials and Methods

2.1. Fabrication Details

The micro hotplates were fabricated on 500 μm double-side oxidized and polished p-type <100> Si wafer; the oxide layer was 2 μm thick. An SiO_2 membrane was used as a supporting layer and a dielectric membrane to thermally isolate the heated area from the silicon substrate. Only two chrome masks were used to fabricate the micro hotplate: One for resistive heater and electrode, the other for the front-side etching window. A specific process flow, including spin-coating the photoresist, lithography, sputtering, lift-off, electroplating, reactive ion etching (RIE), and wet etching, is shown in Figure 3.

Before spin-coating the photoresist, a tackifier layer of HMDS (Hexamethyldisilizane) was spin-coated and dried on a 150 °C hot plate for 5 min to enhance the adhesion of photoresist and substrate. Otherwise, graphics were prone to fall off during lithography. AZ 4330 photoresist was spin-coated at a speed of 6000 rpm to obtain a thickness of 2.5 μm. The Pt resistance wire and a seed layer of Cr/Cu was deposited by magnetron sputtering. A Ni protective layer of about 2.5 μm was electroplated as a mask for RIE to avoid Cu oxidization; a CHF_3 and O_2 gas mixture was used in RIE etching. We used 95 °C 15% KOH solution to perform anisotropic wet etching of the silicon substrate; it took about 2.5 hours to etch about 350 μm, so that the platform could be completely suspended. During the wet etching process, the Ni metal layer effectively protected the front structure and the suspended film. Then, we removed metals Ni, Cu, and Cr to release the device. Finally, the micro hotplate was dipped in ethanol, acetone, and freon (F113) in turn to clean and exchange out water in the cavity. Freon evaporates quickly with low stress, so as to avoid excessive stress and damage in the structure during drying.

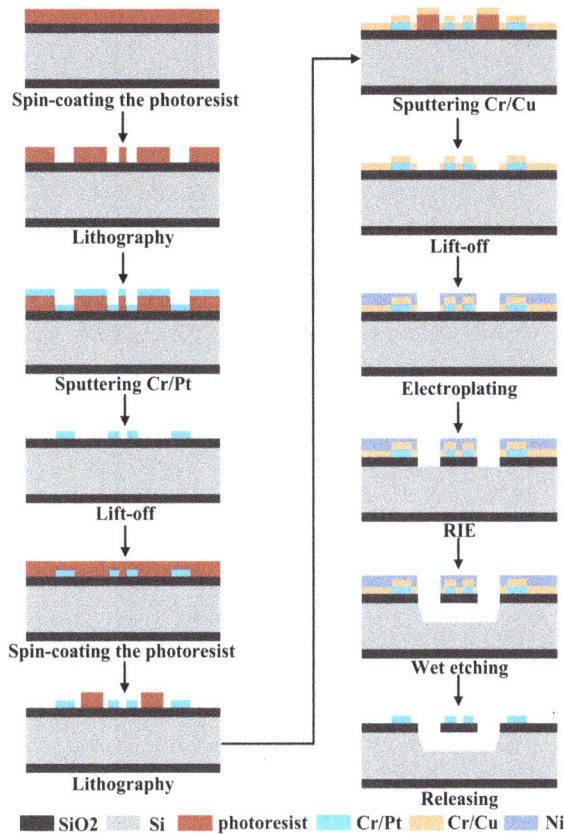

Figure 3. Specific process flow diagram.

2.2. Design

Micromachined metal oxide gas sensors with a micro hotplate often need to be able to operate at a temperature which ranges from 200 °C up to 400 °C, and the temperature distribution needs to be as uniform as possible [19]. Metal platinum is used as a heating resistance wire, as platinum has stable physical and chemical properties, and perfect consistency of processing technology. Additionally, it has a large temperature resistance coefficient and a favorable thermal stability, which is extremely suitable for temperature calibration [20]. For the temperature uniformity of the entire platform, the Pt heating resistance wire evenly bestrewed the heating platform. The number, length, and width of the beam linking platform to the base would affect the conduction of heat, so we designed two types of suspended films, a multibeam structure and reticular structure, to study the effect on thermal performance. After that, serpentine and double spiral heating wires were designed, changing the width and the thickness to study the influence of these parameters on the thermal and electric characteristics.

A stereogram of the micro hotplate and several combinational designs of suspended films and heaters are shown in Figure 4. At the top surface there was a Pt heating resistance wire; designs 1 and 2 were serpentine heaters with a line width of 5 μm, design 3 was a double spiral heater with 5 μm line width, and design 4 was double a spiral heater with a variable line width of 10 μm, 8 μm, 5 μm from inside to outside. The suspended SiO_2 membrane consisted of the heating platform and beams connecting between platform and Si base; the cavity was etched in silicon substrate. Design 1 was named a multibeam structure, whereas the other designs were named reticular structures.

By KOH anisotropic wet etching, the cavity presented an inverted pyramid with a flat bottom surface. The membrane side length was 720 μm and the heater area side length was about 240 μm. Design 1 and 2 were compared to see the impact of the shape of suspended film; design 2 and 3 were compared for the impact of the Pt wire geometry, and design 3 and 4 were compared for the impact of the Pt wire line width.

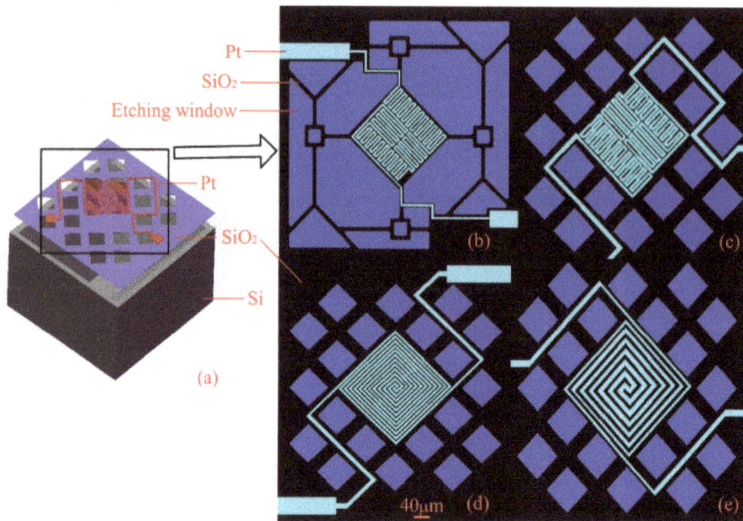

Figure 4. Designs of the micro hotplates: (**a**) Stereogram; (**b**) Design 1: Multibeam structure with a serpentine heater; (**c**) Design 2: Reticular structure with a serpentine heater; (**d**) Design 3: Reticular structure with a double spiral heater; (**e**) Design 4: Reticular structure with a variable line width heater.

2.3. Simulation

Electrothermal-solid mechanics coupling finite element simulation by Comsol Multiphysics 5.2 was applied to calculate surface temperature distribution of the micro hotplates, by which to compare the above four designs, so as to guide an optimum design with more preferable temperature uniformity. The material parameters used in the simulation are shown in Table 2. The temperature coefficient of resistance (TCR) of Pt was set as 2.063e-3; this value was obtained by the temperature-resistance calibration of sputtered Pt wire. The calibration curve is shown in Figure 5.

Table 2. Material properties in simulation.

Material	Density (Kg/m^3)	Young's modulus (Pa)	Poisson's Ratio	Thermal conductivity (W/m*K)	Thermal-expansion coefficient (1/K)
Si	2330	1.9e11	0.2	148	2.5e-6
SiO$_2$	2200	70e9	0.17	1.4	0.5e-6
Pt	21,450	168e9	0.38	71.6	8.8e-6

Primary heat loss is the heat conduction of membrane and convection heat dissipation around the environment. Regarding the research on micro hotplates, there were contradictions regarding the main form of heat loss between convection and conduction in the thermal model. The literature [21] pointed out that convective heat loss was the main factor when the temperature was 50 °C higher than ambient temperature. In our simulation, the setting of the convective heat transfer coefficient directly affected the results. The convective heat transfer coefficient is closely related to the shape and size of the heat transfer surface, the temperature difference between the surface and the fluid, and the flow velocity of the fluid. The convection coefficient under the microscale is much higher than the conventional scale [2,22,23]. The cavity inner surface, the heating platform surface, and the heating wire surface were microscale regions with a high temperature, so the convection coefficient (expressed as h) was set high, while other surfaces of the base with room temperature were set as an air natural convection coefficient. Take design 1 for example, where the convection coefficient h was set as different values to plot the relationship curve between h and the highest temperature on the platform T_{max}. As shown in Figure 6, when h is below 1500 W·m^{-2}·K^{-1}, the value of T_{max} is greatly affected by the setting of h. Thus, it was more reasonable to set h between 1500 to 2500 W·m^{-2}·K^{-1} to reduce error caused by the choice of h value.

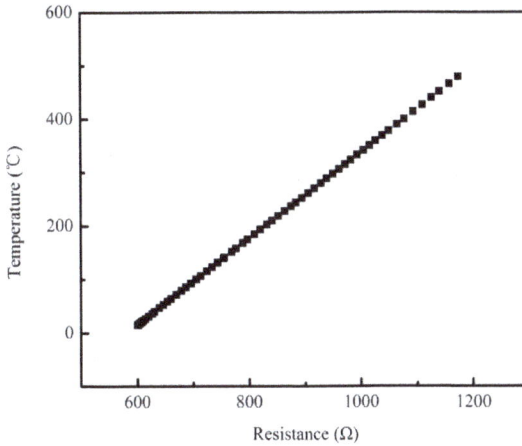

Figure 5. Temperature-resistance calibration curve.

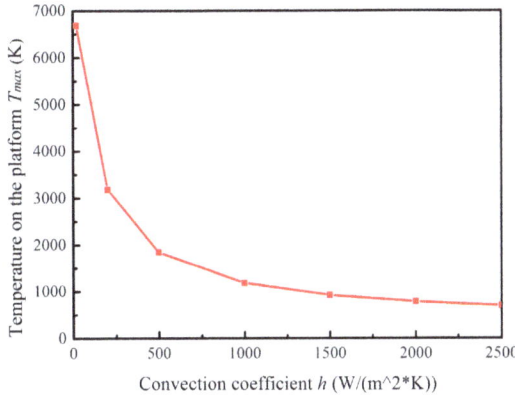

Figure 6. Relationship between convection coefficient h and the highest temperature on the platform T_{max}.

When h was set to 2000 W·m^{-2}·K^{-1}, the Pt resistance wire was regarded as a heat source with a DC voltage on electrodes, and the bottom of the base was fixed. Thermal conduction and convection were both taken into consideration; the temperature distributions of the four designs are shown in Figure 7.

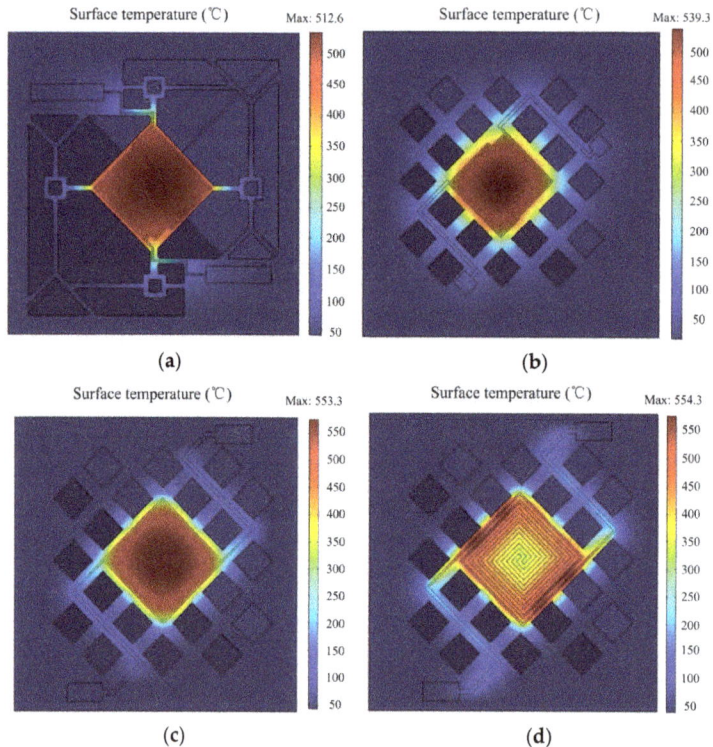

Figure 7. Simulation results: (**a**) Surface temperature of design 1; (**b**) Surface temperature of design 2; (**c**) Surface temperature of design 3; (**d**) Surface temperature of design 4.

Figure 7a indicates that the highest temperature on the platform of design 1 reached 512.6 °C, with a power consumption of 46.8 mW. Temperature decreased gradually outward from the center. The temperature difference on the whole suspended platform was about 120 °C. In addition to the outer ring, temperature difference was within 50 °C in an area of $180 \times 180 \ \mu m^2$.

As shown in Figure 7b, the highest temperature of design 2 reached 539.3 °C with a power consumption of 61.2 mW. The temperature difference on the whole platform was within 200 °C, while the temperature difference was 100 °C in an area of $180 \times 180 \ \mu m^2$. Thus, with a similar maximum temperature, the platform surface temperature difference of design 2 was about two times larger than that of design 1. The reticular membrane had thicker, shorter, and more connecting beams, so temperature conduction was stronger than that of the multibeam structure, leading to faster heat dissipation.

Figure 7c indicated that the highest temperature of design 3 was 553.3 °C, with a power consumption of 65.8 mW; at the moment, the temperature in the most central area was around 550 °C. The temperature in a range of $120 \ \mu m \times \mu m^2$ was uniformly distributed over 525 °C, and temperature difference in the whole platform was up to 180–200 °C. Comparing with Figure 7b, although the temperature difference on the entire suspended heating platform was almost the same, in the center region of $120 \times 120 \ \mu m^2$, the temperature difference was less than half of that of the serpentine heater.

It could be seen from Figure 7d that when the highest temperature of the platform reached 554.3 °C, the temperature difference on the whole platform of the variable width heater was the same as that of the identical width heater in Figure 7c. The biggest difference was that temperature on the center of the platform was lower, whereas the outer ring temperature was relatively high because of the finer resistance wire. The temperature uniformity of the above four designs was much better than the that in the four-beam structure shown in Figure 2b, illustrating that novel suspended films had more uniform temperature distribution. Finer and longer beams helped to slow thermal conduction, and the larger etching window led to rapid convection.

For a better comparison of the temperature difference of the four designs, the temperature distribution on the same position line is shown in Figure 8. In the center of a 120 μm range, the temperature difference of the four designs was 20 °C, 80 °C, 60 °C, and 35 °C, respectively. The smallest temperature difference of design 1 shows that thermal conduction through beams has a great influence on temperature uniformity in the simulation.

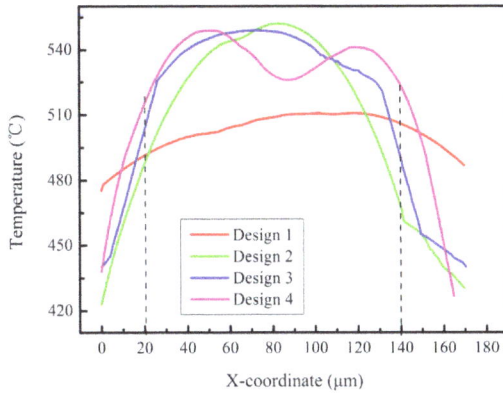

Figure 8. The temperature distribution on the same position line.

2.4. Experiment

The copper wires were soldered on the electrodes and connected to the ammeters through the clamps. In order to facilitate the welding lead, the designs extended the leading wires of the micro hotplates to 4 mm, and expanded the electrodes to 2×2 mm^2. As in the test system diagram shown in Figure 9, DC voltage was applied to the sample, and the heating area temperature distribution was measured using an infrared camera Compix 320, with a pixel size of 6 μm. An amperemeter was used to measure on-state current, while a voltmeter was used to measure the voltage of the sample; thus, different power values were calculated under different DC voltages. Scanning electron microscopy (SEM) of a sample is also shown in Figure 9.

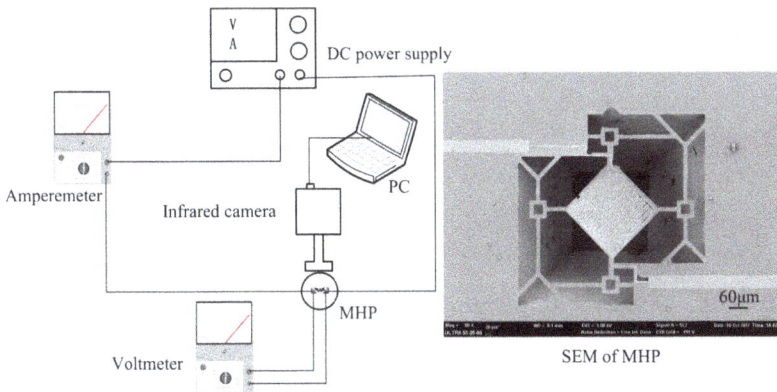

Figure 9. Test system diagram.

3. Results and Discussion

3.1. Temperature Uniformity

After repeated power-on and power-off for several times, a stable infrared thermogram of design 1 was captured (Figure 10). Applying a voltage of 20 V, at the image capture moment, the highest temperature on the central region was 432.6 °C. Figure 10a shows that the temperature was 381.2 °C at the cross point, and the temperature difference on a quarter of an entire micro hotplate (60 × 60 µm²) near the center could be within 50 °C, and in the half area of the whole platform (120 × 120 µm²), the temperature difference range was within 90 °C.

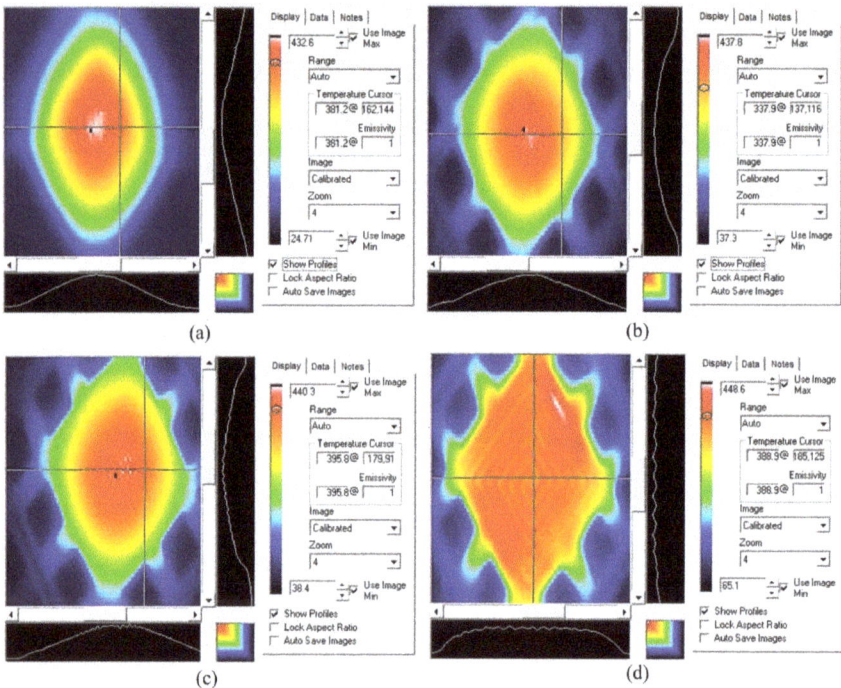

Figure 10. Test results: (**a**) Surface temperature of design 1; (**b**) Surface temperature of design 2; (**c**) Surface temperature of design 3; (**d**) Surface temperature of design 4.

The temperature distribution of design 2 under the voltages of 19 V is shown in Figure 10b. The highest temperature on the center was 437.8 °C. The region with a temperature difference of less than 100 °C did not exceed half of the total area of the platform (120 × 120 µm²). With a similar highest temperature, the platform surface temperature difference of design 2 was larger than that of design 1, and this was consistent with the simulation. The thermal conduction of the reticular membrane was stronger than that of the multibeam structure, for it had thicker, shorter, and more connecting beams.

Figure 10c shows the temperature distribution of design 3 under the voltages of 20 V, where the highest temperature on the center was 440.3 °C. The regional extents of 50 °C and 100 °C temperature difference were almost the same as those in Figure 10b, so only improving heating wire shape from serpentine to double spiral had little effect on temperature uniformity.

Figure 10d shows the temperature distribution of design 4 under 22 V voltage, where the highest temperature was 448.6 °C and the temperature difference on the whole platform was almost within 50 °C. Compared with that of the traditional four-beam micro hotplate listed in Table 1 [13,15], temperature uniformity was increased by about 30–50%. The temperature profiles swept by two orthogonal straight lines on the same position of four designs are also shown in Figure 10. The temperature profile of design 4 was much flatter than that of others, and showed a more uniform temperature distribution. In actual infrared tests, design 4 has the most uniform temperature distribution, which proves that temperature uniformity can be significantly improved by a reasonable layout of line width. Due to the presence of etching windows and the cavity, the convection at the edge of the platform is stronger than that at the center. The resistance wire with a finer width has high heat to compensate for the strong convection of the edge position.

For designs 1–3, temperature decreased from the center to the edge of the platform, and the temperature difference in the experiment was larger than that in the simulation. For design 4, in the simulation, the temperature of periphery was higher due to the thinner resistance wire, while the temperature of the whole platform was more uniform in the experiment. These phenomena were due to the fact that convective heat dissipation was achieved only by setting a constant convection coefficient h in the simulation, but in actual tests, the convection of periphery was more prominent than that of the center because of the existence of the cavity. Previous literature [5,7,10] only had the results of simulations, but infrared temperature images were not given. However, simulation results cannot truly reflect temperature distribution.

3.2. Power Consumption

The powers and the corresponding temperatures at different voltages were measured to draw Figure 11. The relationship of the power consumptions on the applied voltages and the average temperatures for the 120 $\times$ 120 μm^2 active area under these conditions was monitored. As the temperature rose, the power was approximately proportional to the increase. For design 1, when average temperature reached 430 °C, power consumption was only 41.32 mW. For design 2, power consumption was 59.10 mW when average temperature reached 430 °C. For design 3, when average temperature reached 430 °C, power consumption was 63.53 mW. For design 4, with the thickness of 100 nm, when the highest temperature reached 445 °C, power consumption was 115.48 mW, with a thickness of 300 nm, and when the highest temperature reached 450 °C, power consumption was 140.01 mW. These values accorded with the fact that micromachined sensors typically require power consumption which is in the range between 30 mW and 150 mW [18].

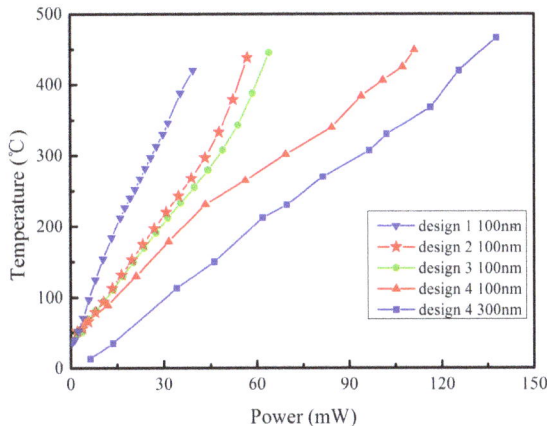

Figure 11. Relationship of power consumption and temperature.

It could be seen that when the temperature of the platform was almost the same, the power consumption of the reticular membrane structure (design 2) was higher than that of the multibeam structure (design 1). This phenomenon contributed to a larger heat dissipation area and multiple short conduction paths of the reticular membrane structure. The power consumption of the double spiral resistance wire (design 3) was slightly higher than that of the serpentine shape (design 2). Comparing design 4 to design 3, with the same shape of resistance wire, power consumption in the high temperature region increased obviously as line width increased. When the width and the shape of the resistive heaters were the same, power consumption increased sharply with the increase of thickness. In conclusion, the main influencing factors of power consumption are the line width and thickness of the resistance wire.

When heated to 500 °C for about 2–3 min, a 100 nm thick resistance wire was broken due to electromigration. By comparison, when voltage was added to heat up 300 nm thick resistance wire to 650 °C, the micro hotplate could still work. Therefore, resistance wire becomes more reliable when increasing the sputtering thickness; meanwhile, triplicating the section of the wire leads to three times more current before reaching the critical current density for electro migration. Increasing the thickness leads to more power consumption. Thus, an appropriate thickness of the resistance wire needs to be chosen to prevent failure and excessive power consumption.

4. Conclusions

Micro hotplates with novel shapes of single-layer SiO_2 suspended films were designed and fabricated through a simplified technological process. Accurate temperature distribution can be obtained by high resolution infrared images. Finer, longer beams can reduce thermal conduction and small etching windows can reduce heat convection, both leading to a more uniform temperature. The shape change of resistive heaters has little effect on power consumption and thermal uniformity. Wider resistive heaters consume more power, but a reasonable variable line width design compensates for the strong convection of the edge position to improve temperature uniformity on the platform. The heaters should not be too thick, because power consumption increases significantly with increasing thickness. Double-spiral resistance wire with variable, thin line width and moderate thickness based on multibeam structure suspended film is the most optimized design. The lowest power consumption and the minimum temperature difference are 43 mW and 50 °C, respectively, when the highest temperature on the suspended platform (240 × 240 μm^2) is 450 °C. Compared to that of the suspended micro hotplates listed in Table 1, the temperature difference of this design drops by about 30–50%.

Author Contributions: Conceptualization, Q.L. and G.D.; Methodology, Q.L. and G.D.; Software, Q.L.; Formal Analysis, Q.L.; Investigation, Q.L. and Y.W.; Resources, J.Y. and G.D.; Writing-Original Draft Preparation, Q.L.; Writing-Review & Editing, Q.L. and G.D.

Funding: The technology development fund of Shanghai Science and Technology Commission: 17DZ2291400.

Acknowledgments: This research was funded by Technology Development Fund of Shanghai Science and Technology Commission grant number No. 17DZ2291400. And the APC was funded by Shanghai Professional Technical Service Platform for Non-Silicon Micro-Nano Integrated Manufacturing.

Conflicts of Interest: The authors declare no conflict of interest.

References

1. Chang, W.Y.; Hsihe, Y.S. Multilayer microheater based on glass substrate using MEMS technology. *Microelectron. Eng.* **2016**, *149*, 25–30.
2. Courbat, J.; Canonica, M.; Teyssieux, D.; Briand, D.; De Rooij, N. Design and fabrication of micro-hotplates made on a polyimide foil: Electrothermal simulation and characterization to achieve power consumption in the low mW range. *J. Micromech. Microeng.* **2010**, *21*, 196–201. [CrossRef]
3. Hwang, W.J.; Shin, K.S.; Roh, J.H.; Lee, D.S.; Choa, S.H. Development of micro-heaters with optimized temperature compensation design for gas sensors. *Sensors* **2011**, *11*, 2580–2591. [CrossRef] [PubMed]
4. Yi, X.; Lai, J.; Liang, H.; Zhai, X. Fabrication of a MEMS micro-hotplate. *JPCS* **2011**, *276*, 12098–12106. [CrossRef]
5. Bhattacharyya, P. Technological journey towards reliable microheater development for mems gas sensors: A review. *IEEE. Trans. Device Mater. Reliab.* **2014**, *14*, 589–599. [CrossRef]

6. Prasad, M. Design, development and reliability testing of a low power bridge-type micromachined hotplate. *Microelectron. Reliab.* **2015**, *55*, 937–944. [CrossRef]

7. Verma, S.; Rajnish; Singhal, A.; Jindal, P. Design and simulation of closed membrane type micro hotplate for gas sensing applications. *Int. J. Eng. Sci. Metal.* **2012**, *2*, 768–771.

8. Rossi, C.; Templeboyer, P.; Esteve, D. Realization and performance of thin SiO$_2$/SiNx membrane for microheater applications. *Sensor. Actuat. A. Phys.* **1998**, *64*, 241–245. [CrossRef]

9. Ali, S.Z.; Udrea, F.; Milne, W.I.; Gardner, J.W. Tungsten-based SOI microhotplates for smart gas sensors. *J. Microelectromech. S.* **2008**, *17*, 1408–1416. [CrossRef]

10. Saxena, G.; Paily, R. Analytical modeling of square microhotplate for gas sensing application. *IEEE. Sens. J.* **2013**, *13*, 4851–4858. [CrossRef]

11. Simon, I.; BaÃrsan, N.; Bauer, M.; Weimar, U. Micromachined metal oxide gas sensors: opportunities to improve sensor performance. *Sensor. Actuat. B. Chem.* **2001**, *73*, 1–26. [CrossRef]

12. Tommasi, A.; Cocuzza, M.; Perrone, D.; Pirri, C.; Mosca, R.; Mosca, R.; Villani, M.; Delmonte, N.; Zappttini, A.; Calestani, D.; et al. Modeling, fabrication and testing of a customizable micromachined hotplate for sensor applications. *Sensors* **2017**, *17*, 62. [CrossRef] [PubMed]

13. Samaeifar, F.; Afifi, A.; Abdollahi, H. Simple fabrication and characterization of a platinum microhotplate based on suspended membrane structure. *Exp. Techniques.* **2014**, *40*, 755–763. [CrossRef]

14. Kaur, M.; Prasad, M. Development of double spiral MEMS hotplate using front-side etching cavity for gas sensors. *NCTP* **2016**, *1724*, 2088–2106.

15. Prasad, M.; Khanna, V.K. A low-power, micromachined, double spiral hotplate for MEMS gas sensors. *Microsyst. Technol.* **2015**, *21*, 2123–2131. [CrossRef]

16. Belmonte, J.C.; Puigcorbe, J.; Arbiol, J.; Vila, A.; Morante, J.R.; Sabate, N.; Gracia, I.; Cane, C. High-temperature low-power performing micromachined suspended micro-hotplate for gas sensing applications. *Sensor. Actuat. B. Chem.* **2006**, *114*, 826–835. [CrossRef]

17. Liu, Q.; Wang, Y.P.; Yao, J.Y.; Ding, G.F. Impact resistance and static strength analysis of an extremely simplified micro hotplate with novel suspended film. *Sensor. Actuat. B. Phys.* **2018**, *280*, 495–504. [CrossRef]

18. Chiou, J.C.; Tsai, S.W.; Lin, C.Y. Liquid phase deposition based SnO$_2$ gas sensor integrated with TaN heater on a micro-hotplate. *IEEE. Sens. J.* **2013**, *13*, 2466–2473. [CrossRef]

19. Marasso, S.L.; Tommasi, A.; Perrone, D.; Cocuzza, M.; Mosca, R.; Villani, M.; Zappettini, A.; Calestani, D. A new method to integrate ZnO nanotetrapods on MEMS micro-hotplates for large scale gas sensor production. *Nanotechnology* **2016**, *27*, 385–503. [CrossRef] [PubMed]

20. Lee, K.N.; Lee, D.S.; Jung, S.W.; Jang, Y.H.; Kim, Y.K.; Seong, W.K. A high-temperature MEMS heater using suspended silicon structures. *J. Micromech. Microeng.* **2009**, *19*, 115011. [CrossRef]

21. Samaeifar, F.; Hajghassem, H.; Afifi, A.; Abdollahi, H. Implementation of high-performance MEMS platinum micro-hotplate. *Sensor Rev.* **2015**, *35*, 116–124. [CrossRef]

22. Peirs, J.; Reynaerts, D.; Van Brussel, H. Scale effects and thermal considerations for micro-actuators. *ICRA* **1998**, *2*, 1516–1521.

23. Hu, X.J.; Jain, A.; Goodson, K.E. Investigation of the natural convection boundary condition in microfabricated structures. *Int. J. Therma. Sci.* **2008**, *47*, 820–824. [CrossRef]

micromachines

MDPI

Article

Automatic Frequency Tuning Technology for Dual-Mass MEMS Gyroscope Based on a Quadrature Modulation Signal

Jia Jia [1,2] iD **, Xukai Ding** [1,2] iD **, Yang Gao** [1,2] iD **and Hongsheng Li** [1,2,*] iD

[1]   School of Instrument Science and Engineering, Southeast University, Nanjing 210096, China;
       230169207@seu.edu.cn (J.J.); dingxukai@126.com (X.D.); gao-yang@seu.edu.cn (Y.G.)
[2]   Key Laboratory of Micro-Inertial Instruments and Advanced Navigation Technology, Ministry of Education,
       Nanjing 210096, China
[*]   Correspondence: hsli@seu.edu.cn; Tel.: +86-25-8379-5920

Received: 17 September 2018; Accepted: 8 October 2018; Published: 10 October 2018

check for updates

Abstract: In order to eliminate the frequency mismatch of MEMS (Microelectromechanical Systems) gyroscopes, this paper proposes a frequency tuning technology based on a quadrature modulation signal. A sinusoidal signal having a frequency greater the gyroscope operating bandwidth is applied to the quadrature stiffness correction combs, and the modulation signal containing the frequency split information is then excited at the gyroscope output. The effects of quadrature correction combs and frequency tuning combs on the resonant frequency of gyroscope are analyzed. The tuning principle based on low frequency input excitation is analyzed, and the tuning system adopting this principle is designed and simulated. The experiments are arranged to verify the theoretical analysis. The wide temperature range test (−20 °C to 60 °C) demonstrates the reliability of the tuning system with a maximum mismatch frequency of less than 0.3 Hz. The scale factor test and static test were carried out at three temperature conditions (−20 °C, room temperature, 60 °C), and the scale factor, zero-bias instability, and angle random walk are improved. Moreover, the closed-loop detection method is adopted, which improves the scale factor nonlinearity and bandwidth under the premise of maintaining the same static performances compared with the open-loop detection by tuning.

Keywords: dual-mass MEMS gyroscope; frequency tuning; frequency split; quadrature modulation signal; frequency mismatch

---

1. Introduction

With the rapid development of MEMS (Microelectromechanical Systems) technology, silicon micromachined gyroscopes have attracted more attention. As a miniature sensor for measuring angular velocity, MEMS gyroscopes have been widely used in military and civilian fields [1–4]. Therefore, the performance requirements of the MEMS gyroscope are also increasing [2,5,6]. When the drive mode and sense mode of the gyroscope have the same resonant frequency (mode-matching), the gyroscope can have a higher signal-to-noise ratio of the output signal without deteriorating the circuit noise. The principle of operation of dual-mass MEMS gyroscopes is based on the Coriolis coupling between the two operating modes (drive and sense modes) when a rotation is applied about the sensitive axis of the device. Efficient energy transfer from the drive mode to the sense mode, which is largely determined by the frequency matching condition, is a principal factor in performance. In practice, however, fabrication imperfections and environmental variations are always present, resulting in a frequency mismatch between the two modes [3,5,7,8]. This frequency mismatch lead to degraded sensitivity, resolution, signal-to-noise ratio, and poor zero bias stability. Therefore, it is necessary to study the method of eliminating frequency split (Δf) so that the gyroscope is in the mode-matching condition.

There are several post-processing frequency tuning technologies to eliminate frequency split, such as local thermal stress technology [9,10], micromachining correction technology [11,12], and electrostatic adjustment technology [1–8,13–19]. In [9,10], the structural stress and material parameters of the gyroscope are changed by the heat generated by loading voltage, and the resonant frequency of the gyroscope is altered to realize the mode-matching. While frequency tuning by micromachining correction technology is achieved by changing the structural parameters of the gyroscope through polysilicon precipitation [11] or laser trimming [12]. The above two techniques share the same drawback of requiring, to some extent, a manual intervention, so that they are not desirable for mass production. Moreover, these two technologies can cause unstable output due to temperature changes and are not suitable for real-time adjustment of the resonant frequency.

A more effective method at present is the electrostatic adjustment technology, which utilizes a structure-specific electrostatic negative stiffness effect to change the stiffness of the structure by adjusting the DC voltage, thereby altering the resonant frequency to achieve the purpose of mode-matching. Complex algorithms are used for parameter fitting [13,14], identification [5,7], and prediction [17] to achieve real-time frequency tuning. These strategies can effectively eliminate the frequency split, but they need a large amount of original data acquisition, and the general applicability is not ideal. Some literature utilize the characteristics of the gyroscope output signal to reduce the frequency mismatch between the two operating modes. When $\Delta f = 0$, the amplitude of the Coriolis signal and the quadrature signal reach the maximum [2,3,16], and the phase difference between the quadrature signal and the drive detection signal is 90° [4,8,18]. However, these frequency tuning strategies do not work properly if the input angular velocity changed. The frequency tuning strategies that can satisfy the normal operation of the gyroscope is to introduce low-frequency oscillation signals into the sense resonator, and realize mode-matching according to the amplitude or phase characteristics of the output signals [1,6,15,19].

This paper presents a real-time automatic tuning technology for dual-mass MEMS gyroscopes. A low frequency oscillation signal (its frequency is greater than the gyroscope's bandwidth) is introduced into quadrature stiffness correction combs, and the degree of the frequency mismatch is then judged according to the output response of the sense mode. This paper is organized as follows. Section 2 gives the structure of the gyroscope. The theory and simulation of frequency tuning are analyzed in Section 3. In Section 4, the relevant experimental results are published to testify the theoretical analysis and contrast the gyroscope's performance. Section 5 concludes this paper with a summary.

2. Dual-mass MEMS Gyroscope Structure

2.1. Gyroscope Overall Structure

In this paper, the structure of the dual-mass MEMS gyroscope is shown in Figure 1. Two fully symmetrical masses perform a simple harmonic vibration of equal amplitude and reverse phase along the X-axis direction under the action of electrostatic driving force. Due to the Coriolis effect, the Coriolis mass drives the sense comb to move along the Y-axis through the U-shaped connecting spring when the angular velocity exists in the sensitive axis (Z-axis), which causes the relative motion between the moveable electrode and the fixed electrode of the sense combs, resulting in the change in differential detection capacitance. By measuring the amount of capacitance change, the corresponding input angular velocity can be obtained. In addition, the gyroscope has another three combs: the sense feedback comb, the quadrature stiffness correction comb, and the frequency tuning comb. The area-changing force rebalance combs are used to suppress the movement by Coriolis force, the angular velocity information is obtained indirectly by the size of the feedback force, and the force rebalance combs will not change the resonant of the two operating modes [6]. The quadrature stiffness correction combs with unequal spacing are designed in Coriolis mass for restraining the quadrature error. The gap-changing frequency tuning combs are for applying a change in the resonant frequency of the sense mode.

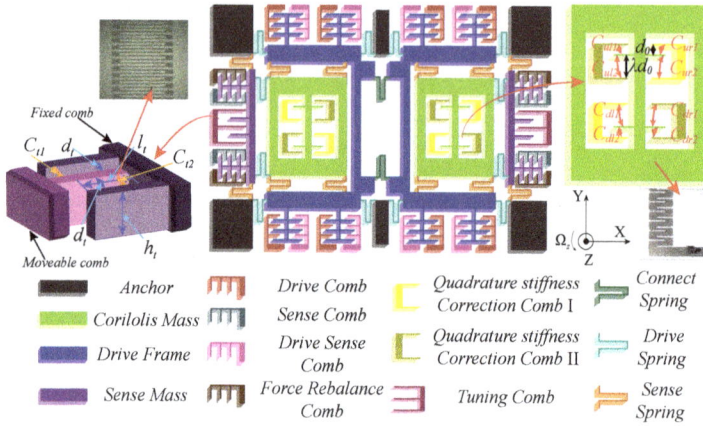

Figure 1. Mechanical model of dual-mass MEMS gyroscope.

The MEMS gyroscope consists of two operating modes—the drive mode and the sense mode. Both modes can be considered as a "spring-mass-damping" second-order system. According to Newton's second law, furthermore, only the effects of quadrature coupling stiffness are considered, simplified dynamics equation for the drive mode and sense mode of MEMS gyroscope are obtained as follows:

$$\begin{bmatrix} m_x & 0 \\ 0 & m_y \end{bmatrix} \begin{bmatrix} \ddot{x} \\ \ddot{y} \end{bmatrix} + \begin{bmatrix} c_x & 0 \\ 0 & c_y \end{bmatrix} \begin{bmatrix} \dot{x} \\ \dot{y} \end{bmatrix} + \begin{bmatrix} k_{xx} & k_{xy} \\ k_{yx} & k_{yy} \end{bmatrix} \begin{bmatrix} x \\ y \end{bmatrix} = \begin{bmatrix} F_x \\ -2m_c\Omega_z\dot{x} + F_y \end{bmatrix} \tag{1}$$

where x and y are the displacement of drive mode and sense mode, respectively; m_x and m_y are the equivalent mass of the two modes, respectively; c_x, k_{xx} and c_y, k_{yy} are the damp coefficients and the stiffness coefficients of the drive and sense modes; k_{xy} and k_{yx} are the coupling stiffness coefficients in each mode; F_x and F_y is the external force applied to drive mode and sense mode, respectively; m_c is the Coriolis mass and $m_c \approx m_y$; Ω_z is the input angular velocity with respect to the Z-axis.

Assume the electrostatic driving force in the drive mode is $F_x = A_F \sin \omega_d t$, where A_F is the amplitude of electrostatic driving force, and ω_d is the frequency of electrostatic driving force. The gyroscope system uses phase-locked loop technology to track the resonant frequency of the driving mode, that is $\omega_d = \omega_x$. Substituting F_x into Equation (1), the stationary solution of x and y can be obtained as

$$x(t) = A_x \sin(\omega_d t + \varphi_x) \tag{2}$$

$$y(t) = \underbrace{A_c \cos(\omega_d t + \varphi_x + \varphi_y)}_{\text{Coriolis Singal}} + \underbrace{A_q \sin(\omega_d t + \varphi_x + \varphi_y)}_{\text{Quadrature Singal}} \tag{3}$$

where A_x, A_c, and A_q can be written as

$$A_x = \frac{A_F/m_x}{\omega_x^2 \sqrt{(1 - \frac{\omega_d^2}{\omega_x^2})^2 + (\frac{\omega_d}{Q_x\omega_x})^2}} \tag{4}$$

$$A_c = \frac{2\Omega_z A_x \omega_d}{\omega_y^2 \sqrt{(1 - \frac{\omega_d^2}{\omega_y^2})^2 + (\frac{\omega_d}{Q_y\omega_y})^2}} \tag{5}$$

$$A_q = \frac{-\Omega_z A_x k_{xy}/m_y}{\omega_y^2 \sqrt{(1 - \frac{\omega_d^2}{\omega_y^2})^2 + (\frac{\omega_d}{Q_y \omega_y})^2}} \tag{6}$$

$$\varphi_x = -\arctan \frac{\omega_x \omega_d}{Q_x(\omega_x^2 - \omega_d^2)} \tag{7}$$

$$\varphi_y = -\arctan \frac{\omega_y \omega_d}{Q_y(\omega_y^2 - \omega_d^2)} \tag{8}$$

where $\omega_x = \sqrt{k_{xx}/m_x}$ $(\omega_x = 2\pi f_x)$ and $\omega_y = \sqrt{k_{yy}/m_y}$ $(\omega_y = 2\pi f_y)$ are the angular frequencies of the drive mode and sense mode, respectively; $Q_x = \frac{\omega_x m_x}{c_x}$ and $Q_y = \frac{\omega_y m_y}{c_y}$ are the quality factors of each mode.

Let $\Delta\omega = |\omega_y - \omega_x|$. The mechanical sensitivity of dual-mass MEMS gyroscope can then be expressed as

$$S = \frac{2A_F Q_x}{m_x \omega_y^2} \cdot \frac{1}{\sqrt{(1 - \frac{\omega_d^2}{\omega_y^2})^2 + (\frac{\omega_d}{Q_y \omega_y})^2}} \tag{9}$$

$$\approx \frac{A_x}{\Delta\omega}.$$

When $\omega_x = \omega_y$, the mechanical sensitivity reaches its maximum value:

$$S_{max} = \frac{2A_F Q_x Q_y}{m_x \omega_d^3} = \frac{2A_x Q_y}{\omega_d}. \tag{10}$$

Based on Equations (9) and (10), the value of ω_x, ω_y, Q_x, and Q_y will affect the size of S and S_{max}. Figure 2 shows the values of S and S_{max} at different temperatures when $A_x = 5$ um. It can be concluded that S does not change monotonically with temperature, but its overall trend increases with temperature and S_{max} decreases with increasing temperature. Furthermore, S_{max} is greater than S at the same temperature. When the amplification factor of the interface circuit is fixed, the scaling factor is proportional to the mechanical sensitivity.

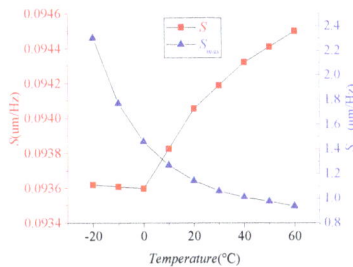

Figure 2. The mechanical sensitivity at different temperature (left red y-axis represents S and right blue y-axis represents S_{max}).

Figure 3 shows the phase relationship between the drive mode and sense mode signals and takes into account changes in the input angular velocity. When the input angular velocity exists or changes, the technologies in [2–4,8,16,18] can not effectively identify the Coriolis signal and the quadrature signal, which leads to frequency tuning failure.

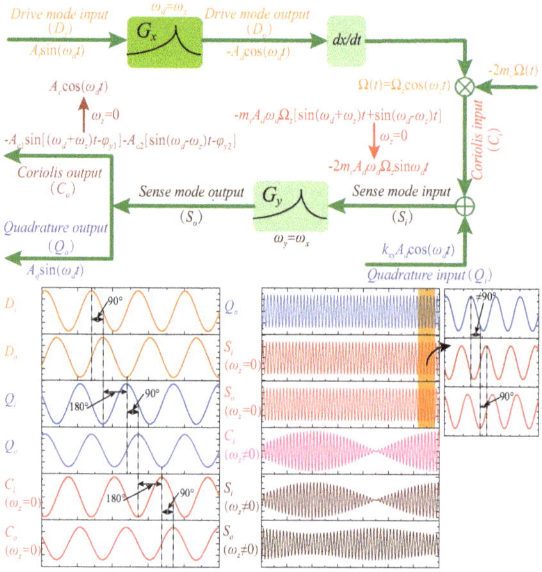

Figure 3. The phase relationship between the drive mode and sense mode signals.

2.2. Quadrature Stiffness Correction Structure

Quadrature stiffness correction structure as shown in Figure 1, these combs have the two degrees of freedom in both X and Y directions, which is located in the Coriolis mass. In addition, they are arranged at unequal intervals, with unequal pitch ratio $\lambda > 1$. The correction voltages V_{q1} and V_{q2} are respectively applied to Quadrature Electrodes 1 and 2. The right mass is used as an example to analyze the working mechanism of the quadrature stiffness correction combs. The stiffness matrix of the correction combs under electrostatic force can be expressed as

$$
\begin{bmatrix} k_{qxx} & k_{qxy} \\ k_{qyx} & k_{qyy} \end{bmatrix} = \begin{bmatrix} -\frac{\partial F_{qx}}{\partial x} & -\frac{\partial F_{qx}}{\partial y} \\ -\frac{\partial F_{qy}}{\partial x} & -\frac{\partial F_{qy}}{\partial y} \end{bmatrix}
\tag{11}
$$

where F_{qx} and F_{qy} can be written as

$$
F_{qx} = \frac{1}{2}V_{q1}^2 \left(\frac{\partial C_{ul1}}{\partial x} + \frac{\partial C_{ul2}}{\partial x} + \frac{\partial C_{dr1}}{\partial x} + \frac{\partial C_{dr2}}{\partial x} \right) + \frac{1}{2}V_{q2}^2 \left(\frac{\partial C_{dl1}}{\partial x} + \frac{\partial C_{dl2}}{\partial x} + \frac{\partial C_{ur1}}{\partial x} + \frac{\partial C_{ur2}}{\partial x} \right)
\tag{12}
$$

$$
F_{qy} = \frac{1}{2}V_{q1}^2 \left(\frac{\partial C_{ul1}}{\partial y} + \frac{\partial C_{ul2}}{\partial y} + \frac{\partial C_{dr1}}{\partial y} + \frac{\partial C_{dr2}}{\partial y} \right) + \frac{1}{2}V_{q2}^2 \left(\frac{\partial C_{dl1}}{\partial y} + \frac{\partial C_{dl2}}{\partial y} + \frac{\partial C_{ur1}}{\partial y} + \frac{\partial C_{ur2}}{\partial y} \right)
\tag{13}
$$

where F_{qx} and F_{qy} are the electrostatic force generated by the quadrature correction combs in the X- and Y-axes, respectively.

Consider the displacement of the sense mode $y \ll d_0$. Moreover, $V_{q1} = V_d + V_q$ and $V_{q2} = V_d - V_q$. Thus, Equation (11) can be simplified as

$$
\begin{bmatrix} k_{qxx} & k_{qxy} \\ k_{qyx} & k_{qyy} \end{bmatrix} = \begin{bmatrix} 0 & -\frac{4n_q\varepsilon_0 h_q}{d_q^2}\left(1 - \frac{1}{\lambda^2}\right)V_d V_q \\ -\frac{4n_q\varepsilon_0 h_q}{d_q^2}\left(1 - \frac{1}{\lambda^2}\right)V_d V_q & -\frac{4n_q\varepsilon_0 h_q l_q}{d_q^3}\left(1 + \frac{1}{\lambda^3}\right)(V_d^2 + V_q^2) \end{bmatrix}
\tag{14}
$$

where ε_0 is the vacuum permittivity; h_q is the thickness of the correct comb; l_q is the initial combing length of fixed comb and movable comb; d_q is the distance between the fixed comb and moveable comb; n_q is the number of quadrature stiffness correction combs; V_d is the preset fixed DC benchmark voltage; V_q is the quadrature adjustment voltage.

From Equation (14), the quadrature stiffness correction structure does not affect the stiffness of drive mode. In the drive mode and the sense mode, it can produce a negative stiffness to counteract the quadrature coupling stiffness. In addition, F_{qy} will generate the negative stiffness in the sense mode, and affected the resonance frequency of sense mode. Figure 4 illustrated the effect of V_q on f_x and f_y. The continuous curves are the theoretical calculation data, and the discrete points are the measured data. When $V_d = 2.048$ V and $V_q = 1$ V, both the test value and calculated value indicate the effect of V_q on f_y is less than 0.06 Hz. This makes it feasible to use the quadrature stiffness correction structure to generate modulation signals for obtaining frequency mismatch information.

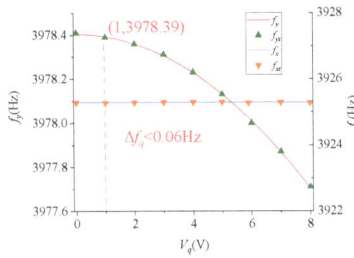

Figure 4. The effect of V_q on f_x and f_y.

2.3. Frequency Tuning Structure

The frequency tuning combs are also shown in Figure 1. The combs are the gap-changing structure and with an equal gap. The movable combs can only move in the Y direction. Taking a single frequency tuning comb as an example, the frequency adjustment voltage V_t is applied to the fixed comb. When the movable comb moves along the Y-axis, the capacitance variation on both sides of the fixed comb can be described as

$$\begin{cases} C_{t1} = \frac{\varepsilon_0 h_t l_t}{d_t - y} \\ C_{t1} = \frac{\varepsilon_0 h_t l_t}{d_t + y} \end{cases} \tag{15}$$

where C_{t1} and C_{t2} are the capacitance between the fixed comb and the movable comb; h_t is the thickness of the comb; l_t is the initial combing length of the fixed comb and the movable comb; d_t is the distance between the fixed comb and the moveable comb.

Furthermore, the stiffness matrix of the frequency tuning comb under electrostatic force can be expressed as

$$\begin{bmatrix} k_{txx} & k_{txy} \\ k_{tyx} & k_{tyy} \end{bmatrix} = \begin{bmatrix} -\frac{\partial F_{tx}}{\partial x} & -\frac{\partial F_{tx}}{\partial y} \\ -\frac{\partial F_{ty}}{\partial x} & -\frac{\partial F_{ty}}{\partial y} \end{bmatrix} \tag{16}$$

where F_{tx} and F_{ty} can be written as

$$F_{tx} = 0, \tag{17}$$

$$F_{ty} = F_{t1y} + F_{t2y} = \frac{1}{2}V_t^2 \left(\frac{\partial C_{t1}}{\partial y} + \frac{\partial C_{t2}}{\partial y} \right) = \frac{\varepsilon_0 h_t l_t V_t^2}{2} \left[\frac{1}{(d_t - y)^2} - \frac{1}{(d_t + y)^2} \right] \tag{18}$$

where F_{tx} and F_{ty} is the electrostatic force generated by the two capacitors in the X and Y directions, respectively.

Consider $y \ll d_t$, Equation (16) can be simplified as

$$\begin{bmatrix} k_{txx} & k_{txy} \\ k_{tyx} & k_{tyy} \end{bmatrix} = \begin{bmatrix} 0 & 0 \\ 0 & -\frac{\varepsilon_0 h_t l_t}{d_t^3}V_t^2 \end{bmatrix}. \tag{19}$$

According to Equation (19), V_t does not cause a change in ω_x. ω_y will monotonically decrease as V_t increases, so that $\omega_x = \omega_y$ at a certain voltage value. Figure 5 shows the variation of f_x and f_y at different V_t values. Similarly, the continuous curves are the theoretical calculation data, and the discrete points are the measured data. Moreover, the frequency adjustment capability of the 8 V DC voltage is 64.72 Hz.

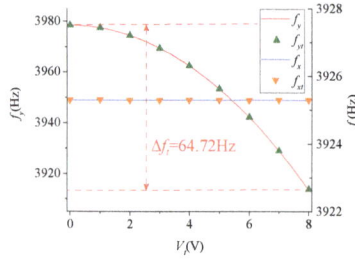

Figure 5. The effect of V_t on f_x and f_y.

3. Automatic Frequency Tuning System

3.1. Frequency Tuning Theory

Combined Equations (14) and (19), the equivalent stiffness of the sense mode can be expressed as

$$k_{eyy} = k_{yy} + k_{tyy} + k_{qyy}. \tag{20}$$

The resonant frequency of the sense mode under V_t, V_d, and V_q can be represented as

$$\omega_{eyy} = \sqrt{\frac{k_{yy}}{m_y} - \frac{n_t \varepsilon_0 h_t l_t}{m_y d_t^3} V_t^2 - \frac{4 n_q \varepsilon_0 h_q l_q}{m_y d_q^3}\left(1 + \frac{1}{\lambda^3}\right)(V_d^2 + V_q^2)}. \tag{21}$$

In this paper, $V_d = 2.048$ V. The influence of V_t and V_q on f_y is shown in Figure 6. f_{qty} and f_{tty} respectively represent the variation curve of f_y when V_q is a sine wave ($V_q = 1 \times \sin(160\pi t)$ V) and a direct current amount ($V_q = 1$ V). When $V_q = 1 \times \sin(\omega_t t)$ V, the resonant frequency of the sense mode will produce sinusoidal fluctuations, but the maximum frequency difference between f_{qty} and f_{tty} is less than 0.011 Hz. Moreover, according to Equation (19) and Figure 5, the voltage of 0.2 mV only causes a deviation of 4×10^{-8} Hz. Therefore, the effect of the sinusoidal signal applied to the quadrature stiffness combs on f_y can be neglected, which demonstrates the feasibility of frequency tuning in the following section.

The automatic frequency tuning loop is shown in Figure 7. The quadrature stiffness correction combs are applied with the low frequency sinusoidal voltage (V_q) and DC benchmark voltage (V_d), which can equivalently produce a modulation excitation signal including ω_t and ω_x. By identifying the output response of the gyroscope under the excitation signal, the frequency matching degree of the two operating modes can be distinguished. The Coriolis signal, the quadrature signal, and the frequency mismatch signal can be obtained by different multiplication demodulations. Figure 7 also shows the configuration of the cut-off frequency for the four low-pass filters. In the frequency tuning loop, a proportional integral controller is used to adjust the frequency tuning voltage to change the resonance frequency of the sense mode.

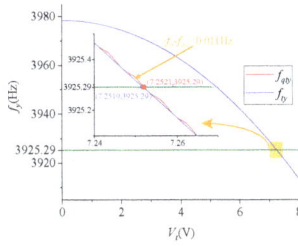

Figure 6. The effect of V_t and V_q on f_y.

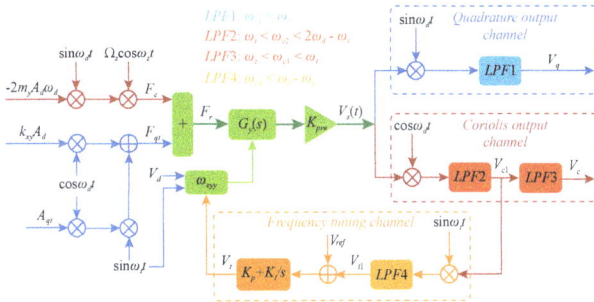

Figure 7. Schematic control loop for the automatic frequency tuning system.

The alternating voltage applied to the quadrature stiffness correction combs can be expressed as

$$V_q = A_t \sin \omega_t t \qquad (22)$$

where A_t and ω_t are the amplitude and frequency of the low frequency sinusoidal signal, respectively.

Combined Equations (14) and (22), the equivalent input force generated by the quadrature channel can be expressed as

$$F_{qt} = A_d \cos \omega_d t * \left[k_{xy} - \frac{4 n_q \varepsilon_0 h_q}{d_0^2} \left(1 - \frac{1}{\lambda^2} \right) V_d A_t \sin \omega_t t \right] \qquad (23)$$

$$= k_{xy} A_d \cos \omega_d t + A_{qt} \cos \omega_d t * \sin \omega_t t.$$

Considering the variation of Ω_z, the input resultant force of the sense mode can be given as

$$F_r = F_c + F_{qt}$$

$$= \underbrace{-2 m_y A_d \omega_d \Omega_z \cos \omega_z t * \sin \omega_d t}_{\text{Coriolis force}} + \underbrace{k_{xy} A_d \cos \omega_d t}_{\text{Quadrature coupling force}} + \underbrace{A_{qt} \cos \omega_d t * \sin \omega_t t}_{\text{Quadrature modulation force}} . \qquad (24)$$

The output of the sense mode should be the sum of the responses of the above three forces,

$$V_s = V_c + V_{qt}$$

$$= \underbrace{A_{c1} \sin[(\omega_d + \omega_z)t + \varphi_{\omega_d + \omega_z}] - A_{c2} \sin[(\omega_d - \omega_z)t + \varphi_{\omega_d - \omega_z}]}_{\text{Coriolis signal}}$$

$$+ \underbrace{A_{t1} \sin[(\omega_d + \omega_t)t + \varphi_{\omega_d + \omega_t}] - A_{t2} \sin[(\omega_d - \omega_t)t + \varphi_{\omega_d - \omega_t}]}_{\text{Frequency tuning signal}} \tag{25}$$

$$+ \underbrace{A_q \cos(\omega_d + \varphi_q)}_{\text{Quadrature signal}}$$

where

$$A_{c1} = \frac{-A_d \omega_d \Omega_z K_{pre}}{\sqrt{[\omega_y^2 - (\omega_d + \omega_z)^2]^2 + [\omega_y(\omega_d + \omega_z)/Q_y]^2}} \tag{26}$$

$$A_{c2} = \frac{-A_d \omega_d \Omega_z K_{pre}}{\sqrt{[\omega_y^2 - (\omega_d - \omega_z)^2]^2 + [\omega_y(\omega_d - \omega_z)/Q_y]^2}} \tag{27}$$

$$A_{t1} = \frac{A_{qt} K_{pre}/2m_y}{\sqrt{[\omega_y^2 - (\omega_d + \omega_t)^2]^2 + [\omega_y(\omega_d + \omega_t)/Q_y]^2}} \tag{28}$$

$$A_{t2} = \frac{A_{qt} K_{pre}/2m_y}{\sqrt{[\omega_y^2 - (\omega_d - \omega_t)^2]^2 + [\omega_y(\omega_d - \omega_t)/Q_y]^2}} \tag{29}$$

$$A_q = \frac{k_{xy} A_d/m_y}{\sqrt{(\omega_y^2 - \omega_d^2)^2 + (\omega_y \omega_d/Q_y)^2}} \tag{30}$$

$$\varphi_{\omega_d + \omega_z} = -\arctan \frac{\omega_y(\omega_d + \omega_z)}{Q_y[\omega_y^2 - (\omega_d + \omega_z)^2]} \tag{31}$$

$$\varphi_{\omega_d - \omega_z} = -\arctan \frac{\omega_y(\omega_d - \omega_z)}{Q_y[\omega_y^2 - (\omega_d - \omega_z)^2]} \tag{32}$$

$$\varphi_{\omega_d + \omega_t} = -\arctan \frac{\omega_y(\omega_d + \omega_t)}{Q_y[\omega_y^2 - (\omega_d + \omega_t)^2]} \tag{33}$$

$$\varphi_{\omega_d - \omega_t} = -\arctan \frac{\omega_y(\omega_d - \omega_t)}{Q_y[\omega_y^2 - (\omega_d - \omega_t)^2]} \tag{34}$$

$$\varphi_q = -\arctan \frac{\omega_y \omega_d}{Q_y(\omega_y^2 - \omega_d^2)}. \tag{35}$$

Therefore, the output of low pass filter *LPF2* can be obtained as

$$V_{c1} = \frac{1}{2} A_{c1} \sin(\omega_z t + \varphi_{\omega_d + \omega_z}) - \frac{1}{2} A_{c2} \sin(\omega_z t - \varphi_{\omega_d - \omega_z})$$

$$+ \frac{1}{2} A_{t1} \sin(\omega_t t + \varphi_{\omega_d + \omega_t}) + \frac{1}{2} A_{t2} \sin(\omega_t t - \varphi_{\omega_d - \omega_t}) \tag{36}$$

$$+ \frac{1}{2} A_q \cos \varphi_q.$$

When $\omega_x = \omega_y$, the value of $\cos \varphi_q$ is zero. The signal V_{c1} pass through the low pass filter *LPF3*, the Coriolis output can be written as

$$V_c = \frac{1}{2} A_{c1} \sin(\omega_z t + \varphi_{\omega_d + \omega_z}) - \frac{1}{2} A_{c2} \sin(\omega_z t - \varphi_{\omega_d - \omega_z}). \tag{37}$$

According to Equation (37), the Coriolis output can eliminate the interference from the quadrature signal and low frequency input signal, and correctly reflect the input angular velocity.

Similarly, V_q and V_{t1} can be expressed as

$$V_q = \frac{1}{2} A_{c1} \cos(\omega_z t + \varphi_{\omega_d + \omega_z}) - \frac{1}{2} A_{c2} \cos(\omega_z t - \varphi_{\omega_d - \omega_z}) - \frac{1}{2} A_q \sin \varphi_q \tag{38}$$

$$
\begin{aligned}
V_{t1} &= \frac{1}{4} A_{t1} \cos \varphi_{\omega_d + \omega_t} + \frac{1}{4} A_{t2} \cos \varphi_{\omega_d - \omega_t} \\
&= \frac{A_{qt} K_{pre}}{8 m_y} (t_{eq1} + t_{eq2})
\end{aligned}
\tag{39}
$$

where t_{eq1} and t_{eq2} can be given as

$$t_{eq1} = \frac{\omega_y^2 - (\omega_d + \omega_t)^2}{[\omega_y^2 - (\omega_d + \omega_t)^2]^2 + [\omega_y(\omega_d + \omega_t)/Q_y]^2} \tag{40}$$

$$t_{eq2} = \frac{\omega_y^2 - (\omega_d - \omega_t)^2}{[\omega_y^2 - (\omega_d - \omega_t)^2]^2 + [\omega_y(\omega_d - \omega_t)/Q_y]^2}. \tag{41}$$

According to Equation (38), the quadrature output contains the momentum associated with ω_z. Since the quadrature quantity is a slow variable, in the actual system, the interference can be filtered out as much as possible by a low pass filter (*LPF1*) with a very low cut-off frequency.

$\omega_y = \omega_d + \Delta\omega$ is substituted into Equation (39), and the expression for $\Delta\omega$ is as follows:

$$V(\Delta\omega) = \frac{A_{qt} K_{pre}}{8 m_y} (V_{\Delta\omega 1} + V_{\Delta\omega 2}) \tag{42}$$

where $V_{\Delta\omega 1}$ and $V_{\Delta\omega 2}$ can be obtained as

$$V_{\Delta\omega 1} = \frac{(\omega_d + \Delta\omega)^2 - (\omega_d + \omega_t)^2}{[(\omega_d + \Delta\omega)^2 - (\omega_d + \omega_t)^2]^2 + [(\omega_d + \Delta\omega)(\omega_d + \omega_t)/Q_y]^2} \tag{43}$$

$$V_{\Delta\omega 2} = \frac{(\omega_d + \Delta\omega)^2 - (\omega_d - \omega_t)^2}{[(\omega_d + \Delta\omega)^2 - (\omega_d - \omega_t)^2]^2 + [(\omega_d + \Delta\omega)(\omega_d - \omega_t)/Q_y]^2}. \tag{44}$$

Let $\Delta\omega = 0$ in Equation (42). The reference voltage for frequency tuning can be obtained as follows:

$$V_{ref} = V(0) = \frac{A_{qt} K_{pre}}{8 m_y} (V_{ref1} + V_{ref2}) \tag{45}$$

where V_{ref1} and V_{ref2} can be written as

$$V_{ref1} = \frac{\omega_d^2 - (\omega_d + \omega_t)^2}{[\omega_d^2 - (\omega_d + \omega_t)^2]^2 + [\omega_d(\omega_d + \omega_t)/Q_y]^2} \tag{46}$$

$$V_{ref2} = \frac{\omega_d^2 - (\omega_d - \omega_t)^2}{[\omega_d^2 - (\omega_d - \omega_t)^2]^2 + [\omega_d(\omega_d - \omega_t)/Q_y]^2}. \tag{47}$$

After the gyroscope structure and circuit parameters are determined, each parameter in Equation (45) is a known quantity except for ω_d and Q_y. ω_d can be acquired by the drive mode control system [20], and Q_y is considered a function of the change with ω_d [21]. Figure 8 shows the relationship between ω_d, Q_y, and V_{ref}. The V_{ref} varies from 0.2597 to 0.2574 mV over a wide temperature range (-20 to $60\,^\circ$C).

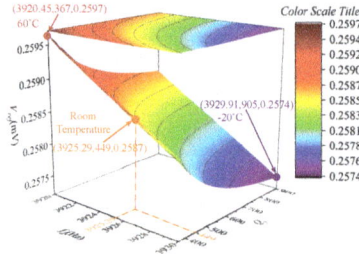

Figure 8. The effect of ω_d and Q_y on V_{ref}.

3.2. Frequency Tuning System Analysis

According to the principle as shown in Figure 7, an automatic frequency tuning system based on low frequency modulation excitation was built on the Simulink simulation platform. The simulation parameters are listed in Table 1. The quadrature stiffness correction comb was originally designed to suppress the quadrature coupling stiffness by applying a slowly varying DC voltage, so the comb cannot respond to a high frequency input. In order to balance the response frequency and the gyroscope's bandwidth, $\omega_t = 80$ Hz is selected. In addition, since the force rebalance comb does not change the f_x and f_y [6], the open-loop and closed-loop detection methods do not interfere with the operation of the frequency tuning system, so the open-loop detection method is adopted in the simulation analysis.

Table 1. Simulation parameters of the dual-mass MEMS gyroscope.

Parameter	Values	Units
Drive mode resonant frequency (ω_x)	$3925.29 \times 2\pi$	rad/s
Drive mode quality factor (Q_x)	4673	
Sense mode resonant frequency (ω_y)	$3978.45 \times 2\pi$	rad/s
Sense mode quality factor (Q_y)	449	
Drive effective mass (m_x)	1.42×10^{-6}	kg
Sense effective mass (m_y)	1.58×10^{-6}	kg
Drive mode capacitance (C_d)	2.88	pF
Sense mode capacitance (C_d)	4.68	pF
Quadrature correction comb number (n_q)	30	
Quadrature correction comb thickness (h_q)	60	um
Quadrature correction comb gap (d_q)	5	um
Comb overlap length (l_q)	10	um
Unequal spacing ratio (λ)	2.5	
Vacuum permittivity (ε_0)	8.85×10^{-12}	F/m
Tuning comb number (n_t)	300	
Tuning comb number thickness (h_t)	60	um
Correction comb gap (d_t)	4	um
Comb overlap length (l_t)	200	um
Low-frequency signal amplitude (A_t)	1	
Input signal frequency (ω_t)	$2\pi \times 80$	rad/s
DC benchmark voltage (V_d)	2.048	V
Interface circuit amplification factor (K_{pre})	7.6159×10^7	
Controller parameters (K_p)	30	
Controller parameters (K_i)	0.0075	
Reference voltage (V_{ref})	0.2587	mV

Figure 9 shows the curves of some observation points when the frequency tuning system is working normally ($\Omega_z = 50\,^\circ$/s, $\Omega_q = 100\,^\circ$/s). The first curve is the low frequency input signal

($\omega_t = 80$ Hz) that applied to the quadrature stiffness correction combs, and its function is to cause the gyroscope to generate a modulated signal containing frequency split information. The second curve is the gyroscope output signal, which is characterized by the modulated signals of ω_t and ω_x. The third and fourth curves represent the changes in the frequency tuning input voltage and the sense mode resonant frequency, respectively. The curves indicate that the system is in a stable state after 0.75 s, the output fluctuation of V_t is less than 0.5 mV, and the fluctuation of the corresponding f_y is less than 0.008 Hz.

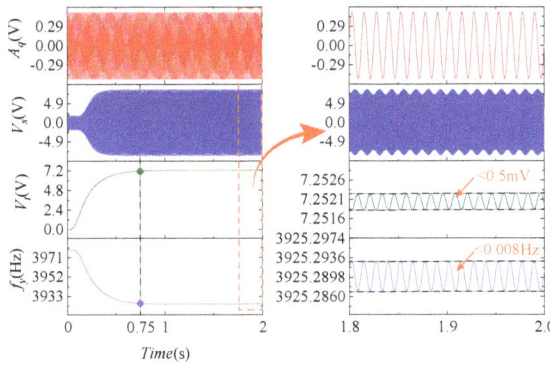

Figure 9. The curves of observation points in the frequency tuning system.

The effects of different Ω_z and Ω_q values (quadrature equivalent input angular velocity [22]) on f_y are shown in Figure 10. The interference fluctuations of Ω_z and Ω_q to f_y are less than 0.005 Hz and 0.0005 Hz, respectively. This illustrates that, when Ω_z and Ω_q exist, the frequency tuning system can still work properly, and eventually can be stabilized at the desired frequency.

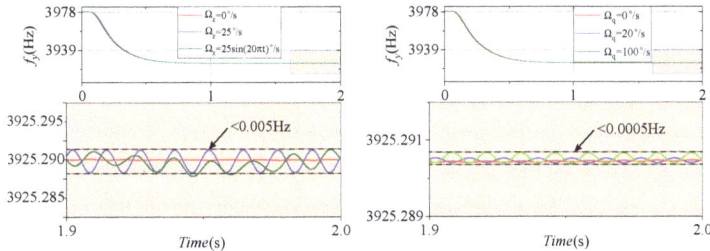

Figure 10. The disturbance of different Ω_z (**left**) and Ω_q (**right**) to f_y.

Consider $\Omega_q = 100°/s$, set Ω_c to $0°/s$, $25°/s$, $50°/s$, and $25 \times \sin(20\pi t)°/s$, and obtain the angular velocity output response curves, as illustrated in Figure 11. This indicates that, when the frequency tuning loop works normally, the system can still detect the input angular velocity.

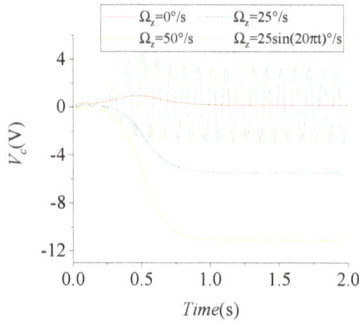

Figure 11. Coriolis path output curves.

4. Experiments

In order to verify the effectiveness of automatic frequency tuning technology in the MEMS gyroscope, the gyroscope control circuit was designed and relevant tests were conducted. Figure 12 shows the experimental test equipment and the gyroscope system circuit. The circuit is mounted on the printed circuit boards (PCBs) and its electrical signals and mechanical structure are connected to each other through metal pins. First, the PCBs are wrapped in rubber pads that protect PCBs and fabric chips from impact and vibration. The test equipment includes the power (GWINSTEK GPS-3303C, GWINSTEK, New Taipei, China) providing ±8 V DC voltage and GND, the oscilloscope (Keysight DSOX2024A, Keysight, Santa Rosa, CA, USA), which is applied to observe the different input and output signals of the gyroscope, the computer, which is devoted to measuring the gyroscope data in a variety of working conditions, and the temperature oven, providing a wide-temperature range environment and a turntable test of the bandwidth of the gyroscope. The experiments were divided into three methods: open-loop detection without frequency tuning (Test 1), open-loop detection with frequency tuning (Test 2), and closed-loop detection with frequency tuning (Test 3).

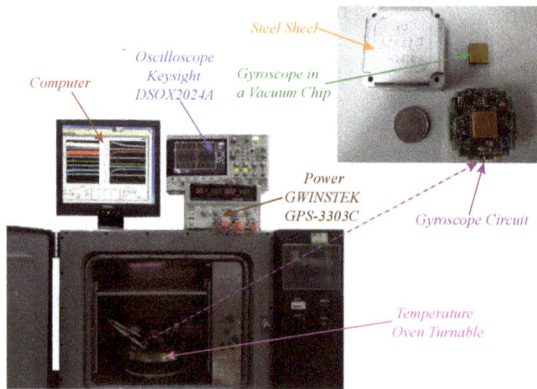

Figure 12. Photos of the MEMS gyroscope circuit and test equipment.

The test curves of the frequency tuning system at room temperature are shown in Figure 13. These four curves are the drive mode detection signal, the drive mode input signal, the quadrature excitation signal (V_q), and the sense mode output signal. The frequency of the drive mode input signal and the drive mode detection signal were both ω_x, and a phase difference of $90°$ was maintained, while the amplitude of the drive mode detection signal remained stable. These curves indicated that the drive mode of the gyroscope was working properly. The frequency of the tuning input signal was 80 Hz, and the sense mode output signal was the modulation signal of 80 Hz and ω_x.

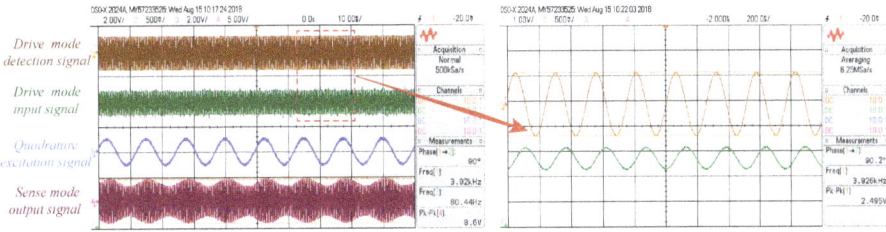

Figure 13. The test curves of the frequency tuning system.

The gyroscope control circuit was placed in the temperature oven for a wide temperature range test, and the temperature range was set from $-20\,°C$ to $60\,°C$. The curve of frequency tuning voltage varying with temperature was obtained under the automatic frequency tuning technology, as shown in Figure 14. V_{t_test} and V_{t_real} are the frequency tuning voltages of the temperature test and the actual temperature conditions, respectively. Among them, V_{t_real} is the frequency tuning voltage at each stable temperature condition under manual adjustment mode, which can be characterized as the real frequency tuning voltage. Δf_{vt} is the frequency split in the wide temperature range test, which represents the mismatch frequency of the gyroscope adopting the frequency tuning technology. In the wide temperature range, the frequency tuning voltage was changed from 7.27021 to 7.24871 V, the maximum difference between the test tuning voltage and the real tuning voltage was 1.986 mV, and the corresponding frequency difference was 0.29326 Hz.

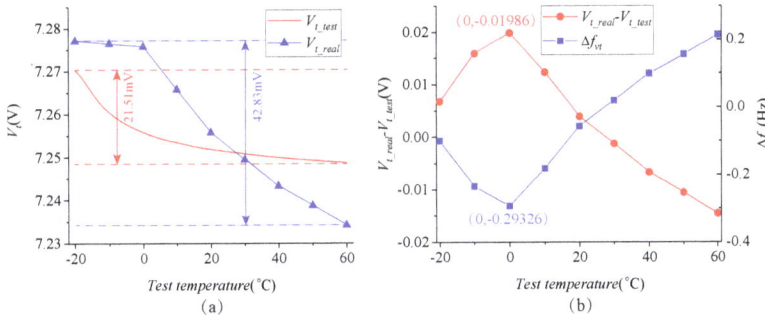

Figure 14. Wide temperature range test curves: (**a**) The variation of frequency tuning voltage, (**b**) The variations of $V_{t\_real} - V_{t\_test}$ and Δf_{vt}.

The scale factor tests (listed in Table 2) at the three different temperatures were arranged with input angular rates Ω_z of $\pm0.1°/s$, $\pm0.2°/s$, $\pm0.5°/s$, $\pm1°/s$, $\pm2°/s$, $\pm5°/s$, $\pm10°/s$, $\pm20°/s$, $\pm50°/s$, and $\pm100°/s$, and Figure 15 shows the residuals of the fit. According to Equation (10), Q_x and Q_y change with temperature, resulting in a large change in the scale factor value of Test 2. The scale factor of Test 2 was greater than that of Test 1, which is theoretically demonstrated in Figure 2, but its scale factor nonlinearly was degraded. When the input angular velocity is $\pm100°/s$, the residual error of Test 1 and Test 2 between the measured data and fitting data reached the maximum. Test 3 adopted a close-loop detection method that made the scale factor independent of the mechanical sensitivity, and the nonlinearity was improved.

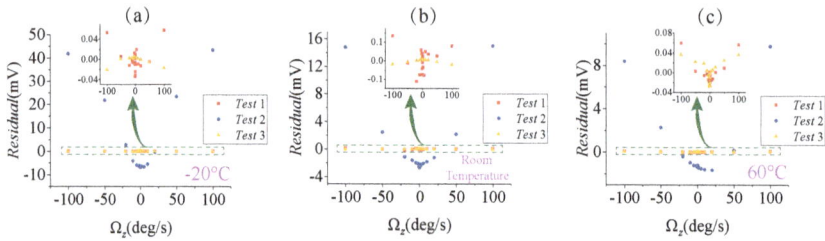

Figure 15. The residual errors of the scale factor under different temperatures: (**a**) −20 °C condition, (**b**) Room temperature condition, (**c**) 60 °C condition.

Table 2. Scale factor performance of the tested gyroscope.

Temperature	Test Type	Scale Factor (mV/°/s)	Scale Factor Nonlinearity (ppm)	Scale Factor Asymmetry (ppm)
−20 °C	Test 1	−2.039	142	546
	Test 2	40.453	107,155	46,978
	Test 3	3.857	53	212
60 °C	Test 1	−2.048	144	596
	Test 2	19.324	2526	10,074
	Test 3	3.780	96	568
Room temperature	Test 1	−2.051	322	1009
	Test 2	22.024	3398	14,792
	Test 3	3.786	28	128

The static output performance of the three tests at three different temperature was also tested, as listed in Table 3. The angle random walk (ARW) of the same test mode remained stable under different temperature conditions. However, as the temperature rose, the damping coupling increased, resulting in deterioration of the zero bias stability. Comparing Test 1 and Test 2, the static performance of Test 2 was improved due to the adoption of frequency tuning technology. The performance of Test 3 and Test 2 was basically the same.

Table 3. Static performance of the three test methods at different temperatures.

Temperature	Test Type	Zero Bias (°/s)	Zero Bias Stability (°/h)	ARW (°/√h)
−20 °C	Test 1	1.138	24.746	12.780
	Test 2	5.554	23.964	4.966
	Test 3	5.381	24.580	4.862
60 °C	Test 1	5.991	70.187	12.536
	Test 2	13.127	67.429	5.085
	Test 3	13.526	64.926	5.058
Room temperature	Test 1	1.458	43.439	12.229
	Test 2	2.602	37.648	4.784
	Test 3	2.601	39.545	4.956

The bandwidth of the gyroscope at room temperature is shown in Figure 16. The bandwidth of Test 1 was 31 Hz, which is approximately equal to $0.54\Delta f$ [20], while the detection transfer function of Test 2 can be approximated as a low-pass filter with a cut-off frequency of $\omega_y/2Q_y$ [17], so the bandwidth of Test 2 was 5 Hz. Because of the previous bandwidth expansion technology [23], the bandwidth of Test 3 reached 15 Hz.

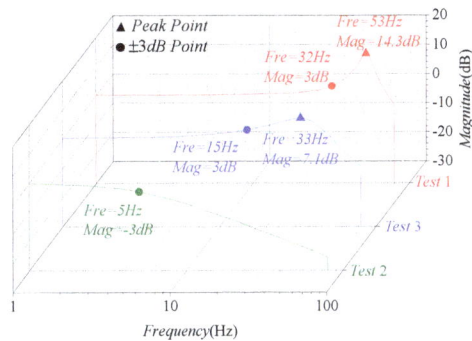

Figure 16. Gyroscope bandwidth under the three test methods.

5. Conclusions

This paper focuses on the automatic frequency tuning technology based on a quadrature modulation signal. The quadrature stiffness correction combs are applied to a DC benchmark voltage and a low frequency sinusoidal signal whose frequency is higher than the gyroscope's bandwidth, which can equivalently produce a modulation excitation signal acting on the input of gyroscope. By identifying the output response of the gyroscope under this excitation signal, the frequency mismatch degree of the two operating modes can be distinguished. In order to obtain a frequency tuning signal, a Coriolis signal, and a quadrature signal, a low pass filter with different cut-off frequencies were configured for demodulation. Simulation analysis and experimental results demonstrate the feasibility of the automatic frequency tuning system. The wide temperature range test demonstrates the reliability of the frequency tuning system with a maximum mismatch frequency of less than 0.3 Hz in the range of −20 °C to 60 °C. The scale factor test and static test of the gyroscope at three different temperatures (−20 °C, room temperature, 60 °C) prove that the performance of the gyroscope under a mode-matching condition is improved. When the method of open-loop detection with frequency tuning, compared with the method of open-loop detection without frequency tuning, was employed, the scale factors were increased by 19.8 times, 10.7 times, and 9.4 times, the ARW was improved by 157%, 147% and 156%, and the zero bias stability was promoted by 3.26%, 4.09% and 17.08% at −20 °C, room temperature, and 60 °C, respectively. In addition, the method of closed-loop detection by frequency tuning was adopted, and, compared with the method of open-loop detection with frequency tuning, the scale factor nonlinearity and bandwidth under the premise of maintaining the same static performance improved. However, the large damping coupling and the small quality factor resulted in a large drift of the gyroscope static output, which made the improvement of the zero bias stability in the mode-matching condition not obvious. Moreover, the structure of quadrature correction combs cannot respond to higher frequency sinusoidal signals, thereby limiting the working bandwidth of the gyroscope. Therefore, it is necessary to design a quadrature stiffness correction comb that can respond to a higher frequency sinusoidal input signal that improves the operating bandwidth limitations of the gyroscope. The quality factor of the gyroscope needs to be improved, and an effective compensation method can be adopted to suppress the zero bias drift caused by the small quality factor and damping coupling, thereby improving the static performances of the gyroscope.

Author Contributions: Conceptualization: J.J. and H.L.; data curation: J.J. and X.D.; formal analysis: J.J., Y.G., and H.L.; methodology: X.D.; writing—review & editing: J.J.

Conflicts of Interest: The authors declare no conflict of interest.

References

1. Ezekwe, C.D.; Boser, B.E. A Mode-Matching ΔΣ Closed-Loop Vibratory Gyroscope Readout Interface with a 0.004°/s/√Hz Noise Floor Over a 50 Hz Band. *IEEE J. Solid State Circ.* **2008**, *43*, 3039–3048. [CrossRef]
2. Sharma, A.; Zaman, M.F.; Ayazi, F. A Sub-0.2°/hr Bias Drift Micromechanical Silicon Gyroscope with Automatic CMOS Mode-Matching. *IEEE J. Solid State Circ.* **2009**, *44*, 1593–1608. [CrossRef]
3. Antonello, R.; Oboe, R.; Prandi, L.; Biganzoli, F. Automatic mode matching in MEMS vibrating gyroscopes using extremum-seeking control. *IEEE Trans. Ind. Electron.* **2009**, *56*, 3880–3891. [CrossRef]
4. Sonmezoglu, S.; Alper, S.E.; Akin, T. An automatically mode-matched MEMS gyroscope with wide and tunable bandwidth. *J. Microelectromech. Syst.* **2014**, *23*, 284–297. [CrossRef]
5. Painter, C.C.; Shkel, A.M. Active structural error suppression in MEMS vibratory rate integrating gyroscopes. *IEEE Sens. J.* **2003**, *3*, 595–606. [CrossRef]
6. Xu, L.; Li, H.; Ni, Y.; Liu, J.; Huang, L. Frequency tuning of work modes in z-axis dual-mass silicon microgyroscope. *J. Sens.* **2014**, *2014*, 891735. [CrossRef]
7. Park, S.; Horowitz, R.; Hong, S.K.; Nam, Y. Trajectory-switching algorithm for a MEMS gyroscope. *IEEE Trans. Instrum. Meas.* **2007**, *56*, 2561–2569. [CrossRef]
8. Sung, S.; Sung, W.T.; Kim, C.; Yun, S.; Lee, Y.J. On the mode-matched control of MEMS vibratory gyroscope via phase-domain analysis and design. *IEEE/ASME Trans. Mechatron.* **2009**, *14*, 446–455. [CrossRef]
9. Wang, K.; Wong, A.C.; Hsu, W.T.; Nguyen, C.C. Frequency trimming and Q-factor enhancement of micromechanical resonators via localized filament annealing. In Proceedings of the International Solid State Sensors and Actuators Conference (Transducers '97), Chicago, IL, USA, 19 June 1997; Volume 1, pp. 109–112.
10. Remtema, T.; Lin, L. Active frequency tuning for micro resonators by localized thermal stressing effects. *Sens. Actuators A Phys.* **2001**, *91*, 326–332. [CrossRef]
11. Joachim, D.; Lin, L. Characterization of selective polysilicon deposition for MEMS resonator tuning. *J. Microelectromech. Syst.* **2003**, *12*, 193–200. [CrossRef]
12. Abdelmoneum, M.A.; Demirci, M.M.; Lin, Y.W.; Nguyen, C.C. Location-dependent frequency tuning of vibrating micromechanical resonators via laser trimming. In Proceedings of the 2004 IEEE International Frequency Control Symposium and Exposition, Montreal, QC, Canada, 23–27 August 2004; pp. 272–279.
13. Keymeulen, D.; Fink, W.; Ferguson, M.I.; Peay, C.; Oks, B.; Terrile, R.; Yee, K. Tuning of MEMS devices using evolutionary computation and open-loop frequency response. In Proceedings of the 2005 IEEE Aerospace Conference, Big Sky, MT, USA, 5–12 March 2005; pp. 1–8.
14. Kim, D.; M'Closkey, R. Real-time tuning of MEMS gyro dynamics. In Proceedings of the American Control Conference, Portland, OR, USA, 8–10 Jun 2005; pp. 3598–3603.
15. Geiger, W.; Bartholomeyczik, J.; Breng, U.; Gutmann, W.; Hafen, M.; Handrich, E.; Huber, M.; Jackle, A.; Kempfer, U.; Kopmann, H.; et al. MEMS IMU for ahrs applications. In Proceedings of the 2008 IEEE/ION Position, Location and Navigation Symposium, Monterey, CA, USA, 5–8 May 2008; pp. 225–231.
16. Hu, Z.; Gallacher, B.; Burdess, J.; Fell, C.; Townsend, K. Precision mode matching of MEMS gyroscope by feedback control. In Proceedings of the 2011 IEEE Sensors, Limerick, Ireland, 28–31 October 2011; pp. 16–19.
17. He, C.; Zhao, Q.; Huang, Q.; Liu, D.; Yang, Z.; Zhang, D.; Yan, G. A MEMS vibratory gyroscope with real-time mode-matching and robust control for the sense mode. *IEEE Sens. J.* **2015**, *15*, 2069–2077. [CrossRef]
18. Xu, L.; Li, H.; Yang, C.; Huang, L. Comparison of three automatic mode-matching methods for silicon micro-gyroscopes based on phase characteristic. *IEEE Sens. J.* **2016**, *16*, 610–619. [CrossRef]
19. Prikhodko, I.P.; Gregory, J.A.; Clark, W.A.; Geen, J.A.; Judy, M.W.; Ahn, C.H.; Kenny, T.W. Mode-matched MEMS Coriolis vibratory gyroscopes: Myth or reality? In Proceedings of the 2016 IEEE/ION Position, Location and Navigation Symposium (PLANS), Savannah, GA, USA, 11–14 April 2016; pp. 1–4.
20. Yang, C.; Li, H. Digital control system for the MEMS tuning fork gyroscope based on synchronous integral demodulator. *IEEE Sens. J.* **2015**, *15*, 5755–5764. [CrossRef]
21. Cao, H.L.; Li, H.S.; Lu, X.; Ni, Y.F. Temperature Model for a Vacuum Packaged MEMS Gyroscope Structure. In *Key Engineering Materials*; Trans Tech Publications: Zurich, Switzerland, 2013; Volume 562, pp. 280–285.

22. Cao, H.; Li, H.; Kou, Z.; Shi, Y.; Tang, J.; Ma, Z.; Shen, C.; Liu, J. Optimization and experimentation of dual-mass MEMS gyroscope quadrature error correction methods. *Sensors* **2016**, *16*, 71. [CrossRef] [PubMed]

23. Cao, H.; Li, H.; Shao, X.; Liu, Z.; Kou, Z.; Shan, Y.; Shi, Y.; Shen, C.; Liu, J. Sensing mode coupling analysis for dual-mass MEMS gyroscope and bandwidth expansion within wide-temperature range. *Mech. Syst. Signal Process.* **2018**, *98*, 448–464. [CrossRef]

micromachines

MDPI

Article

Miniaturized NIR Spectrometer Based on Novel MOEMS Scanning Tilted Grating

Jian Huang [1,2,3,*], Quan Wen [1,4,*], Qiuyu Nie [1,5], Fei Chang [1,5], Ying Zhou [1,3] and Zhiyu Wen [1,3]

[1] Key Laboratory of Fundamental Science of Micro/Nano-Device and System Technology,
    Chongqing University, Chongqing 400044, China; qiu30@163.com (Q.N.);
    changfei0602@gmail.com (F.C.); yzhou@cqu.edu.cn (Y.Z.); wzy@cqu.edu.cn (Z.W.)
[2] College of Information Engineering, Qujing Normal University, Qujing 655000, China
[3] Microsystem Research Center, College of Optoelectronic Engineering, Chongqing University,
    Chongqing 400044, China
[4] Fraunhofer Institute for Electronic Nano Systems (ENAS), 09131 Chemnitz, Germany
[5] College of Electronic and Information Engineering, Southwest University, Chongqing 400715, China
*   Correspondence: huangjian7@gmail.com (J.H.); Quan.Wen@enas.fraunhofer.de (Q.W.);
    Tel.: +86-023-65111010 (J.H.); +49(0)-371-4500-1252 (Q.W.)

Received: 15 August 2018; Accepted: 18 September 2018; Published: 20 September 2018

check for updates

Abstract: This paper presents a dispersive near-infrared spectrometer with features of miniaturization, portability and low cost. The application of a resonantly-driven scanning grating mirror (SGM) as a dispersive element in a crossed Czerny–Turner configuration enables the design of a miniaturized spectrometer that can detect the full spectra using only one single InGaAs diode. In addition, a high accuracy recalculation is realized, which can convert time-dependent measurements to spectrum information by utilizing the deflection position detector integrated on SGM and its associated closed-loop control circuit. Finally, the spectrometer prototype is subjected to a series of tests to characterize the instrument's performance fully. The results of the experiment show that the spectrometer works in a spectral range of 800 nm–1800 nm with a resolution of less than 10 nm, a size of $9 \times 7 \times 7$ cm^3, a wavelength stability better than ± 1 nm and a measuring time of less than 1 ms. Furthermore, the power consumption of the instrument is 3 W at 5 V DC, and the signal-to-noise ratio is 3267 at full scale. Therefore, this spectrometer could be a potential alternative to classical spectrometers in process control applications or could be used as a portable or airborne spectroscopic sensor.

Keywords: micro-NIR spectrometer; scanning grating mirror; deflection position detector

1. Introduction

Because of their ability to detect the composition and content of substances non-destructively, near-infrared spectrometers have a wide range of applications [1]. However, due to their disadvantages in terms of volume, power consumption, portability and price, classical spectrometers have limited application in some fields, like remote sensing, in-field spectroscopy, astronomy, analytical chemistry and process control. Therefore, the miniaturization of the NIR spectral instruments has become a development trend and has an urgent application requirement.

MEMS and MOEMS technology have experienced rapid progress in recent decades. In the field of NIR spectroscopy, MEMS- and MOEMS-based NIR spectrometers have made important contributions to the process of instrument miniaturization. The incorporation of MEMS- or MOEMS-based devices into NIR spectrometers has become one of the research hotspots in the field of spectroscopic instruments, and such devices are characterized by cost effectiveness, portability, low power consumption, high speed and small size [2,3].

Since linear array detectors are expensive in the near-infrared band, in order to reduce the cost and size of the spectrometer, MOEMS-based NIR spectral instruments are usually designed with integrated movable scanning components so that a single-tube detector can be used to obtain the entire spectrum. This type of spectrometer can be classified into two categories based on spectroscopic principles: (1) nondispersive methods and (2) dispersive methods. The typical representatives of the former are the Fourier transform spectrometer (FTS) [4], the Fabry–Perot interferometer (FPI) [5,6], the grating light modulator spectrometer [7] and the quantum dot spectrometer [8]. Sandner proposed a Fourier transform spectrometer that utilizes an out-of-plane MEMS mirror in 2007 [9]. An FTS has several basic advantages, including a high signal-to-noise ratio (SNR) and linearity and a broad wavelength range. It also has disadvantages, such as light fluctuation, which creates noise at all wavelengths, and the need for significant amount of data analysis. A major hurdle, especially for MEMS, is that the FTS requires significant movement of reflecting surfaces, which miniaturized devices cannot easily achieve. Malinen proposed a piezo-actuated NIR FPI in 2010 [10]. This kind of instrument needs to be mass manufactured in high volumes to drive down the cost of individual FPI chips.

Kraft proposed a single-detector micro-electro-mechanical scanning grating spectrometer in 2006 [11], which combines the features of a spectrometer monochromator with the advantages of optical MEMS components. Spectral scanning can be accomplished by the rotation of an oscillating reflection grating, which belongs to the grating dispersion type. Because scanning grating is not integrated with any position detecting device, a laser projection system is used to detect the deflection angle of the dispersive element, which would enlarge and complicate the entire system. Furthermore, an additional deflection position detection device results in an increase in instrument volume, which cannot meet the special application requirements of the micro-NIR spectrometer based on the MOEMS scanning grating mirror. In addition, the signal-to-noise ratio is relatively low because blazed grating is not introduced.

The spectrometer presented in this paper was designed and implemented based on an MOEMS scanning grating micro-mirror, integrated with a deflection position detector. The angle signal is generated synchronously with the movement of the mirror, which serves as a trigger signal of the spectrum A/D acquisition and a reference signal of the mirror closed loop control. Therefore, the scanning range can be dynamically adjusted according to the respective requirements. In addition, the spectral changes can be monitored with a high time resolution (less than 1 ms), and the driving voltage is lower than 1 V.

The optical design and spectral reconstruction of the spectrometer are introduced in Section 2. After this, the system characterization is described in Section 3. Finally, a brief conclusion is given in Section 4.

2. System Design

The spectrometer based on MOEMS technology consists of four main parts, including the fabrication of MOEMS core components, optical design, circuit design and spectral reconstruction.

This paper focuses mainly on optical design, system integration and spectral reconstruction.

2.1. MEMS Component

The scanning dispersive element is the key component in the spectrometer system. It is vital to have a small and robust movable diffraction grating device.

Two kinds of devices were fabricated by Chongqing University Microsystem Research Center based on different materials, including single crystalline silicon [12–14] and FR4 [15]. To take thermal sensitivity into account, the Si MEMS electromagnetic scanning grating mirror (SGM) was selected as the scanning tilted grating. The movable mirror plate is suspended inside a frame by two rectangular torsion bars. One side of the movable plate is integrated with the blazed grating, and the other is integrated with the driving and sensing coils, as shown in Figure 1a. The specific parameters of the SGM are shown in Table 1. The corresponding signal processing circuits were also prepared [16].

When the component was driven at a voltage of 650 mV and a frequency slightly above the resonance frequency, e.g., at 620 Hz, its mechanical deflection angle was shown to reach ±4°. As a result of the combination of the optical layout and mirror blazed grating (period of 4 μm), the scanning wavelength range can cover 800–1800 nm.

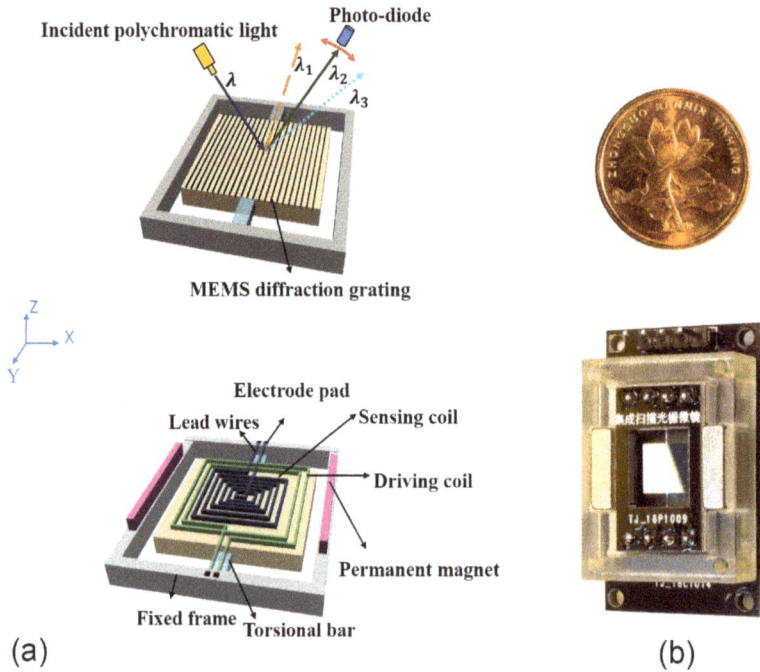

(a)   (b)

Figure 1. Schematic of the electromagnetic MEMS scanning mirror. (**a**) The integrated blazed grating (front) and driving and sensing coils (back); (**b**) scanning grating micro-mirror component.

Table 1. Scanning tilted grating parameters.

Parameter	Value
Resonance frequency (Hz)	620 Hz
Mechanical deflection angle (°)	±4 (driving voltage: 650 mV)
Mirror plate (mm^2)	6
Quality factor (Q) (a.u.)	125
Grating groove density (lines/mm)	250
Blazed angle (°)	7.9

2.2. Optical Design

Due to its compact size and ability to suppress stray light effectively, the Czerny–Turner system has become the classical structure of commercial spectrometers. This paper presents a modified version of the Czerny–Turner system. First, there is light crossing in this layout, which further reduces the instrument's volume. In addition, the focal lengths of the focusing mirror and the collimation mirror are not equal. This eliminates coma aberration at the specified wavelength. It is an asymmetrical crossed Czerny–Turner system with a 48° total deflection angle between the incident and refracted beam.

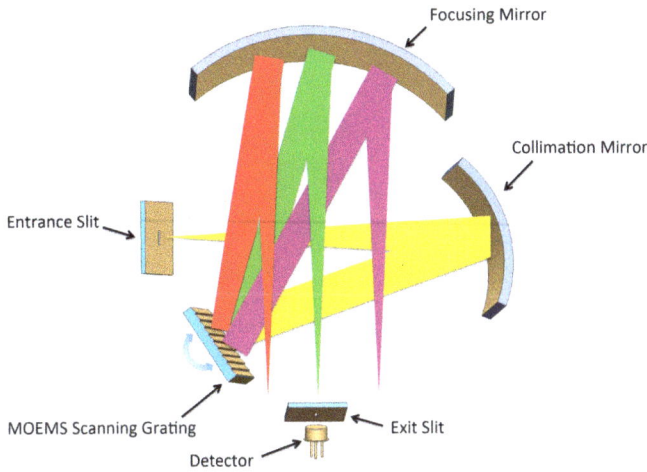

Figure 2. Principal drawing of the spectrometer showing the asymmetrical crossed Czerny–Turner setup.

The 3D optical layout is shown in Figure 2. The optics system includes an entrance slit, collimating mirror, scanning grating mirror (SGM), focusing mirror, exit slit and detector. In this optics system, the light radiation to be analyzed passes through a vertical 50-μm entrance slit aperture, collimated by a spherical mirror (focal length of 50 mm). It is diffracted by the scanning grating and focused again by a second spherical mirror (focal length of 75 mm) to the exit slit, and then, it reaches the detector (Hamamatsu G12181-05). Between the exit slit and focusing mirror, the spectrometer is equipped with an 800-nm long-pass filter as the second-order filter. The specific design parameters of the spectrometer are shown in Table 2.

Table 2. Spectrometer parameters.

Parameter	Value
Object distance (mm)	50
Wavelength range (nm)	800–1800
Entrance slit (μm)	50
Exit slit (μm)	85
NA (a.u.)	0.22
Image distance (mm)	75
Center wavelength (nm)	1300
Angle of incidence (°)	14
Diffraction order (a.u.)	+1

In addition to the portability and compactness of the spectrometer, another important feature is the scanning speed. When the scanning grating micromirror oscillates at a resonant frequency of 620 Hz, the time taken for a single spectral scan is only 0.83 ms. The fast scanning capability can be applied to either monitor the spectral signal changes at a high speed or to reduce the effection of random noise by co-adding multiple times in a short time period.

2.3. The Spectra Reconstruction

There are two full scans within one mirror oscillation, forwards and backwards, as shown in Figure 3a. Therefore, it is enough to acquire one spectrum of the forward scan. The return point of the mirror is the correct starting point for sampling a new spectrum.

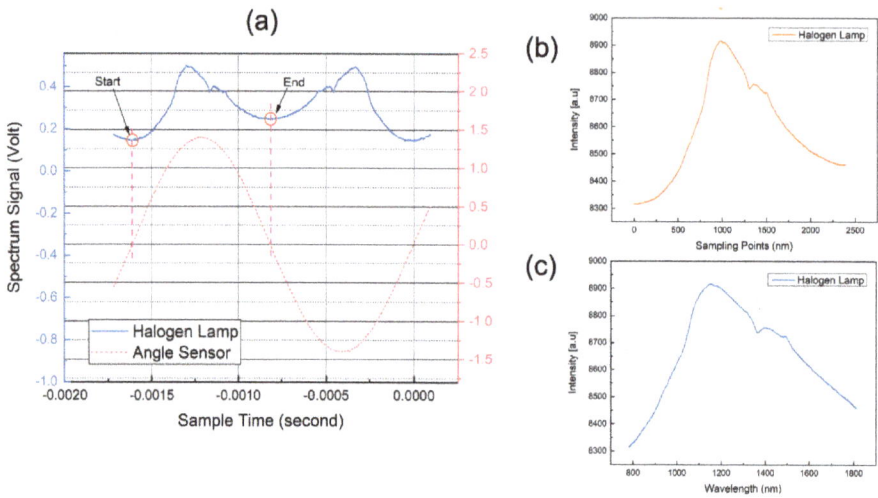

Figure 3. (a) Two full scans within one mirror oscillation; (b) time-dependent spectrum of a halogen lamp and (c) wavelength-dependent spectrum of a halogen lamp using reconstruction.

Spectrum acquisition with one single detector is always a time-dependent measurement. When the spectrometer circuit is working, A/D sampling will be performed on the single-tube detector to obtain the relationship between light intensity and time; it is $I(t)$, as shown in Figure 3b.

During the movement of the scanning grating mirror, the change in the deflection angle with time is nonlinear. Therefore, it is necessary to unfold the spectrum in the wavelength domain by means of the deflection position signal and the basic grating equation.

The maximum deflection angle of the mirror can be obtained from the deflection position detector. Together with the known sinusoidal movement of the mirror over time, $\phi(t)$ can be obtained with:

$$\phi(t) = \phi_{max} \cdot \cos(2 \cdot \pi \cdot f_{mirror} \cdot t). \tag{1}$$

The deflection position detector shows the mark of the spectra. There is a linear relationship between the deflection angle and the wavelength. By using the combined grating equation, the wavelength sliding over the detector at a certain timestamp can be determined. It is $\lambda(t)$ and is described as:

$$\lambda(t) = g \cdot (\sin \beta(t) - \sin \alpha(t)) \tag{2}$$

where:

$$\alpha(t) = \alpha_{middle} + \phi(t) \tag{3}$$

$$\beta(t) = \beta_{middle} - \phi(t) \tag{4}$$

$$\beta_{middle} = arcsin(\frac{\lambda_{middle}}{g} + \sin \alpha_{middle}) \tag{5}$$

where α_{middle}, β_{middle}, λ_{middle} and g represent the angle of incidence, the angle of diffraction, the wavelength when the grating mirror in its non-deflected position and the grating constant.

The final spectrum $I(\lambda)$ can be calculated by $I(t)$ and $\lambda(t)$, as shown in Figure 3c.

3. System Characterization

The experimental program followed the classical arrangement (light source-sample-fiber optics-spectrometer), with a 400 μm ($NA = 0.22$) low-OH silica fiber. All tests were performed at room temperature, i.e., $22 \pm 2\,°C$, unless otherwise specified.

3.1. Wavelength Range

While the mechanical deflection angle reaches ±4°, the spectral scanning range can cover 800–1800 nm, as shown in Figure 4a. Since the photosensitivity of InGaAs detector varies at each wavelength, the light intensities of the 800-nm filter and the 1800-nm filter are different.

As the prominent bands are 1446 nm for water and 1650 nm–1800 nm for carbon compounds, the 800-nm–1800-nm range is appropriate. Even though some interesting bands can be detected between 1900 nm and 2500 nm, the additional effort required for detector cooling increases system complexity and is not suitable for portable applications.

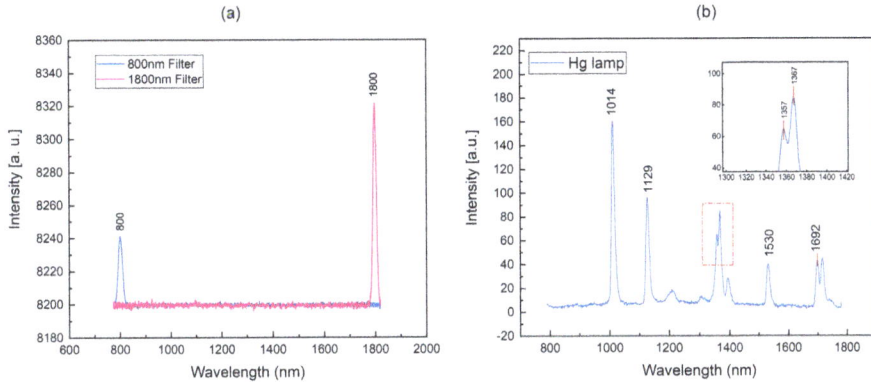

Figure 4. (**a**) Wavelength range; (**b**) spectral resolution.

3.2. Spectral Resolution

To evaluate the effective spectral resolution of the spectrometer prototype, a special test setup was established in the laboratory. Light from a standard high pressure mercury lamp was coupled to the fiber of the spectrometer. The result of the measurement is depicted in Figure 4b. It shows the whole spectral range from 800 nm–1800 nm and a magnified section of two adjacent peaks spaced 10 nm apart [17]. Obviously, the spectrometer prototype can distinguish two peaks at 1357 nm and 1367 nm. According to Rayleigh's criterion, the resolution of the spectrometer is better than 10 nm.

In addition, the theoretical spectral resolution can be calculated by Equation (6).

$$\delta\lambda = \frac{W_{slit} \cdot g \cdot cos\alpha}{m \cdot L_c} \tag{6}$$

where W_{slit} is the width of the entrance slit; m is the spectral order; L_c is the focal length of the collimating lens; g is the grating constant; and α is the incidence of grating. The theoretical resolution and the measured resolution are shown in Table 3.

Table 3. Comparison of the theoretical resolution with the measured resolution.

Parameter	Theoretical	Measured
Resolution (nm)	3.9	10

Because of assembly errors, part machining errors and system aberrations, the measured resolution is larger than the theoretical value. In practical applications of NIR spectroscopy, the requirements for spectral resolutions in 800–1800 nm are intermediate due to the broad structure of the over-tone and combination bands. Therefore, for most applications of an NIR spectrometer as a pocket-sized or handheld spectra analyzer, this resolution (10 nm) is at a good enough level.

3.3. Wavelength Stability

From the perspective of chemometrics, 1 nm is the minimum requirement for long-term stability.

In order to obtain accurate measurement data in terms of wavelength stability, a 1714-nm interference band-pass filter illuminated by a tungsten halogen lamp was tested 10 times every 15 min, as shown in Figure 5a. By analyzing the degree of deviation of the 10 measurements from the center wavelength (1714 nm), it was concluded that the wavelength stability was better than ±1 nm, meeting the requirement for long-term stability, as shown in Figure 5b.

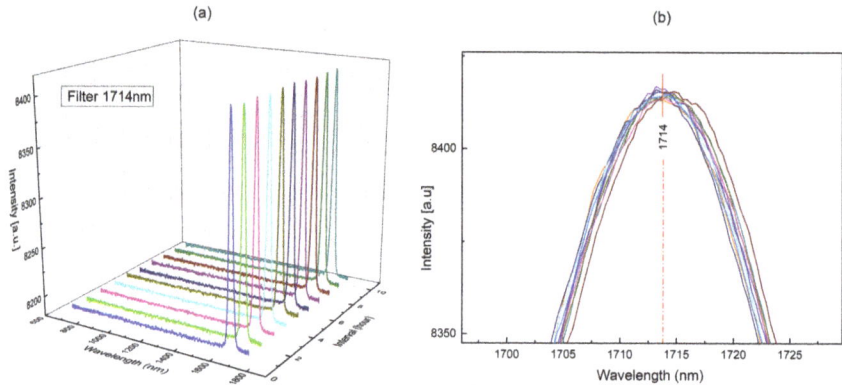

Figure 5. (**a**) Three-dimensional schematic of stability testing; (**b**) 10 measurements around the central wavelength of a 1714-nm filter.

3.4. Signal-to-Noise Ratio Characteristics

For the reliable evaluation of spectra, it is important to ensure a sufficient signal level and low noise. Independent of the spectral setup, the signal-to-noise ratio (SNR), typically given for the lowest signal level that can be detected, contributes to the performance description of a spectrometer.

The SNR improves with an increasing number of co-added scans [11]. For the spectrometer prototype described in this paper, when sampling 50 times for averaging, the signal-to-noise ratio is 1700, and a single scan acquired within 0.8 ms is noisier, about 200. When the light source is strong enough to make the spectrometer work at full scale and the co-added scans exceed 100 times, the signal-to-noise ratio can reach 3267, as shown in Table 4. As a handheld instrument, this is satisfactory for most NIR applications.

Table 4. Standard deviation, peak intensity and SNR of the spectrum of a tungsten halogen lamp source with 1, 50 and 100 averaging.

Parameter	1×	50×	100×
Baseline intensity SD (a.u.)	0.0049	0.0006	0.0003
Peak intensity (a.u.)	0.9913	0.9874	0.9802
SNR (a.u.)	200:1	1700:1	3267:1

Packaged in an aluminum shell with a three-dimensional size of $90 \times 70 \times 70$ mm^3, the resulting spectrometer prototype has a weight of approximately 0.65 kg and a power consumption of less than 3 W at 5 V DC. It is comparable in size to a tennis ball (Figure 6b).

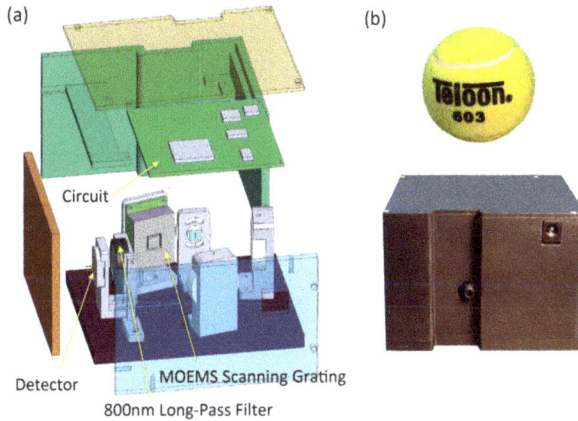

Figure 6. (a) Expanded assembly drawing of the spectrometer; (b) a single-detector NIR microspectrometer encased in an aluminum housing, which is comparable in size to a tennis ball.

The power consumption and weight of the MEMS spectrometer in comparison with the SGS 1900 [11] and irSys E2.1 [18,19] are listed in Table 5.

Table 5. Important parameters of the microspectrometer prototype and the corresponding values for the Hiperscan SGS 1900 and irSys E2.1 for comparison.

Parameter	Microspectrometer	SGS 1900	irSys E2.1
Wavelength range (nm)	800–1800	1200–1900	910–2100
Spectral resolution (nm) *	10	10	11
Volume (cm^3)	441	600	810.6
SNR (a.u.) [†]	3267:1	1700:1	2500:1
Power consumption (W) [‡]	3	5	5
Scan time (ms)	0.83	4	4
Power supply (V)	5	7.5	24

* 50-μm entrance slit; [†] measured at 1650 nm with 100 averaging; [‡] including complete readout electronics.

4. Conclusions

In summary, a prototype of a miniaturized NIR spectrometer based on novel MOEMS scanning grating integrated with a deflection position detector was presented in this paper. As a scanning grating Czerny–Turner structure spectrometer, it has the characteristics of miniaturization, low power consumption and low cost when introducing the MOEMS device as the core device. A series of test results showed that the spectrometer's scanning range can cover 800–1800 nm with a single scan time of less than 1 ms, a spectral resolution of better than 10 nm and a wavelength stability of better than ±1 nm. In addition, the signal-to-noise ratio was shown to be sufficient to satisfy most NIR analyses. In the future, the performance of the instrument could be further improved by employing a parabolic collimation mirror while controlling the magnification of the optical system so that the exit slit can be removed to improve the signal-to-noise ratio.

Author Contributions: J.H., Q.W. and Z.W. conceived of and designed the spectrometer. Q.N., Y.Z. and Z.W. fabricated the SGM device. J.H. and F.C. designed and implemented the spectrometer software system. J.H. and Q.W. wrote and revised the paper respectively.

Acknowledgments: This work is supported by the Special Funds of the National Natural Science Foundation of China (Grant No. 61327002) and the Fundamental Research Funds for Central Universities (Grant No. 106112017CDJXY120006).

Conflicts of Interest: The authors declare no conflict of interest.

References

1. Antila, J.; Tuohiniemi, M.; Rissanen, A.; Kantojärvi, U.; Lahti, M.; Viherkanto, K.; Kaarre, M.; Malinen, J. MEMS-and MOEMS-Based Near-Infrared Spectrometers. *Encycl. Anal. Chem.* **2014**. [CrossRef]
2. Schuler, L.P.; Milne, J.S.; Dell, J.; Faraone, L. MEMS-based microspectrometer technologies for NIR and MIR wavelengths. *J. Phys. D* **2009**, *42*, 133001. [CrossRef]
3. Hong, L.; Sengupta, K. Fully Integrated Optical Spectrometer in Visible and Near-IR in CMOS. *IEEE Trans. Biomed. Circuits Syst.* **2017**, *11*, 1176–1191. [CrossRef] [PubMed]
4. Xie, H.; Lan, S.; Wang, D.; Wang, W.; Sun, J.; Liu, H.; Cheng, J.; Ding, J.; Qin, Z.; Chen, Q.; et al. Miniature fourier transform spectrometers based on electrothermal MEMS mirrors with large piston scan range. In Proceedings of the IEEE Sensors, Busan, Korea, 1–4 November 2015; pp. 1–4.
5. Akujarvi, A.; Guo, B.; Mannila, R.; Rissanen, A. MOEMS FPI sensors for NIR-MIR microspectrometer applications. In Proceedings of the SPIE, San Francisco, CA, USA, 13–18 February 2016; Volume 9760.
6. Rissanen, A.; Akujarvi, A.; Antila, J.; Blomberg, M.; Saari, H. MOEMS miniature spectrometers using tuneable Fabry-Perot interferometers. *J. Micro-Nanolithog. MEMS MOEMS* **2012**, *11*, 023003. [CrossRef]
7. Xu, J.; Liu, H.; Lin, C.; Sun, Q. SNR analysis and Hadamard mask modification of DMD Hadamard Transform Near-Infrared spectrometer. *Opt. Commun.* **2017**, *383*, 250–254. [CrossRef]
8. Bao, J.; Bawendi, M.G. A colloidal quantum dot spectrometer. *Nature* **2015**, *523*, 67–70. [CrossRef] [PubMed]
9. Sandner, T.; Kenda, A.; Drabe, C.; Schenk, H.; Scherf, W. Miniaturized FTIR-spectrometer based on optical MEMS translatory actuator. In Proceedings of the MOEMS and Miniaturized Systems VI, San Jose, CA, USA, 20–25 January 2007; International Society for Optics and Photonics: Bellingham, WA, USA, 2007; Volume 6466, p. 646602. [CrossRef]
10. Malinen, J.; Saari, H.; Kemeny, G.; Shi, Z.; Anderson, C. Comparative performance studies between tunable filter and push-broom chemical imaging systems. In Proceedings of the Next-Generation Spectroscopic Technologies III, Orlando, FL, USA, 5–9 April 2010; International Society for Optics and Photonics: Bellingham, WA, USA, 2010; Volume 7680, p. 76800E. [CrossRef]
11. Kraft, M.; Kenda, A.; Frank, A.; Scherf, W.; Heberer, A.; Sandner, T.; Schenk, H.; Zimmer, F. Single-detector micro-electro-mechanical scanning grating spectrometer. *Anal. Bioanal. Chem.* **2006**, *386*, 1259–1266. [CrossRef] [PubMed]
12. Nie, Q.; Wen, Z.; Huang, J. Design and fabrication of a MEMS high-efficiency NIR-scanning grating based on tilted (1 1 1) silicon wafer. *Eur. Phys. J. Appl. Phys.* **2015**, *72*, 10702. [CrossRef]
13. Zhou, Y.; Wen, Q.; Wen, Z.; Yang, T. Modeling of MOEMS electromagnetic scanning grating mirror for NIR micro-spectrometer. *AIP Adv.* **2016**, *6*, 025025. [CrossRef]
14. Zhou, Y.; Wen, Q.; Wen, Z.; Huang, J.; Chang, F. An electromagnetic scanning mirror integrated with blazed grating and angle sensor for a near infrared micro spectrometer. *J. Micromech. Microeng.* **2017**, *27*, 125009. [CrossRef]
15. Lei, H.; Wen, Q.; Yu, F.; Zhou, Y.; Wen, Z. FR4-Based Electromagnetic Scanning Micromirror Integrated with Angle Sensor. *J. Micromech. Microeng.* **2018**, *9*, 214. [CrossRef]
16. Liu, H.; Wen, Z.; Li, D.; Huang, J.; Zhou, Y.; Guo, P. A Control and Detecting System of Micro-Near-Infrared Spectrometer Based on a MOEMS Scanning Grating Mirror. *J. Micromech. Microeng.* **2018**, *9*, 152. [CrossRef]
17. Kramida, A.; Ralchenko, Y.; Reader, J.; NIST ASD Team. *NIST Atomic Spectra Database (ver. 5.5.4)*; National Institute of Standards and Technology: Gaithersburg, MD, USA, 2017. Available online: https://physics.nist.gov/asd/ (accessed on 14 August 2017).
18. Otto, T.; Saupe, R.; Bruch, R. A novel dual-detector micro-spectrometer. In Proceedings of the SPIE, San Jose, CA, USA, 22–27 January 2005; The International Society for Optical Engineering: Bellingham, WA, USA, 2005; Volume 5719, pp. 76–82.
19. TQ-Systems. Technology Report, TQ-Systems Gmbh, Blankenburgstr. 81, Chemnitz, Germany. 2012. Available online: https://www.tq-group.com (accessed on 12 August 2018).

micromachines

MDPI

Article

Design, Simulation and Experimental Study of the Linear Magnetic Microactuator

Hanlin Feng [1], Xiaodan Miao [1,*] and Zhuoqing Yang [2]

1   College of Mechanical Engineering, Shanghai University of Engineering Science, Shanghai 201620, China; m010216123@sues.edu.cn
2   National Key Laboratory of Science and Technology on Micro/Nano Fabrication, School of Electronics Information and Electrical Engineering, Shanghai Jiao Tong University, Shanghai 200240, China; yzhuoqing@sjtu.edu.cn
*   Correspondence: 01120003@sues.edu.cn; Tel.: +86-021-6779-1413

Received: 3 August 2018; Accepted: 6 September 2018; Published: 11 September 2018

check for updates

Abstract: This paper reports the design, simulation and experimental study of a linear magnetic microactuator for portable electronic equipment and microsatellite high resolution remote sensing technology. The linear magnetic microactuator consists of a planar microcoil, a supporter and a microspring. Its bistable mechanism can be kept without current by external permanent magnetic force, and can be switched by the bidirectional electromagnetic force. The linearization and threshold of the bistable mechanism was optimized by topology structure design of the microspring. The linear microactuator was then fabricated based on non-silicon technology and the prototype was tested. The testing results indicated that the bistable mechanism was realized with a fast response of 0.96 ms, which verified the simulation and analysis.

Keywords: MEMS; microactuator; magnetic

1. Introduction

"Microactuator" generally refers to a driver with small size, high positioning accuracy and low energy consumption, and is an important part of microelectromechanical systems (MEMS) [1]. Its main function is to realize the transformation and output force or displacement (including displacement and angle), which forms the operation and power for MEMS devices [2]. With the development of portable electronic equipment and microsatellite high resolution remote sensing technology, linear microactuators with fast response and high precision characteristics have been attracting more attention [3–6]. In recent years, the experimental study and modeling of the voice coil motors (VCM) actuator [3,4] and permanent rotatory linear actuator [7,8] have been presented. However, the VCM actuator comprises two or more permanent magnets, a yoke, a fixed base, a moving part and a coil. These parts were fabricated separately, and then were assembled. The componentry was complex, and as a result, the precision of the actuator was affected, as well as its volume. Closed-loop control was used to improve the precision control of the MEMS microactuator [9–12]. Sensors were used in the control system, but this lead to increased complexity of the system, and the system will be influenced by environment. Since the displacement of the microactuator is always limited in micrometers, the sensor precision may have a negative influence on the precision control. In addition, the electrostatic, electrothermal and piezoelectric actuators were also studied for high precision control [13,14]. However, the driving voltage of the electrostatic microactuator is above 12 V, which is not compatible with integrated circuit (IC) technology. The electrothermal miroactuator usually responds at a slow speed, and the displacement of the piezoelectric microactuator is limited. Compared with the actuators mentioned above, the linear electromagnetic microactuator can provide large displacement, and high precision control in a small volume with a low driving voltage [15–19]. The linear microactuator also has higher control accuracy and faster response [20–23]. However, for linear electromagnetic microactuators, the planar microcoil and the permanent magnet were fabricated on two separate wafers, and the prototype was formed using a bonding process. In current study, although the volume was smaller, the fabrication process was simplified, and the characteristics was obviously improved, the linear microactuator's high integrity, large displacement and high precision can be improved further.

2. Structural Design of the Linear Magnetic Microactuator

The linear magnetic microactuator consists of a planar microcoil, a supporter and a microspring, as shown in Figure 1a. The microcoil can provide bidirectional electromagnetic force, which can increase or decrease the magnetic force in combination with the external permanent magnetic force on the microspring. The dimension and material of the microactuator is shown in Table 1. The microactuator is usually in either open or closed state, and switching between the two states is achieved by the balance between the magnetic force and the elastic force. If the microactuator is in the open state, the microspring stays flat, as shown in Figure 1b. When a positive current is fed into the microcoil, an electromagnetic force is generated, and the magnetic force on the microspring increases. When the magnetic force is larger than the elastic force, the microspring will be attracted down to the bottom, and the microactuator is switched to closed state. At the same time, the current is decreased, and the microactuator remains in closed state because of the permanent magnetic force, as shown in Figure 1c. This means the microactuator can work in a closed state without current, which reduces its power consumption greatly. When a negative current is fed into the planar microcoil, a negative electromagnetic force is generated, and the magnetic force is reduced. When the magnetic force is smaller than elastic force, the microspring will be pushed up into the open position as a result of the elastic force, as shown in Figure 1a.

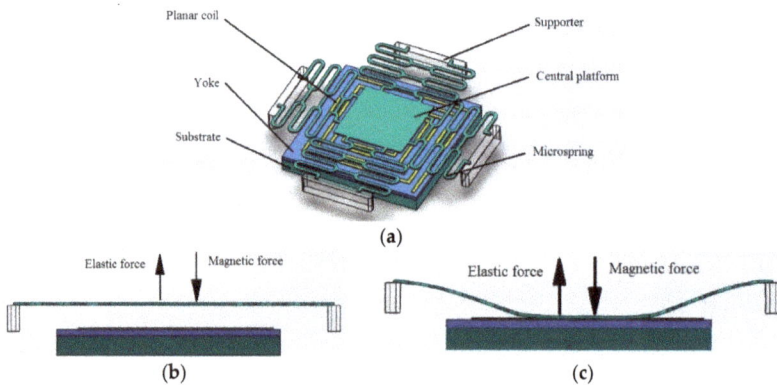

Figure 1. The structure and working mechanism of the microactuator. (**a**) Structure; (**b**) Open state; (**c**) Closed state.

Table 1. Properties of the microactuator.

Part	Material	Value
Central Platform	Permalloy	1 mm × 1 mm × 15 μm
Planar Microcoil	Copper	2.5 mm
Microspring	Nickel	3 mm × 3 mm × 12 μm
Yoke	Permalloy	2.8 mm × 2.8 mm × 50 μm
Substrate	Ferrite	2.8 mm × 2.8 mm × 200 μm
Supporter	Nickel	1.2 mm × 0.2 mm × 160 μm

The working principle of the microactuator is dependent on the matching between the magnetic force and the elastic force, as shown in Figure 2. There are two intersections of the magnetic force curve and the elastic curve; that is, they enclose the working threshold. By changing the direction of the current in the microcoil, the electromagnetic force becomes positive or negative, the magnetic force is increased or decreased, and two stable states can be obtained. It is assumed that the initial state corresponds to the disconnected state, that is, point A in Figure 2, which indicates the first stable state. When a positive current is applied, the magnetic force increases, which becomes larger than the elastic force. Then the microactuator is closed at B point indicated in Figure 2. In this state, the closing magnetic force is provided only by external permanent magnet without extra power consumption. When the reverse current is applied, the electromagnetic force is gradually reduced. The microactuator will then return to the disconnected state.

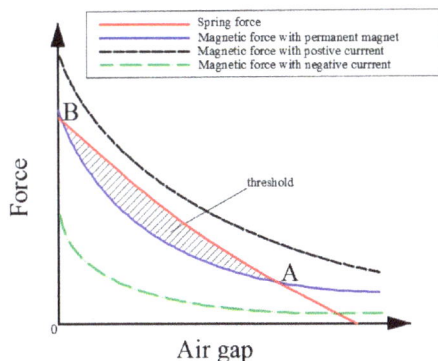

Figure 2. The relationship between the electromagnetic force, the elastic force and the air gap.

3. Topology Design and Simulation of the Microspring

The switching of states in the microactuator is obtained by the elastic deformation of the micro spring, so the spring's linearity has a great influence on the working mechanism. In order to ensure the precise movement of microactuator, the microspring needs to provide a compliant linear motion. Through our previous experimental research combined with other researchers' experimental conclusions in [24,25], it is found that the rectangular cantilever beam in the traditional process can provide both translation motion and rotation error, which causes a torsion pendulum and affects the accuracy. Thus, a symmetrical structure can reduce the rotation error and achieve precise control of pure translational motion. As a result, a spring with four topology structures was designed, as shown in Figure 3. In order to analyze the elastic force, ANSYS (12.0 version, Canonsburg, PA, USA) was used to carry out a three-dimensional simulation experiment of our MEMS microactuator.

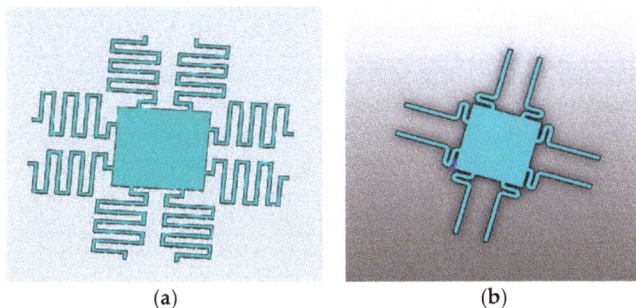

(a)   (b)

Figure 3. *Cont.*

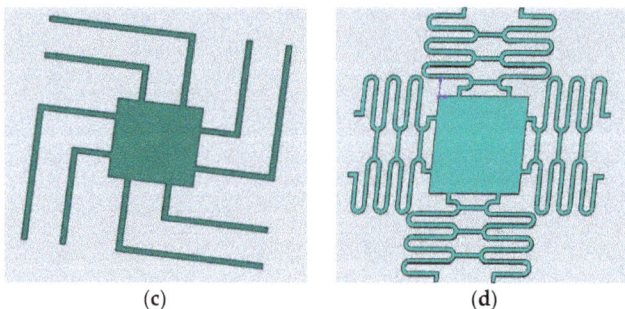

(c)   (d)

Figure 3. Four kinds of cantilever beam structure. (**a**) U-type; (**b**) N-type; (**c**) L-type; (**d**) W-type.

According to the traditional mechanism, the elastic force is proportional to the deformation. However, in natural conditions, nonlinear phenomena are common. The nonlinearity of the structure will lead to a nonlinear relationship between the external load and the strain generated by the structure, which leads to the microactuator being uncontrollable. Therefore, ANSYS static simulation of each spring model was carried out on four different topology structures, and the load-carrying capacity increased from 2 to 15 mN. The maximum spring deformation value was obtained under the nonlinear calculation conditions at each load, and compared with the linear value as the characterization of the nonlinear degree. As shown in Figure 4, the abscissa is the load value, the ordinate is the spring deformation, and the thick black line is the linear result curve. By comparing the nonlinear deformation of four cantilever structures under different load forces, it can be obtained that different structure has different linearization.

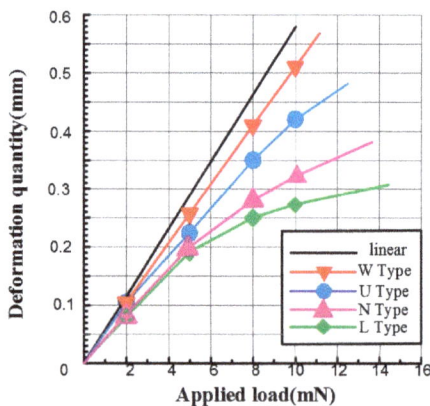

Figure 4. Nonlinear simulation curves of four kinds of structural springs.

The nonlinear curves of the four springs with the same linear stiffness and different plane geometry gradually deviate from the linear deformation curve, and the offset increases sequentially. The nonlinear results corresponding to the U and W-type springs are close to the linear results, and the elastic coefficient curves are close to the straight line, which shows that the nonlinear degree of the two shapes and sizes of the springs is low, and the d spring is the lowest. The deformation of W spring under an 8 mN load is close to that of a-type spring, but the deformation increases after a 10 mN load. It can be seen that for different topological structures, different nonlinear characteristics will be generated under nonlinear conditions.

4. Simulation of the Bistable Mechanism of the Microactuator

The above results provided a basis for further reducing the nonlinear characteristics of springs under large deformation conditions through topology design. It is also proved that the structure topology optimization design method using ANSYS can effectively analyze the influence of the nonlinear degree of the microspring from the geometric nonlinear angle, and reduce the influence of the nonlinearity on the bistability of the device through structural topology optimization.

ANSYS was used to conduct the magnetic simulation of the whole magnetic circuit structure. Due to the symmetry of the whole structure, only a quarter of the structure needed to be modeled and analyzed. First, the overall parameters of the structure were determined: The number of the coil is 30 turns; the relative permeability of the permanent magnet is 50,000; the coercive force is 1×10^5 A/m and the air gap size is 160 µm. Under the condition of these constant structural parameters, analysis of the bistable mechanism of microactuators with four different topology structures of the microspring was carried out with 300 mA negative current, 400 mA positive current, 500 mA positive current, and permanent magnet force. The results are shown in Figure 5.

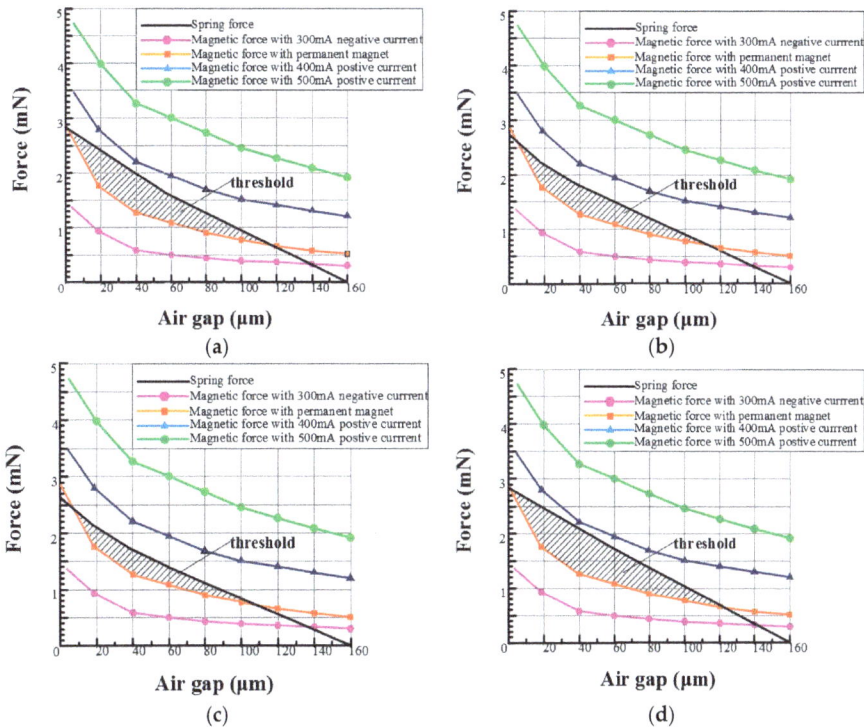

Figure 5. Matching diagram of elastic force and magnetic force for microsprings with four topology structures. (**a**) U-type; (**b**) N-type; (**c**) L-type; (**d**) W-type.

When a 300 mA negative current is applied, the resultant electromagnetic force is less than the spring force, and the microactuator is returned to its original state. When a 400 mA and 500 mA positive current are applied, the resultant electromagnetic force is greater than the spring force, and the microactuator is closed. In addition, the spring force curve has two intersection points with the magnetic force produced by the permanent magnet, which indicates the bistable mechanism of microactuator.

In order to compare the threshold formed by the two curves for the four topological structures, the curve-fitting method is used to fit the data of the curves generated by permanent magnet and the four curves. First, the polynomial fitting method is applied to the electromagnetic force curve. Assuming that the air gap x is an independent variable, the electromagnetic force F is a dependent variable. In order to ensure the accuracy of the curve, 5th-order polynomials are used to fit the data. It can be obtained that:

$$F = p_1 \times x^5 + p_2 \times x^4 + p_3 \times x^3 + p_4 \times x^2 + p_5 \times x + p_6 \tag{1}$$

It can be seen that the determinable coefficient of the fitting curve R-square is 0.9996 and the correction factor adjusted R-square is 0.9989, so the fitting accuracy is higher and the error is smaller, and it can be used for the calculation of the threshold. The results of the polynomial are in Table 2.

Table 2. Results of the polynomial.

$p_1 = -1.551 \times e^{-10}$	$p_2 = 8.225 \times e^{-8}$
$p_3 = -1.691 \times e^{-5}$	$p_4 = 0.0017$
$p_5 = -0.08869$	$p_6 = 3.003$

Secondly, four groups of elastic force curves were fitted: Elastic curve of the U-type spring:

$$F_a = 2.5e^{-5} \times x^2 - 0.0215x + 2.8 \tag{2}$$

Elastic curve of the N-type spring:

$$F_b = 3.081e^{-5} \times x^2 - 0.0212x + 2.604 \tag{3}$$

Elastic curve of L-type spring:

$$F_c = 3.556e^{-5} \times x^2 - 0.0219x + 2.604 \tag{4}$$

Elastic curve of W-type spring:

$$F_d = -0.0175x + 2.8 \tag{5}$$

The integral method was used to calculate the threshold area of the force curve and the electromagnetic force curve. The solution is shown in Table 3.

Table 3. Threshold area of four types of microactuator spring.

Type	Threshold Area
D_a	40.2
D_b	23.45
D_c	22.74
D_d	53.48

From the calculated threshold area, it can be obtained that the threshold area is smaller with the stronger nonlinearity. The threshold area of N-type (D_b) and L-type (D_c) springs with the worst linearity is 23.45 and 22.74 respectively, which is obviously smaller than the threshold area formed by the U-type (D_a) and W-type (D_d) springs. In the matching curve between U-type and W-type springs and electromagnetic force, the degree of similarity of the two curves is similar when the air gap spacing is less than 60 μm. When the gap spacing is greater than 60 μm, it can be seen that the nonlinear degree of the elastic resilience curve of the U-type spring is enhanced, which leads to the reduction of the threshold area. Therefore, the threshold area of the W spring structure is larger, which indicates better reliability of the bistable mechanism.

In addition, the electromagnetic force analysis of the above simulation was based on the analysis of the permanent magnet substrate. By analyzing the electromagnetic force produced by different substrate materials, the effects of different substrate materials on the size of the electromagnetic force were analyzed in the case of the same overall structure. With the same 300 mA current and overall structure size, two materials of permanent magnet and glass were used as the substrates of the microactuator, as shown in Figure 6. Because the permanent magnetic material can generate extra magnetic force on the microspring, a larger magnetic force can be produced for the same current by using a permanent magnet substrate to ensure a higher driving speed.

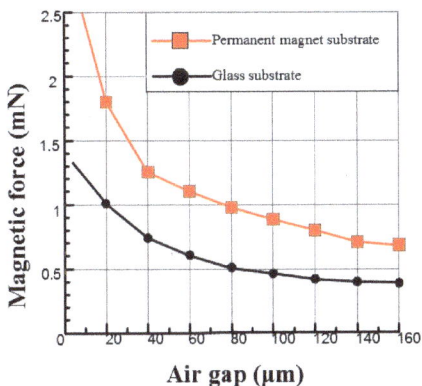

Figure 6. Variation of magnetic force with current on different substrates.

Finally, through the simulation and analysis, it can be seen that a permanent magnet substrate and the W-type microspring can reduce the influence of nonlinearity on the bistable mechanism at 300 mA of current. Thus, the response speed and working stability of the microactuator is improved.

5. Fabrication and Testing of the Bistable Microactuator

The bistable magnetic microactuator was fabricated based on the non-silicon micro-micromachining process on a single wafer. The fabrication is described as follows: (a) The chromium/copper seed layer was sputtered on to the ferrite wafer, and then the photoresist was spin coated, followed by electroplating the permalloy yoke and copper planar microcoil. (b) After forming polyimide as an insulation layer for the planar microcoil, the photoresist was coated on two layers thick, then heat treating as sacrificial layer for the air space between the microactuator and microspring. The supporter was then electroplated, followed by the microspring. (c) Last, the thick photoresist and Cr/Cu seed layer were etched smoothly layer by layer, then the suspended structure was released and the spring could be moved in a space as shown in Figure 7.

The bistable mechanism testing system was established, and 5 V, generated by a B&K 2706 power amplifier (Agilent 6813B, Agilent company, Santa Clara, CA, USA) incorporated with a GW waveform generator (GFG-8016G, RIGOL, Beijing, China), was applied to the microcoil. The results were observed by the oscilloscope (Agilent MSO6034, Agilent company). By comparing the input signal (lower level) and output signal (upper level), the bistable mechanism could be observed as shown in Figure 8. The microactuator is in the first stable state, then, when the positive current is fed into the coil, the microactuator is switched into the second stable state. When the current was reduced, the microactuator could stay in the second stable state without power consumption until the next switching current. The difference between the driving voltage and switching voltage at the higher level means a response time of 0.96 ms. The bistable mechanism verified the reliability of the design and simulation described above.

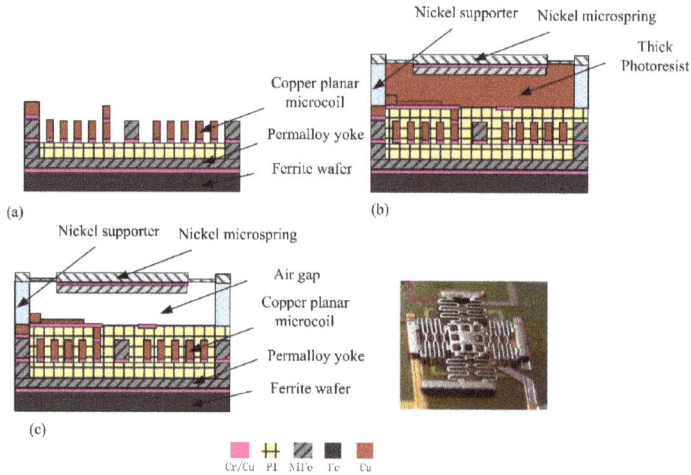

Figure 7. The fabrication process and the prototype.

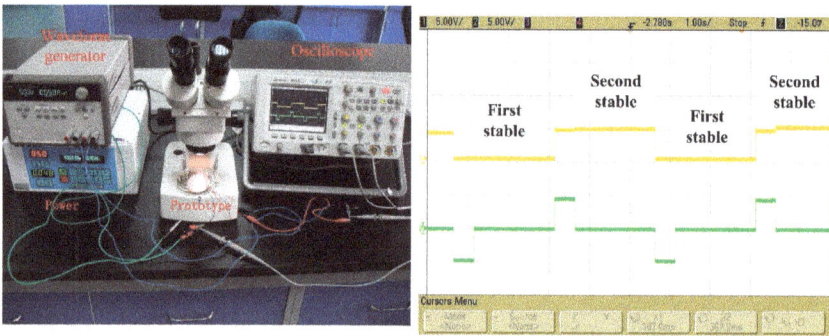

Figure 8. Bistable mechanism of the prototype.

6. Conclusions

This paper presents a linear microactuator, which has high response, large displacement and high precision for use in portable electronic equipment and microsatellite high resolution remote sensing technology. The linear microactuator comprises a planar microcoil, a supporter and a microspring. In order to optimize the linearity of the microactuator, microsprings with four topology structures were simulated and analyzed. By comparing the simulation results, the best linear spring, d-type, could improve the threshold of the bistable mechanism. The redundancy of the fabrication process of the linear microactuator is increased, which has a positive influence on the precision of the microactuator. The linear microactuator was fabricated based on nonsilicon process, and the bistable mechanism of the prototype was tested. The results showed that the prototype could realize a bistable mechanism with a response time of 0.96 ms, which verified the reliability of the topology design and simulation.

Author Contributions: Formal analysis, H.F.; project administration, X.M.; validation, Z.Y.; writing—original draft, X.M.; Writing—review and editing, X.M.

Acknowledgments: This work is supported by the Natural Science Foundation of China (No. 51605277, 11702168, 561703268).

Micromachines **2018**, *9*, 454

Conflicts of Interest: The authors declare no conflict of interest.

References

1. Judy, J.W. Microelectromechanical systems (MEMS): Fabrication, design and applications. *Smart Mater. Struct.* **2001**, *10*, 1115–1134. [CrossRef]
2. Tanaka, M. An industrial and applied review of new MEMS devices features. *Microelectron. Eng.* **2007**, *84*, 1341–1344. [CrossRef]
3. Chung, M.J.; Yee, Y.H.; Ahn, W.H. Development of compact camera module having auto focus actuator and mechanical shutter system for mobile phone. In Proceedings of the 2007 International Conference on Control, Automation and Systems, Seoul, Korea, 17–20 October 2007.
4. Liu, C.S.; Ko, S.S.; Lin, P.D. Experimental characterization of high-performance miniature auto-focusing VCM actuator. *IEEE Trans. Magn.* **2011**, *47*, 738–745. [CrossRef]
5. Tao, K.; Tang, L.; Wu, J.; Lye, S.W.; Chang, H.; Miao, J. Investigation of multimodal electret-based MEMS energy harvester with impact-induced nonlinearity. *J. Microelectromech. Syst.* **2018**, *27*, 276–288. [CrossRef]
6. Wang, P.; Liu, R.; Ding, W.; Zhang, P.; Pan, L.; Dai, G.; Zou, H.; Dong, K.; Xu, C.; Wang, Z.L. Complementary electromagnetic-triboelectric active sensor for detecting multiple mechanical triggering. *Adv. Funct. Mater.* **2018**, *28*, 1705808. [CrossRef]
7. Krebs, G.; Tounzi, A.; Pauwels, B.; Willemot, D.; Piriou, F. Modeling of a linear and rotary permanent magnet actuator. *IEEE Trans. Magn.* **2008**, *44*, 4357–4360. [CrossRef]
8. Jin, P.; Yuan, Y.; Jian, G.; Lin, H.; Fang, S.; Yang, H. Static characteristics of novel air-cored linear and rotary Halbach permanent magnet actuator. *IEEE Trans. Magn.* **2014**, *50*, 977–980. [CrossRef]
9. Chen, H.; Pallapa, M.; Sun, W.J.; Sun, Z.D.; Yeow, J.T.W. Nonlinear control of an electromagnetic polymer MEMS hard-magnetic micromirror and its imaging application. *J. Micromech. Microeng.* **2014**, *24*, 045004. [CrossRef]
10. Bryzek, J.; Abbott, H.; Flannery, A.; Cagle, D.; Maitan, J. Control issues for MEMS. In Proceedings of the 42nd IEEE conference on Decision and Control, Maui, HI, USA, 9–12 December 2003.
11. Peng, Y.; Cao, J.; Guo, Z.; Yu, H. A linear actuator for precision position of dual objects. *Smart Mater. Struct.* **2015**, *24*, 125039. [CrossRef]
12. Borovic, B.; Hong, C.; Zhang, X.M.; Liu, A.Q.; Lewis, F.L. Open vs. closed-loop control of the MEMS electrostatic comb drive. In Proceedings of the 2005 IEEE International Symposium on Mediterrean Conference on Control and Automation Intelligent Control, Limassol, Cyprus, 27–29 June 2005.
13. Yu, S.; Jang, K.; Cha, S.; Lee, Y.; Kwon, O.; Kwon, K.; Choi, J. A piezoelectric actuator driver circuit for automatic focusing of mobile phone cameras. In Proceedings of the2008 IEEE International Symposium on Circuits and Systems, Seattle, WA, USA, 18–21 May 2008.
14. Park, C.; Cha, S.; Lee, Y.; Kwon, O.; Park, D.; Kwon, K.; Lee, J. A highly accurate piezoelectric actuator driver IC for auto-focus in camera module of mobile phone. In Proceedings of the 2010 IEEE International Symposium on Circuits and Systems, Paris, France, 30 May–2 June 2010.
15. Watanabe, Y.; Kobayashi, S.; Iwamatsu, S.; Yahagi, T.; Sato, M.; Oizumi, N. Motion monitoring of MEMS actuator with electromagnetic induction. *Electron. Commun. Jpn.* **2014**, *97*, 52–57. [CrossRef]
16. Cheng, H.D.; Hsiao, S.Y.; Wu, M.; Fang, W. Monolithic Bi-directional linear microactuator for light beam manipulation. In Proceedings of the IEEE/LEOS International Conference on Optical MEMS and Their Applications Conference, Big Sky, MT, USA, 21–24 August 2006.
17. Choi, Y.M.; Gorman, J.J.; Dagalakis, N.G.; Yang, S.H.; Kim, Y.; Yoo, J.M. A high-bandwidth electromagnetic MEMS motion stage for scanning applications. *J. Micromech. Microeng.* **2012**, *22*, 105012. [CrossRef]
18. Braune, S.; Liu, S. A novel linear actuator for variable valve actuation. In Proceedings of the 2005 IEEE International Conference on Industrial Technology, Hong Kong, China, 14–17 December 2005; pp. 377–382.
19. Ruffert, C.; Gehrking, R.; Ponick, B.; Gatzen, H.H. Magnetic levitation assisted guide for a linear micro-actuator. *IEEE Trans. Magn.* **2006**, *42*, 3785–3787. [CrossRef]
20. Sun, S.; Dai, X.; Wang, K.; Xiang, X.; Ding, G.; Zhao, X. Nonlinear electromagnetic vibration energy harvester with closed magnetic circuit. *IEEE Magn. Lett.* **2018**, *9*, 1–4. [CrossRef]
21. Dai, X.; Miao, X.; Sui, L.; Zhou, H.; Zhao, X.; Ding, G. Tuning of nonlinear vibration via topology variation and its application in energy harvesting. *Appl. Phys. Lett.* **2012**, *100*, 031902. [CrossRef]

22. Miao, X.; Dai, X.; Huang, Y.; Ding, G.; Zhao, X. Segmented magnetic circuit simulation of the large displacement planar micro-coil actuator with enclosed magnetic yokes. *Microelectron. Eng.* **2014**, *129*, 38–45. [CrossRef]

23. Sun, S.; Dai, X.; Sun, Y.; Xiang, X.; Ding, G.; Zhao, X. MEMS-based wide-bandwidth electromagnetic energy harvester with electroplated nickel structure. *J. Micromech. Microeng.* **2017**, *27*, 115007. [CrossRef]

24. Luharuka, R.; LeBlanc, S.; Bintoro, J.S.; Berthelot, Y.H.; Hesketh, P.J. Simulated and experimental dynamic response characterization of an electromagnetic valve. *Sens. Actuators A* **2008**, *143*, 399–408. [CrossRef]

25. Wagner, B.; Benecke, W. Microfabricated actuator with moving permanent magnet. In Proceedings of the 1991 IEEE Micro Electro Mechanical Systems, Nara, Japan, 30 December 1990–2 January 1991.

micromachines

Article

Frequency Characteristic of Resonant Micro Fluidic Chip for Oil Detection Based on Resistance Parameter

Zilei Yu, Lin Zeng, Hongpeng Zhang *, Guogang Yang, Wenqi Wang and Wanheng Zhang

Marine Engineering College, Dalian Maritime University, Dalian 110621, China;
ray_yzl417@dlmu.edu.cn or yzl950417@126.com (Z.Y.); bobzl@dlmu.edu.cn (L.Z.);
yanggg@dlmu.edu.cn (G.Y.); sarawang@dlmu.edu.cn (W.W.); zwanheng123@dlmu.edu.cn (W.Z.)
* Correspondence: zhppeter@dlmu.edu.cn; Tel.: +86-0411-8472-9565

Received: 11 May 2018; Accepted: 6 July 2018; Published: 9 July 2018

check for
updates

Abstract: Monitoring the working condition of hydraulic equipment is significance in industrial fields. The abnormal wear of the hydraulic system can be revealed by detecting the variety and size of micro metal debris in the hydraulic oil. We thus present the design and implementation of a micro detection system of hydraulic oil metal debris based on inductor capacitor (LC) resonant circuit in this paper. By changing the resonant frequency of the micro fluidic chip, we can detect the metal debris of hydraulic oil and analyze the sensitivity of the micro fluidic chip at different resonant frequencies. We then obtained the most suitable resonant frequency. The chip would generate a positive resistance pulse when the iron particles pass through the detection area and the sensitivity of the chip decreased with resonant frequency. The chip would generate a negative resistance pulse when the copper particles pass through the detection area and the sensitivity of the chip increased with resonant frequency. The experimental results show that the change of resonant frequency has a great effect on the copper particles and little on the iron particles. Thus, a relatively big resonant frequency can be selected for chip designing and testing. In practice, we can choose a relatively big resonant frequency in this micro fluidic chip designing. The resonant micro fluidic chip is capable of detecting 20–30 μm iron particles and 70–80 μm copper particles at 0.9 MHz resonant frequency.

Keywords: resonant frequency; resistance parameter; micro fluidic; oil detection

1. Introduction

Hydraulic technology is widely applied in the country's core industries, for example, aerospace, energy, and manufacturing industry due to its advantages of a large transmission force, flexible layout, and high efficiency. In the ship engineering field, hydraulic technology is always applied in major devices such as steering gears, propulsion systems, and lifting equipment. Hydraulic oil as the hydraulic system transmission medium is a kind of high cleanliness fluid, and more than 75% of hydraulic system failures are caused by hydraulic oil pollution [1]. Thus, predicting and diagnosing the contamination of hydraulic oil is the key to protecting the hydraulic system.

In hydraulic systems, hydraulic oil is inevitably contaminated due to internal generation and external intrusion, among which solid particle contamination is the main source of pollution. Some of these solid particle contaminants are external dust, and metal abrasive particles produced by internal mechanical wear. The latter is the most important cause of mechanical failure in hydraulic systems. The reference data [2–4] shows that the metal abrasive particles in the hydraulic oil have a constant concentration and the size of the abrasive particles is usually 10–20 μm when the hydraulic systems are under normal working conditions. On the contrary, when abnormal wear occurs in the hydraulic systems, the concentration of the metal abrasive particles in the hydraulic oil increases and the size of the particles becomes 50–100 μm. If the hydraulic system continues to work under this working

Micromachines **2018**, *9*, 344; doi:10.3390/mi9070344   189   www.mdpi.com/journal/micromachines

condition, the concentration and size of the metal abrasive particles in the hydraulic oil will still increase until the hydraulic system fails. Thus, how to achieve the detection of particulate contaminants in hydraulic oil is an important content for predicting and diagnosing hydraulic oil contamination.

The main methods currently used for the detection of hydraulic oil include ferrous analysis, spectroscopic analysis, magnetic plug analysis, screen damping, particle counting, etc. [5]. With the exception of the particle counting method, several other methods can only estimate the pollution degree of the oil and require professional laboratory personnel to carry out analysis and measurement. Furthermore, these methods are difficult to achieve in oil on-line monitoring. The particle counting method is a common method for detecting fluid contamination, through the analysis of the amplitude and quantity of the pulses generated by the contaminants passing through the detection area, the size and number of the contaminants can be obtained and the accurate measurement of the oil contaminants can be truly achieved [6,7]. Further, the particle counting method can achieve on-line monitoring, so that it can better predict and diagnose the operation conditions of mechanical equipment and reduce time-consuming and labor-intensive shutdown inspections. Thus, this method has broad application prospects.

Currently, the combination of the micro fluidic electromagnetic technique and particle counting method can achieve some effects, mainly in the three following ways: (1) resistive pulse sensor (RPS), measuring the change of electrolyte resistance when a particle passes through micro channel [8–10], using multiple pores on a single chip as the detection area, detecting the resistance change when particles pass through the area; (2) capacitive counter sensor, measuring the difference in relativity permittivity between oil and metal particles [11,12], using opposite placement metal rods as the electrodes and detecting the changes when particles pass through the electrodes; and (3) inductive counter sensor, measuring the relative different permeability between ferrous particles, nonferrous metal particles, and non-metallic particles [13] using the inductance coil at high frequency condition to detect the particles. However, resistive pulse sensor detection is insensitive due to the fact that oil is non-conductive and non-isothermal. Capacitive counter sensor detection is interfered by a few water droplets in lubricant oil (the relativity permittivity of water and metal particles are both much larger than the relativity permittivity of oil), and it is also incapable of distinguishing the nonferrous metal particles and ferrous particles [14,15]. Du et al. demonstrated the feasibility and proposed a sensor using inductor capacitor (LC) resonant method based on the principle of such an inductive counter sensor and detected 32 μm iron particles and 75 μm copper particles with different excitation frequency by experiment [16,17]. Different from Du detection based on inductance parameter, in this paper, a resonant micro fluidic chip based on electromagnetic theory for detecting resistance parameter is designed and the frequency characteristic of a micro resistance sensor is explored. The theoretical calculations using MATLAB (MATLAB R2015a, The MathWorks, Natick, MA, USA) and experimental verification are performed that the change of resonant frequency has effect on detection. The resonant micro fluidic chip is capable of detecting 20–30 μm iron particles and 70–80 μm copper particles when the resonant frequency is 0.9 MHz, which improves the detection accuracy of ferromagnetic particles to some extent

2. Chip Design and Fabrication

The micro fluidic resonant oil detection chip is designed to detect the metal debris in oil based on electromagnetic theory as shown in Figure 1. The chip is composited of a plane coil, a chip capacitor, and a micro fluidic channel.

The micro fluidic channel, which is close to the inner wall, goes across the inner hole of the coil, and the coil is connected with the capacitor in parallel. The diameter of micro-channel D_1 is 300 μm (see Figure 2a). The diameter of the coil wire core D_2 is 70 μm. There is a coat of insulating paint with a thickness of 10 μm around the coil wire core and the diameter of the coil inner hole D_3 is 900 μm (see Figure 2b).

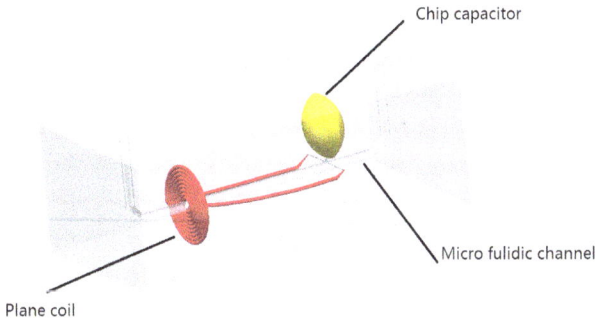

Figure 1. Micro fluidic resonant oil detection chip model figure.

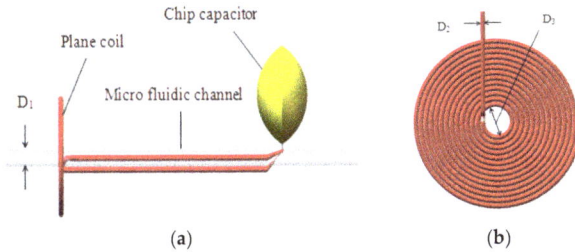

Figure 2. The design of micro fluidic resonant oil detection chip: (**a**) cross-section of the chip: D_1 is the diameter of the micro-channel and (**b**) a sketch of a single-layer coil: D_2 is the diameter of the coil wire core and D_3 is the inner diameter of the single-layer coil.

To fabricate the detection chip, a model of a coil connected with capacitors and a straight micro-channel was made first. The single-layer coil was made of enamel copper wire and wound by a winding machine (SRDZ23-1B, Zhongshan Shili Wire Winder Equipment Co., Ltd., Zhongshan, Guangdong, China); the number of turns is varied according to different experiments. The straight micro-channel model was made of an iron rod with a diameter of 300 µm and a length of 7 cm. It was put into the coil inner hole close to the inner wall (see Figure 2a). After that, both the micro-channel model, the capacitor, and the coil were fixed to a glass substrate using glue. The micro-channel model, the capacitor, the coil, and glass substrate formed a chip mold, upon which the liquid polydimethylsiloxane (PDMS) was poured. The chip mold was then placed in a thermostat with a temperature of 80 °C for 1 h. Finally, after the liquid PDMS was solidified, the straight micro-channel model was removed using pliers and the resonant oil detection chip fabrication was completed. The diameter of the micro-channel model equals to the diameter of the straight micro-channel.

3. Theoretical Analysis

The micro fluidic resonant oil detection chip presented in this paper is a resistance sensor, the change of the inductor coil causes the impedance of the entire circuit to change, and the chip detects the real part of the impedance change.

When an excitation voltage is applied to the coil, an alternating current is formed in the coil. Such a coil can be simplified as a circular current-carrying conductor and the magnetic field is shown in Figure 3. It is easy to know that the transverse components of the magnetic field generated by the coils cancel each other out and only have a magnetic field with an axial component. According to the Biot-Savart Law, the center of the coil has the highest magnetic induction. When metal particles pass through this position, the particles are magnetized and the coils generate increased magnetic flux;

at the same time, particles will also generate eddy current effects at this position which will reduce the magnetic induction of the original magnetic field. For ferromagnetic particles, their relative magnetic permeability is much larger than 1, so when they pass through the detection area, the magnetization field inside them is much larger than the weakened magnetic field generated by the inner eddy current, and the apparent inductance of the coil increase. For non-ferromagnetic metal particles, the relative magnetic permeability is about equal to 1, so there is no magnetization effect. When the particles pass through the detection area, the magnetization field inside the particle is smaller than the weakened magnetic field generated by the inner eddy current, and the apparent inductance value of the coil reduce.

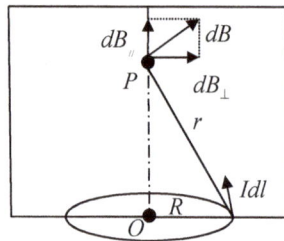

Figure 3. Magnetic field on the circular current-carrying conductor.

3.1. Particle Differentiate Detection

The equivalent circuit diagram of the micro fluidic oil detection chip described herein is shown in Figure 4. Among the figure, the inductor coil is equivalent to a pair of series inductance L_0 and resistor R_0, the entire circuit is composed of a series branch and a capacitor C_0 in parallel. The resistance, inductance, and capacitance are expressed by a complex representation, the impedance of the entire circuit is Z_0. According to the character of the parallel circuit, the branch connected in series with a resistor and an inductor corresponds to a single inductor coil with excitation voltage. Under a constant excitation, a time-harmonic magnetic field is generated and the magnetic induction intensity is constant. When the particles pass through the center of the coil, the particles are magnetized and generate an eddy current effect, resulting in a change in the apparent inductance value of the coil [18,19].

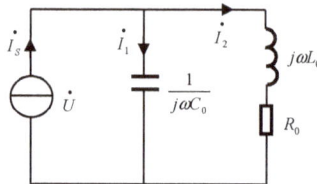

Figure 4. The equivalent circuit of micro fluidic resonant oil detection chip.

When the resonant angular frequency exists in a parallel resonant circuit, it is satisfied by $1 - \frac{C_0 R_0}{L_0} > 0$ and it can be expressed as:

$$\omega_0 = \frac{1}{\sqrt{L_0 C_0}} \sqrt{1 - \frac{C_0 R_0}{L_0}} \qquad (1)$$

When the excitation frequency is the resonant frequency, the entire circuit presents a purely resistive state. When no particles pass through the detection area, the initial equation of the equivalent circuit can be expressed as:

$$Z_0 = \frac{R_0 + j\omega_0 L_0}{(1 - \omega_0^2 L_0 C_0) + j\omega_0 R_0 C_0} = \frac{R_0 + j\omega_0(L_0 - \omega_0^2 L_0^2 C_0 - \omega_0 R_0^2 C_0)}{(1 - \omega_0^2 L_0 C_0)^2 + \omega_0^2 R_0^2 C_0^2} \tag{2}$$

The real part of the impedance can be expressed as:

$$\mathrm{Re}(Z_0) = \frac{R_0}{(1 - \omega_0^2 L_0 C_0)^2 + \omega_0^2 R_0^2 C_0^2} \tag{3}$$

When there are metal particles passing through the detection area, the apparent inductance value of the coil changes as described above, so that the changed value produced is the inductance ΔL, and the impedance change value of the entire circuit is ΔZ. At this time, the real part of the changes can be expressed as:

$$\mathrm{Re}(Z_0 + \Delta Z) = \frac{R_0}{X_1^2 + X_2^2} \tag{4}$$

Among them:

$$\begin{cases} X_1 = [1 - \omega_0^2 (L_0 + \Delta L) C_0]^2 \\ X_2 = \omega_0^2 R_0^2 C_0^2 \end{cases} \tag{5}$$

Because only the inductance value has changed, only X_1 has changed in Formula (5). As can be seen, the resonance angular frequency from the previous Formula (1), X_1 can be expressed as:

$$X_1 = [1 - \frac{1}{L_0}(1 - \frac{C_0 R_0^2}{L_0})(L_0 + \Delta L)]^2 \tag{6}$$

It can be seen from the beginning that when ferromagnetic particles pass through the detection area, the apparent inductance of the coil increases, that is, $\Delta L > 0$; and when the non-ferromagnetic metal particles pass through the detection area, the apparent inductance of the coil decreases, that is, $\Delta L < 0$. The Formulas (3) and (6) show that when the ferromagnetic particles pass, X_1 decreases and the real part of the circuit impedance increases. When non-ferromagnetic metal particles pass, X_1 increases and the real part of the circuit impedance decreases. Therefore, ferromagnetic particles and non-ferromagnetic particles can be distinguished by detecting the real part of the entire circuit impedance.

3.2. Effect of Excitation Frequency on Detection

For a single coil micro fluidic chip, the frequency characteristics have been studied. Through the research, it has found that when the excitation frequency is lower than 2 MHz, the frequency change has little effect on the chip detection capability [20]. The inductor coil used in this paper has an inductance of approximately 10 μH and a resistance of approximately 0.8 Ω. In the previous study, our research team has found that a single coil chip was used for detection the 80–90 μm iron particles, the apparent inductance value generated by the inductor coil was changed approximately 4×10^{-5} μH. And for the 150–160 μm copper particles, the apparent inductance value generated by the inductor coil was changed approximately -3×10^{-5} μH. In this paper, the resonant frequency of the detection chip is changed by changing the value of the chip capacitor in parallel, and the frequency of the excitation voltage is set to the resonant frequency. From Formulas (3)–(5), the real part impedance signal amplitude can be expressed as:

$$\Delta Z = \frac{R_0}{[1 - \omega^2(L_0 + \Delta L)C]^2 + \omega^2 R_0^2 C^2} - \frac{R_0}{(1 - \omega^2 L_0 C)^2 + \omega^2 R_0^2 C^2} \tag{7}$$

Among them, the capacitor and the excitation angular frequency are variables, the resistance and the inductance of the inductor coil are constants. And the relationship between resonant frequency and resonant angular frequency is:

$$f = \frac{\omega}{2\pi} \tag{8}$$

By MATLAB calculation, we can get the curve of the real part impedance signal amplitude value of the entire circuit as shown in Figure 5 with the excitation frequency f changes.

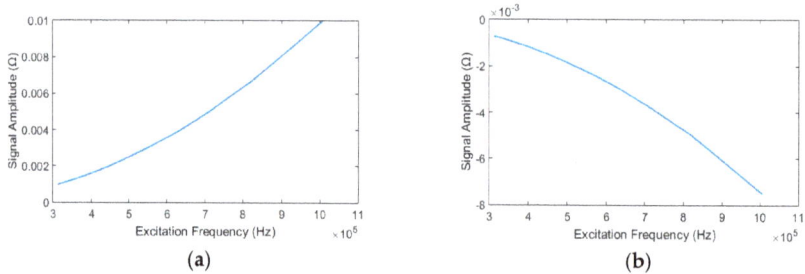

(a)

(b)

Figure 5. The relationship between excitation frequency and impedance signal amplitude value: (a) impedance signal value produced by ferromagnetic particle and (b) impedance signal value produced by non-ferromagnetic metal particle.

From the curve in Figure 5, we can see that as the excitation frequency increases, for ferromagnetic particles, the resulting change in the real part of the impedance generated after passing through the detection area increases; for non-ferromagnetic metal particles, the resulting change in the real part of the impedance generated after passing through the detection area decreases. From the foregoing description, it can be seen that the real part of the impedance of the entire circuit increases as the ferromagnetic particles pass through the detection area, and the real part of the impedance of the entire circuit decreases as the non-ferromagnetic particles pass through the detection area. Thence, for the ferromagnetic particles and the non-ferromagnetic particles, the absolute value of the real part of the impedance change produced by the detection area increases with the increase of the excitation frequency.

4. Experiments and Discussion

The impedance detection system is shown in Figure 6. It consists of a micro-injection pump (Harvard Apparatus B-85259, Harvard Apparatus, Holliston, MA, USA), a microscope (Nikon AZ100, Nikon, Tokyo, Japan), a micro fluidic chip, an inductance (L), capacitance (C), and resistance (R) meter (Agilent E4980A, Agilent Technologies Inc., Bayan Lepas, Malaysia) and a computer with LabVIEW software (LabVIEW 2011, National Instruments, Austin, TX, USA).

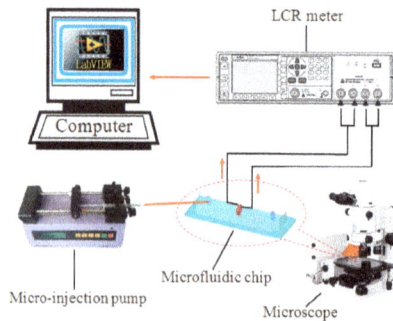

Figure 6. Schematic diagram of impedance detection system.

4.1. Experiments Preparations

In the experiment to investigate the effect of excitation frequency on detection, we used iron particles with sizes of 70–80 µm for ferromagnetic particles, and we used copper particles with sizes of 130–140 µm for non-ferromagnetic particles (Hefei Shatai Mechanical and Electrical Technology Co., Ltd., Hefei, Anhui, China). We weighed 4 mg for each particle using a precision balance (Precisa XS255A, Precisa Gravimetrics AG, Luzern, Switzerland), and mixed the particles with 100 mL of Marine hydraulic oil (marine hydraulic oil (The Great Wall L-HM 46, Sinopec Lubricant Co., Ltd., Beijing, China) by oscillator (IKA S25, IKA, Staufen, Germany), then put them into plastic test tube as experiment material.

In the experiment of exploring the lowest limit of chip detection, we used iron particles with size of 20–30 µm, 30–40 µm, 40–50 µm for ferromagnetic particles, and we used copper particles with size of 70–80 µm, 80–90 µm, 90–100 µm for non-ferromagnetic particles.

In the experiments, in order to ensure that unrelated variables are consistent when detecting particles, we designed the detection chip as shown in Figure 7. This chip contains 1 inductor and 5 capacitors of different sizes. When the excitation frequency is changed, the coil is connected with different capacitor for detection, and the excitation frequency is keeping unchanged. In this way, the particles can be detected in a chip having the same flow channel and the same inductance coil and the external factors interferes can be relatively reduced.

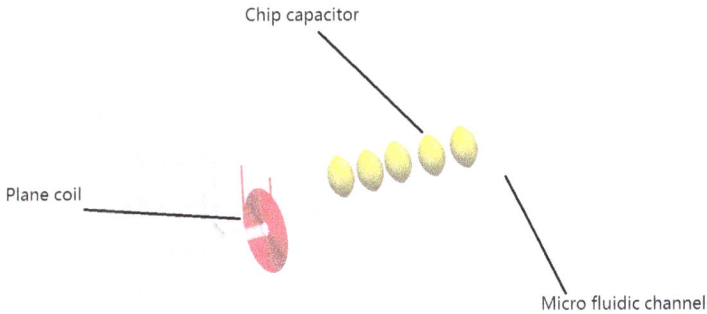

Figure 7. Micro fluidic resonant oil detection chip model figure in the experiments.

4.2. The Influence of Excitation Frequency on Detection

In the experiment, we set the excitation voltage of the chip to 2 V by LCR meter and the injection plastic of the microinjection pump to 40 µL/min. Then put 70–80 µm iron particles and 140–150 µm copper particles into the chip and connected micro-injector pump and finally started the experiment.

Among them, the detection results of 70–80 µm iron particles and 140–150 µm copper particles are shown in Figure 8, respectively, at the same voltage and frequency. From Figure 8a,b, we can see that an upward pulse is generated when iron particles pass through the detection area and a downward pulse is generated by copper particles. Thus, we can distinguish the ferromagnetic particles and the non-ferromagnetic particles by the pulse direction judgment.

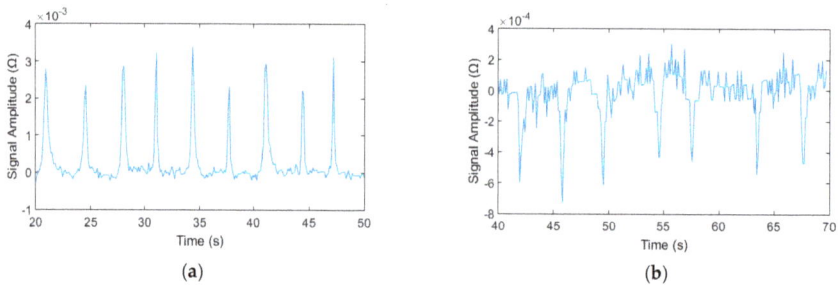

Figure 8. Detection results of iron and copper particles, the excitation voltage was 2 V and the excitation frequency was 0.9 MHz: (**a**) iron particles with sizes ranging from 70–80 μm and (**b**) copper particles with sizes ranging from 140–150 μm.

In the experiments, we used 70–80 μm iron particles and 140–150 μm copper particles to investigate the influence of excitation frequency on detection precision. The size of the chip capacitors and the excitation frequency are shown in Table 1, the excitation frequency is the resonant frequency corresponding to the connected capacitor. And the highest resonant frequency of the chip is 0.9 MHz in laboratory conditions because the minimum capacitor in our laboratory is 0.010 μF.

Table 1. Different excitation frequencies in the experiments.

Chip Capacitors	Excitation Frequency [1]
0.10 μF	0.30 MHz
0.068 μF	0.35 MHz
0.047 μF	0.50 MHz
0.022 μF	0.65 MHz
0.010 μF	0.90 MHz

[1] Excitation frequency $f = \omega/2\pi$, ω is excitation angular frequency.

The experimental results are shown in Figure 9a,b, each average data point and its error bar were evaluated by 10 measurements. From the Figure 9a, we can see that for the same size of iron particles, with the increase of the excitation frequency, the generated pulse size increase, but the signal-to-noise ratio change negatively with the frequency change. From the Figure 9b, we can find that for the same size copper particles, with the increase of the excitation frequency, the generated pulse size increase, and the signal-to-noise ratio (SNR) also change in proportion to the frequency. Furthermore, the amplitude curve of experiments is consistent with theoretical calculations in Figure 5.

4.3. Detection Limit of the Detection Chip

From Figure 9 we can also know that the SNR of copper particle detection signal is more effective than the SNR of iron particle detection signal with excitation frequency change. Thus, we have chosen 0.9 MHz as the excitation frequency to investigate the detection lowest limit of the detection chip in the experiments. Moreover, we used iron particles with size of 20–30 μm, 30–40 μm, and 40–50 μm and copper particles with size of 70–80 μm, 80–90 μm, and 90–100 μm as the experiment's materials.

Through the experiments, we found that the limitation of detection chip is iron particle 20–30 μm and copper particle 70–80 μm when the excitation voltage is 2 V, the excitation frequency is 0.9 MHz and the injection flow rate is 40 μL/min. The detection signal and the particle figure as shown in Figure 10.

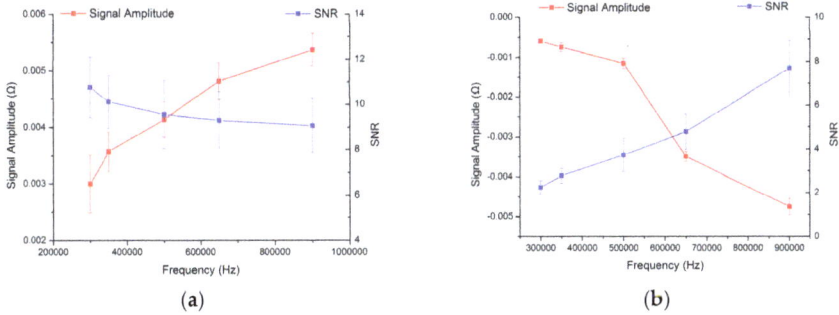

Figure 9. Influence of particle amplitude and signal-to-noise ratio (SNR) with excitation frequency: (**a**) 70–80 μm Iron particles detection results, the blue line was SNR, and the red line was signal amplitude and (**b**) 140–150 μm copper particles detection results, the blue line was SNR, and the red line was signal amplitude.

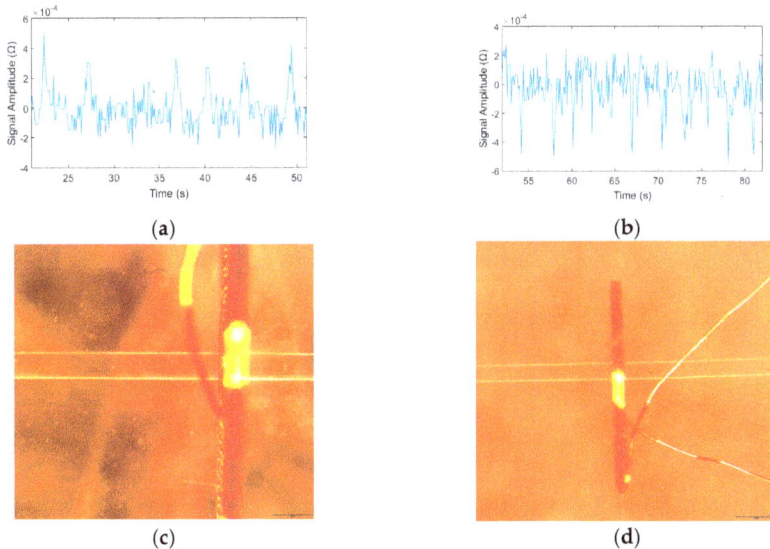

Figure 10. The limitation detection results of iron and copper particles, the excitation voltage was 2 V and the excitation frequency was 0.9 MHz: (**a**) iron particles with sizes ranging from 20–30 μm; (**b**) copper particles with sizes ranging from 70–80 μm; (**c**) 20–30 μm iron particle under microscope; and (**d**) 70–80 μm copper particle under microscope.

4.4. Mixtures Detection of the Detection Chip

The mixtures of 70–80 μm iron particles and 140–150 μm copper particles and the mixtures of 20–30 μm and 70–80 μm iron particles were detected at 0.9 MHz and 2 V condition in experiments to verify that the detection chip can distinguish different size ferromagnetic and non-ferromagnetic particles. And the detection results were shown in Figure 11. The positive signals were generated by iron particles and the negative signals were generated by copper particles in the Figure 10a. The larger positive signals were generated by 70–80 μm iron particles and the smaller were generated by 20–30 μm particles in the Figure 10b.

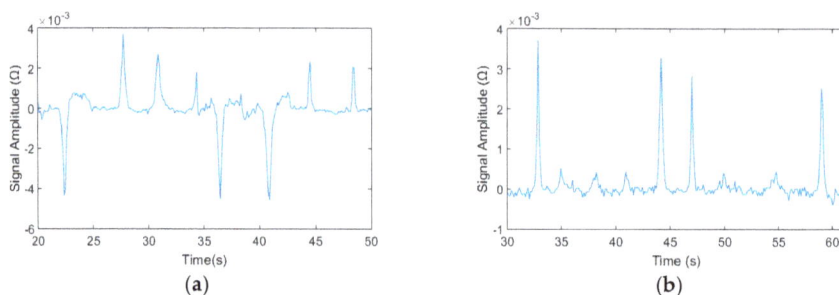

Figure 11. The mixtures detection result of the chip: (**a**) the mixtures of 70–80 µm iron particles and 140–150 µm copper particles results, the positive signals were iron particles, the negative signals were copper particles and (**b**) the mixtures of 20–30 µm and 70–80 µm iron particles detection results, the larger signals were 70–80 µm iron particles, the smaller signals were 20–30 µm iron particles.

5. Conclusions

A resistance micro fluidic detection chip based on resonant circuit for detecting metal particles in hydraulic oil has been designed in this paper. The experimental results are consistent with the theoretical calculations. The chip would generate a positive resistance pulse when the iron particles pass through the detection area, and the chip would generate a negative resistance pulse when the copper particles pass through the detection area, and the detection signal amplitude increased with resonant frequency. With excitation frequency increased, the signal-to-noise ratio decreased for ferromagnetic particles and increased for non-ferromagnetic particles. Since the highest resonant frequency of the chip is 0.9 MHz in laboratory conditions, we used it as the chip's excitation frequency to investigate the limitation of detection and detected 20–30 µm iron particles and 70–80 µm copper particles. We can effectively predict the hydraulic system fault by the size of the contaminants and its location to some extent by distinguishing the contaminants. The condition monitoring and fault diagnosis is significance to the ship's hydraulic system.

The detection accuracy and the throughput of this chip are still limited. In our future works, we will improve the detection accuracy by increase the turns of the coil and improve the throughput by increase the diameter of the fluid channel.

Author Contributions: Z.Y. and L.Z. conceived and designed the experiments; Z.Y. performed the experiments and wrote the first draft of this manuscript; H.Z. and G.Y. aided in theoretical; W.W. and W.Z. aided in harvest, finished writing, and editing of the manuscript.

Funding: This work was supported by the Natural Science Foundation of China (51679022), the Fundamental Research Funds for the Central Universities (3132017013).

Conflicts of Interest: The authors declare no conflict of interest.

References

1. Zhang, X.; Zhang, H.; Sun, Y.; Chen, H.; Guo, L. Effects of eddy current within particles on the 3D solenoid microfluidic detection chip. *Appl. Mech. Mater.* **2013**, *385*, 546–549. [CrossRef]
2. Tucker, J.E.; Schultz, A.; Lu, C.; Sebok, T.; Holloway, C.; Tankersley, L.L.; Reintjes, J. Lasernet fines optical wear debris monitor. In Proceedings of the International Conference on Condition Monitoring, Swansea, Wales, 12–15 April 1999; pp. 445–452.
3. Zhang, H.; Huang, W.; Zhang, Y.; Shen, Y.; Li, D. Design of the microfluidic chip of oil detection. *Appl. Mech. Mater.* **2012**, *117*, 517–520. [CrossRef]
4. Du, L.; Zhu, X.; Han, Y.; Zhe, J. High throughput wear debris detection in lubricants using a resonance frequency division multiplexed sensor. *Tribol. Lett.* **2013**, *51*, 453–460. [CrossRef]

5. Du, L.; Zhe, J. On-line wear debris detection in lubricating oil for condition based health monitoring of rotary machinery. *Electr. Eng.* **2011**, *4*, 1–9. [CrossRef]
6. Kumar, M.; Mukherjee, P.S.; Misra, N.M. Advancement and current status of wear debris analysis for machine condition monitoring: A review. *Ind. Lubr. Tribol.* **2013**, *65*, 3–11. [CrossRef]
7. Wu, T.; Peng, Y.; Wu, H.; Zhang, X.; Wang, J. Full-life dynamic identification of wear state based on on-line wear debris image features. *Mech. Syst. Signal Process.* **2014**, *42*, 404–414. [CrossRef]
8. Deblois, R.W.; Bean, C.P. Counting and sizing of submicron particles by the resistive pulse technique. *Rev. Sci. Instrum.* **1970**, *41*, 909–916. [CrossRef]
9. Carbonaro, A.; Sohn, L.L. A resistive-pulse sensor chip for multianalyte immunoassays. *Lab Chip* **2005**, *5*, 1155–1160. [CrossRef] [PubMed]
10. Wu, X.; Kang, Y.; Wang, Y.; Xu, D.; Li, D.; Li, D. Microfluidic differential resistive pulse sensors. *Electrophoresis* **2008**, *29*, 2754–2759. [CrossRef] [PubMed]
11. Sohn, L.L.; Saleh, O.A.; Facer, G.R.; Beavis, A.J.; Allan, R.S.; Notterman, D.A. Capacitance cytometry: Measuring biological cells one by one. *Proc. Natl. Acad. Sci. USA* **2000**, *97*, 10786–10790. [CrossRef] [PubMed]
12. Murali, S.; Jagtiani, A.V.; Xia, X.; Carletta, J.; Zhe, J. A microfluidic coulter counting device for metal wear detection in lubrication oil. *Rev. Sci. Instrum.* **2009**, *80*, 016105. [CrossRef] [PubMed]
13. Du, L.; Zhe, J.; Carletta, J.; Veillette, R.; Choy, F. Real-time monitoring of wear debris in lubrication oil using a microfluidic inductive coulter counting device. *Microfluid. Nanofluid.* **2010**, *9*, 1241–1245. [CrossRef]
14. Flanagan, I.M.; Jordan, J.R.; Whittington, H.W. An inductive method for estimating the composition and size of metal particles. *Meas. Sci. Technol.* **1990**, *1*, 381. [CrossRef]
15. Hong, W.; Wang, S.; Tomovic, M.; Han, L.; Shi, J. Radial inductive debris detection sensor and performance analysis. *Meas. Sci. Technol.* **2013**, *24*, 125103. [CrossRef]
16. Du, L.; Zhe, J. A high throughput inductive pulse sensor for online oil debris monitoring. *Tribol. Int.* **2011**, *44*, 175–179. [CrossRef]
17. Du, L.; Zhu, X.; Han, Y.; Zhan, L.; Zhe, J. Improving sensitivity of an inductive pulse sensor for detection of metallic wear debris in lubricants using parallel LC resonance method. *Meas. Sci. Technol.* **2013**, *24*, 075106. [CrossRef]
18. Zhang, X.; Zhang, H.; Sun, Y.; Chen, H.; Zhang, Y. Research on the output characteristics of microfluidic inductive sensor. *J. Nanomater.* **2014**, *2014*, 725246. [CrossRef]
19. Yu, Z.; Zhang, H.; Zeng, L.; Teng, H. Detection of metal particles based on micro-fluidic resonant chip. *J. Electron. Meas. Instrum.* **2017**, *31*, 1627–1632.
20. Zhang, X. Study on Metal Particle Magnetization in Harmonic Field. Ph.D. Thesis, Dalian Maritime University, Dalian, China, 2014.

micromachines

MDPI

Article

Adaptive Backstepping Design of a Microgyroscope

Yunmei Fang, Juntao Fei * and Yuzheng Yang

College of Electrical and Mechanical Engineering, Hohai University, Changzhou 213022, China;
yunmeif@163.com (Y.F.); smithcopy@163.com (Y.Y.)
* Correspondence: jtfei@hhu.edu.cn; Tel.: +86-519-8519-2023

Received: 14 May 2018; Accepted: 29 June 2018; Published: 3 July 2018

check for
updates

Abstract: This paper presents a novel algorithm for the design and analysis of an adaptive backstepping controller (ABC) for a microgyroscope. Firstly, Lagrange–Maxwell electromechanical equations are established to derive the dynamic model of a z-axis microgyroscope. Secondly, a nonlinear controller as a backstepping design approach is introduced and deployed in order to drive the trajectory tracking errors to converge to zero with asymptotic stability. Meanwhile, an adaptive estimator is developed and implemented with the backstepping controller to update the value of the parameter estimates in the Lyapunov framework in real-time. In addition, the unknown system parameters including the angular velocity may be estimated online if the persistent excitation (PE) requirement is met. A robust compensator is incorporated in the adaptive backstepping algorithm to suppress the parameter variations and external disturbances. Finally, simulation studies are conducted to prove the validity of the proposed ABC scheme with guaranteed asymptotic stability and excellent tracking performance, as well as consistent parameter estimates in the presence of model uncertainties and disturbances.

Keywords: adaptive control; backstepping approach; tracking performance; microgyroscope

---

1. Introduction

As primary information sensors, microgyroscopes have a large potential for several types of applications in navigation, control, and guidance systems. Fabrication imperfections in microgyroscopes always generate some coupling between oscillation modes. Meanwhile, the performance of the microgyroscope is subject to quadrature errors, time-varying parameters, and external disturbances. Nevertheless, recent applications require sensors with improved performance. The incorporation of advanced control systems into their existing dynamics seems to be an effective way to improve the microgyroscope performance.

During the past decades, many researchers have spent great deal of effort in the design of microgyroscope structures and control systems [1–17]. The conventional controller for a microgyroscope is to force the drive mode into a known oscillatory motion and then detect the Coriolis effect coupling along the orthogonal sense mode, which provides the information about the applied angular velocity. However, the conventional controllers are immanently sensitive to some typical types of fabrication imperfections, such as the cross-damping term, which produces zero-rate output. To solve these problems, advanced control schemes such as adaptive controller [2–5], sliding mode controller [6], compound robust controller [7], adaptive neural controller [8–10], and adaptive fuzzy controller [11–13] have been applied to microgyroscopes. A mode-matched force-rebalance control for a microgyroscope was investigated in [14]. Adaptive dynamic surface control for a triaxial microgyroscope with nonlinear inputs was developed in [15]. Flatness-based adaptive fuzzy control of an electrostatically actuated micro-electro-mechanical system (MEMS) and self-adaptive nonlinear stops for mechanical shock protection of MEMS were discussed in [16,17], respectively.

A backstepping controller [18] that can achieve the goals of tracking and stabilization is a recursive design procedure based on a Lyapunov framework, breaking a full system design into a sequence of lower-order systems. Nevertheless, compared with sliding mode control, the backstepping algorithm has two merits: the first is that it can relax the matching condition for a class of systems which can satisfy the strict feedback form; the second is that it can refrain from cancellation of the useful nonlinearities existing in the nonlinear system. The fundamental rule of backstepping is to recursively design a controller and step back out of the subsystem progressively, guaranteeing stability at each step, until reaching the final external control step. In [19,20], adaptive backstepping controllers were deployed for an air-breathing hypersonic vehicle and a fuel cell/boost converter system. A backstepping controller was applied to a linear 2 × 2 hyperbolic system in [21]. Adaptive intelligent control with backstepping design for dynamic systems were developed in [22–26]. Adaptive command-filtered backstepping control of robot arms with compliant actuators was introduced in [27]. However, so far, an adaptive backstepping controller has not been deployed to a microgyroscope. Based on our preliminary work in [28], our work will explore an adaptive backstepping scheme with a parameter estimator for a microgyroscope. Compared with existing works, the main contributions of the proposed backstepping approach are emphasized as:

(1) Backstepping is a nonlinear control approach based on Lyapunov stability theorem by means of recursion process. Backstepping design is a powerful tool for dynamic systems with pure or strict feedback forms. A major advantage of backstepping is that it has the flexibility to avoid cancellations of useful nonlinearities and achieve regulation and tracking properties. However, the vibratory microgyroscope is neither of these two forms. Therefore, the microgyroscope motion equations should be transformed into a cascade-like system to be suitable for the backstepping approach.

(2) An adaptive control strategy is deployed in the backstepping procedure to deal with parameter uncertainties and external disturbances. The Lyapunov-based adaptive controller is obtained to guarantee the asymptotic stability of the closed-loop system and the consistent parameter estimates, including the external angular velocity if the persistent excitation (PE) condition is satisfied. In addition, a robust term is incorporated in the adaptive backstepping algorithm to suppress the lumped disturbances.

2. Microgyroscope Dynamics

A z-axis vibratory microgyroscope mainly consists of three components: the sensitive element; electrostatic actuations and sensing mechanisms; and the rigid frame rotating along the rotation z-axis. Figure 1 shows a schematic diagram of a microgyroscope. The motion equations of the microgyroscope are developed from the Lagrange–Maxwell equation [1,2]:

$$\frac{\mathrm{d}}{\mathrm{d}t}\left(\frac{\partial L}{\partial \dot{x}_i}\right) - \frac{\partial L}{\partial x_i} + \frac{\partial F}{\partial \dot{x}_i} = Q_i, \tag{1}$$

where $L = E_K - E_P$ is Lagrange's function, E_K and E_P are kinetic and potential energies of the sensitive element, respectively, F is the generalized damping force, Q_i are generalized forces acting on the sensitive element, and i ranges from 1 corresponding to the number of considered degrees of freedom (2 in our system).

The motion equations can be obtained according to (1) and coordinate transformation knowledge. Assuming that the angular velocity is almost constant over a sufficiently long time interval, $\Omega_x \approx \Omega_y \approx 0$, only the component of the angular velocity Ω_z causes a dynamic coupling between the x-y axes. Considering fabrication imperfections, which cause extra coupling, the motion equations are obtained as:

$$\begin{cases} m\ddot{x} + d_{xx}\dot{x} + d_{xy}\dot{y} + k_{xx}x + k_{xy}y = u_x + d_x + 2m\Omega_z\dot{y} \\ m\ddot{y} + d_{xy}\dot{x} + d_{yy}\dot{y} + k_{xy}x + k_{yy}y = u_y + d_y - 2m\Omega_z\dot{x} \end{cases}, \tag{2}$$

where x and y are the coordinates regarding the gyro frame existing in Cartesian coordinates; m is the mass; $d_{xx}, d_{yy}, k_{xx}, k_{yy}$ are called the damping and spring coefficients; d_{xy}, k_{xy} are called quadrature errors, which are coupled damping and spring terms, respectively; u_x, u_y are called control forces; and d_x, d_y represent bounded unknown disturbances (note that the lumped disturbances d_x and d_y could also contain the effects of the time-varying unknown but bounded parameter uncertainties); and $2m\Omega_z\dot{y}, 2m\Omega_z\dot{x}$ are the Coriolis forces used to reconstruct the information of the unknown angular velocity Ω_z.

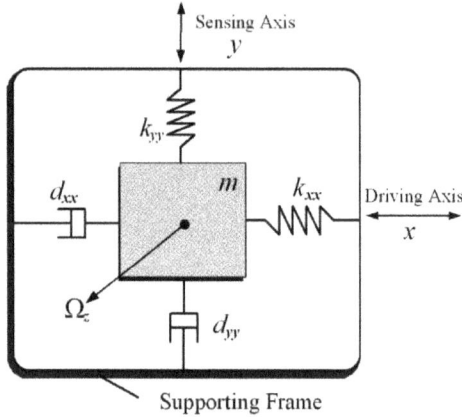

Figure 1. Schematic model of a z-axis MEMS vibratory gyroscope.

Dividing both sides of the motion Equation in (2) by reference mass m, reference length q_0, and natural resonance frequency ω_0^2, we get the non-dimensional equation as:

$$\ddot{x} + d_{xx}\dot{x} + d_{xy}\dot{y} + \omega_x^2 x + \omega_{xy}y = u_x + 2\Omega_z\dot{y} + d_x$$
$$\ddot{y} + d_{xy}\dot{x} + d_{yy}\dot{y} + \omega_{xy}x + \omega_y^2 y = u_y - 2\Omega_z\dot{x} + d_y \tag{3}$$

where $\dfrac{d_{xx}}{m\omega_0} \to d_{xx}$, $\dfrac{d_{xy}}{m\omega_0} \to d_{xy}$, $\dfrac{d_{yy}}{m\omega_0} \to d_{yxy}$, $\dfrac{\Omega_z}{\omega_0} \to \Omega_z$, $\sqrt{\dfrac{k_{xx}}{m\omega_0^2}} \to \omega_x$, $\sqrt{\dfrac{k_{yy}}{m\omega_0^2}} \to \omega_y$, $\dfrac{k_{xy}}{m\omega_0^2} \to \omega_{xy}$.

Equation (3) can be transformed into the vector form equation as:

$$\ddot{q} + D\dot{q} + Kq = u - 2\Omega\dot{q} + d, \tag{4}$$

where $q = \begin{bmatrix} x \\ y \end{bmatrix}$, $u = \begin{bmatrix} u_x \\ u_y \end{bmatrix}$, $d = \begin{bmatrix} d_x \\ d_y \end{bmatrix}$, $D = \begin{bmatrix} d_{xx} & d_{xy} \\ d_{xy} & d_{yy} \end{bmatrix}$, $K = \begin{bmatrix} k_{xx} & k_{xy} \\ k_{xy} & k_{yy} \end{bmatrix}$,

$\Omega = \begin{bmatrix} 0 & -\Omega_z \\ \Omega_z & 0 \end{bmatrix}$. Note that $D = D^T, K = K^T, \Omega = -\Omega^T$ and the input disturbances are assumed to be bounded by $\|d\| \le \rho$, where ρ is a scalar.

Considering a system with parametric uncertainties and external disturbances, the dynamics of the microgyroscope (4) can be represented as:

$$\ddot{q} + (D + 2\Omega + \Delta D)\dot{q} + (K + \Delta K)q = u + d, \tag{5}$$

where ΔD is the unknown parameter uncertainties of $D + 2\Omega$, and ΔK is the unknown parameter uncertainties of K.

Rewriting Equation (5) as

$$\ddot{q} + (D + 2\Omega)\dot{q} + Kq = u + d_f,\tag{6}$$

where $d_f = d - \Delta D\dot{q} - \Delta Kq$, representing the matched, lumped parametric uncertainties and external disturbances.

Despite these difficulties, an adaptive backstepping control (ABC) algorithm is deployed to guarantee the tracking performance, asymptotic stability, and parameter estimations of the microgyroscope system in the following section.

3. Adaptive Backstepping Control Design

Motivated by the research results in [18–22], a backstepping controller was to achieve the goals of tracking and stabilization by a recursive design procedure. We firstly show that if the parameters of the microgyroscope are known, the backstepping controller guarantees zero tracking error and asymptotic stability. Then, we will utilize an adaptive backstepping scheme to deal with the case of the unknown parameters. Figure 2 describes the block diagram of the proposed ABC approach of a microgyroscope.

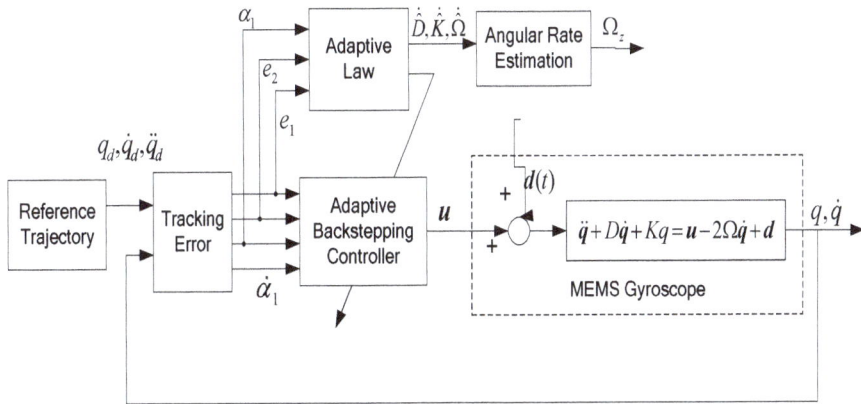

Figure 2. Block diagram of the proposed adaptive backstepping control of a microgyroscope.

As seen from Equation (3), since the coupled microgyroscope motion equation is not formulated in "strict-feedback" form, it should to be transformed into a form which could make backstepping design approach available. We define $X_1 = q$, $X_2 = \dot{q}$.

The dynamics in (3) can be transformed as the following cascade form:

$$\begin{cases} \dot{X}_1 = X_2 \\ \dot{X}_2 = -(D + 2\Omega)X_2 - KX_1 + u + d_f \end{cases}.\tag{7}$$

The control objective for a z-axis microgyroscope is to track a reference oscillation trajectory q_d as closely as possible and make all the signals in the closed-loop system be uniformly bounded. For the microgyroscope in (5), the backstepping control design can be synthesized in two steps.

Step 1: Treat X_2 as a virtual control force and design a control law for it to make X_1 follow the reference trajectory.

Firstly, the tracking error is defined as $e_1 = q - q_d = X_1 - q_d$, where q_d is the reference trajectory of q. Assume the first and second derivatives of the reference trajectory q_d are all bounded. Considering D, K, Ω are known, we treat X_2 as a control input and design a virtual controller α_1 for it such that

$\lim_{t \to \infty} q = q_d$ (i.e., $\lim_{t \to \infty} e_1(t) = 0$). To make the tracking error e_1 converge to zero, we study the dynamics of e_1 derived by differentiating the both sides of $e_1 = X_1 - q_d$, then we obtain $\dot{e}_1 = X_2 - \dot{q}_d$.

Now that X_2 is treated as a control input, we naturally design the following simple virtual control law for X_2 to make e_1 converge to zero exponentially:

$$X_2 = \alpha_1 \equiv -c_1 e_1 + \dot{q}_d, \tag{8}$$

where c_1 is a positive definite symmetric matrix.

With virtual control law (8), the dynamics of $\dot{e}_1 = X_2 - \dot{q}_d$ become

$$\dot{e}_1 = -c_1 e_1. \tag{9}$$

Due to the positive property of c_1, tracking error e_1 will approach zero exponentially. Roughly speaking, X_1 rapidly approximates to q_d.

Step 2: However, X_2 is not the actual control input, but a state variable. We cannot operate X_2 directly. So, let us move on to the second line of (5), which reveals the dynamics of X_2. We design the real control force to make X_2 converge to α_1.

Define e_2 as an error variable that is the deviation between X_2 and its virtual control law α_1, that is, $e_2 = X_2 - \alpha_1$.

We derive the dynamics of e_2 as

$$
\begin{aligned}
\dot{e}_2 &= \dot{X}_2 - \dot{\alpha}_1 \\
&= -(D + 2\Omega)(e_2 + \alpha_1) - K(e_1 + q_d) + u + d_f - \dot{\alpha}_1 \\
&= -(D + 2\Omega)e_2 - K(e_1 + q_d) - (D + 2\Omega)\alpha_1 - \dot{\alpha}_1 + u + d_f
\end{aligned}
\tag{10}
$$

In (10), the actual control u appears. Our target is to design u such that e_1, e_2 converge to zero. Select a Lyapunov function V for the whole system as:

$$V = \frac{1}{2}e_1{}^\mathrm{T}e_1 + \frac{1}{2}e_2{}^\mathrm{T}e_2. \tag{11}$$

Its first time derivative is given by:

$$
\begin{aligned}
\dot{V} &= e_1{}^\mathrm{T}\dot{e}_1 + e_2{}^\mathrm{T}\dot{e}_2 = e_1{}^\mathrm{T}(X_2 - \dot{q}_d) + e_2{}^\mathrm{T}\dot{e}_2 \\
&= e_1{}^\mathrm{T}(-c_1 e_1 + e_2) + e_2{}^\mathrm{T}[-(D + 2\Omega)e_2 - K(e_1 + q_d) - (D + 2\Omega)\alpha_1 - \dot{\alpha}_1 + u + d_f]
\end{aligned}
\tag{12}
$$

We finally derive and design the real controller u. $\dot{V}$ must satisfy $\dot{V} \leq 0$. Some terms in (13) are definitely negative, and we shall keep them. Some terms are positive or indefinite, and we will use the control force to cancel them. Thus, we design the control effort as:

$$u = -c_2 e_2 - e_1 + (D + 2\Omega)e_2 + K(e_1 + q_d) + (D + 2\Omega)\alpha_1 + \dot{\alpha}_1 - \rho\mathrm{sgn}(e_2), \tag{13}$$

where c_2 is a positive, definite, and symmetric matrix. The last term $-\rho\mathrm{sgn}(e_2)$ in (15) is a robust compensator for the parameter variations and external disturbances.

Substituting Equation (13) into Equation (12) generates

$$\dot{V} = -e_1{}^\mathrm{T}c_1 e_1 - e_2{}^\mathrm{T}c_2 e_2 + e_2{}^\mathrm{T}d_f - \rho e_2{}^\mathrm{T}\mathrm{sgn}(e_2) \leq 0. \tag{14}$$

Because $-e_1{}^\mathrm{T}c_1 e_1 \leq 0$, $-e_2{}^\mathrm{T}c_2 e_2 \leq 0$, and $e_2{}^\mathrm{T}d_f - \rho e_2{}^\mathrm{T}\mathrm{sgn}(e_2) \leq \|e_2\|_1\|d_f\|_1 - \rho\|e_2\|_1 \leq 0$, $\dot{V}$ coincides with zero if and only if the three terms are simultaneously equal to zero. Because of c_1 and c_2 being symmetric positive definite matrices, both $-e_1{}^\mathrm{T}c_1 e_1$ and $-e_2{}^\mathrm{T}c_2 e_2$ equal to zero if and only if $e_1 = 0$ and $e_2 = 0$. Therefore, $\dot{V} = 0$ contains no trajectories other than $[e_1^\mathrm{T}, e_2^\mathrm{T}]^\mathrm{T} = 0$. According to

Lasalle's invariance principle, the origin zero is globally asymptotically stable. Then, $e_1, e_2 \to 0$ as $t \to \infty$.

4. Adaptive Estimator

In the following, we will develop the procedure to deal with unknown system dynamics, lumped parametric uncertainties, and disturbances. The modified controller in (13) is

$$u = -c_2 e_2 - e_1 + \hat{D}(e_2 + \alpha_1) + \hat{K}(e_1 + q_d) + \hat{\Omega}(2e_2 + 2\alpha_1) + \dot{\alpha}_1 - \rho \text{sgn}(e_2), \tag{15}$$

where $\hat{D}, \hat{K}$ and $\hat{\Omega}$ are the estimates of D, K and Ω, respectively. Regarding the characteristics and performance of the proposed ABC strategy, we state the following theorem.

Theorem 1. *In the presence of lumped disturbances d_f, the adaptive controller (15) with the adaptive estimator (16) applied to the microgyroscope model (3) guarantees that all the closed-loop signals are bounded and that state tracking errors converge to zero asymptotically.*

$$\begin{aligned}
\dot{\hat{D}}^T &= -\frac{1}{2}\gamma_D[(e_2 + \alpha_1)e_2^T + e_2(e_2 + \alpha_1)^T] \\
\dot{\hat{K}}^T &= -\frac{1}{2}\gamma_K[(e_1 + q_d)e_2^T + e_2(e_1 + q_d)^T] \\
\dot{\hat{\Omega}}^T &= \gamma_\Omega[e_2(e_2 + \alpha_1)^T - (e_2 + \alpha_1)e_2^T]
\end{aligned} \tag{16}$$

where $\gamma_D > 0, \gamma_K > 0, \gamma_\Omega > 0$.

Proof. Substituting (16) into (5) yields

$$\begin{cases} \dot{e}_1 = e_2 + \alpha_1 - \dot{q}_d \\ \dot{e}_2 = [-c_2 e_2 - e_1 + d_f - \rho\text{sgn}(e_2)] + \tilde{D}(e_2 + \alpha_1) + \tilde{K}(e_1 + q_d) + \tilde{\Omega}(2e_2 + 2\alpha_1) \end{cases}, \tag{17}$$

where $\tilde{D} = \hat{D} - D, \tilde{K} = \hat{K} - K, \tilde{\Omega} = \hat{\Omega} - \Omega$, represent the estimation errors.

Consider the Lyapunov function candidate as the form of (18):

$$V = \frac{1}{2}e_1^T e_1 + \frac{1}{2}e_2^T e_2 + \frac{1}{2}\text{tr}\{\gamma_D^{-1}\tilde{D}\tilde{D}^T\} + \frac{1}{2}\text{tr}\{\gamma_K^{-1}\tilde{K}\tilde{K}^T\} + \frac{1}{2}\text{tr}\{\gamma_\Omega^{-1}\tilde{\Omega}\tilde{\Omega}^T\}, \tag{18}$$

where $\text{tr}\{\cdot\}$ is the matrix trace operator.
Differentiating (18) generates

$$\begin{aligned}
\dot{V} &= [-e_1^T c_1 e_1 - e_2^T c_2 e_2 + e_2^T d_f - \rho e_2^T \text{sgn}(e_2)] \\
&+ e_2^T[\tilde{D}(e_2 + \alpha_1) + \tilde{K}(e_1 + q_d) + \tilde{\Omega}(2e_2 + 2\alpha_1)] \\
&+ \text{tr}\{\gamma_D^{-1}\tilde{D}\dot{\tilde{D}}^T\} + \text{tr}\{\gamma_K^{-1}\tilde{K}\dot{\tilde{K}}^T\} + \text{tr}\{\gamma_\Omega^{-1}\tilde{\Omega}\dot{\tilde{\Omega}}^T\}
\end{aligned} \tag{19}$$

Substituting the adaptive estimator (16) into (19), and $\dot{\tilde{D}} = \dot{\hat{D}}^T, \dot{\tilde{K}} = \dot{\hat{K}}^T, \dot{\tilde{\Omega}} = -\dot{\hat{\Omega}}^T$, we obtain

$$\dot{V} = -e_1^T c_1 e_1 - e_2^T c_2 e_2 + e_2^T d_f - \rho e_2^T \text{sgn}(e_2) \leq 0. \tag{20}$$

Note that (20) and (14) are identical. Thus, e_1 and e_2 converge to zero asymptotically. The adaptive laws that guarantee the tracking error converges to zero do not mean the parameter estimates are consistent only if the PE condition can be satisfied. Since the reference trajectories contain two distinct nonzero frequencies, the PE condition is satisfied, and the microgyroscope has sufficient persistence of excitation to permit the accurate identification of major fabrication imperfections and all the unknown system parameters.

5. Simulation Study

The proposed ABC scheme was evaluated on a lumped z-axis microgyroscope sensor [1,2]. The physical parameters are described as:

$$m = 1.8 \times 10^{-7} \text{kg}, k_{xx} = 63.955 \frac{\text{N}}{\text{m}}, k_{yy} = 95.92 \frac{\text{N}}{\text{m}}, k_{xy} = 12.776 \frac{\text{N}}{\text{m}}$$

$$d_{xx} = 1.8 \times 10^{-6} \frac{\text{N} \cdot \text{s}}{\text{m}}, d_{yy} = 1.8 \times 10^{-6} \frac{\text{N} \cdot \text{s}}{\text{m}}, d_{xy} = 3.6 \times 10^{-7} \frac{\text{N} \cdot \text{s}}{\text{m}}$$

We chose 1 μm as the reference length q_0. It is known that the usual natural frequency of a microgyroscope is in the kHz range, so chose the ω_0 as 1 kHz. Assume the unknown angular velocity is $\Omega_z = 10 \, \text{rad/s}$. Non-dimensionalizing the physical parameters, we obtained the following nondimensional parameter matrices defined in (3):

$$D = \begin{bmatrix} 0.01 & 0.002 \\ 0.002 & 0.01 \end{bmatrix}, K = \begin{bmatrix} 355.3 & 70.99 \\ 70.99 & 532.9 \end{bmatrix}, \Omega = \begin{bmatrix} 0 & -0.01 \\ 0.01 & 0 \end{bmatrix}.$$

The desired trajectory should be the resonance of vibration modes. The reference trajectories were selected as $x_d = \cos(\omega_1 t), y_d = \cos(\omega_2 t)$, where $\omega_1 = 6.17, \omega_2 = 5.11$. Here ω_1, ω_2 were chosen to be the resonance frequencies of the z-axis MEMS vibratory gyroscope. We assumed that ω_1, ω_2 were fixed in the simulation period.

The lumped parametric uncertainties and external disturbances are given by $d_f = d - \Delta D\dot{q} - \Delta Kq$. As for model uncertainties, there were $\pm 20\%$ parameter variations for the spring and damping coefficients and $\pm 20\%$ magnitude changes in the coupling terms. Random signal $d = \begin{bmatrix} randn(1,1) & randn(1,1) \end{bmatrix}$ was considered as disturbance.

Let D_0, K_0 and Ω_0 to be the nominal values of D, K and Ω, respectively. Figure 3 shows the tracking error using a "dull" controller without any adaptation strategies by solely replacing D, K, Ω in (15) with D_0, K_0, Ω_0. The control parameters are $c_1 = c_2 = 20I$, where I is the unit matrix. For the moment, there is no disturbance. It must be noted that all of the system parameters, including the gyroscope, controller, and disturbance parameters are nondimensional herein, meaning that all of the parameters on vertical axes in the following figures are unitless. The simulation time was nondimensional, as were the simulation positions. Though they were nondimensional, the same class of parameters could be compared with each other, due to the unified reference physical quantity.

From Figure 3, due to the modeling error, the "dull" controller which relied on the nominal parameters led to a stable system, but the tracking errors were obvious. For comparison, Figure 4 depicts the tracking error using the proposed ABC approach, and Figure 5 shows the adaptation procedure of the parameter estimates. Figure 6 plots the control forces for the microgyroscope.

Obviously different from the result depicted in Figure 3, tracking errors approached zero quickly when using the proposed ABC scheme. Since the reference trajectories contained two different nonzero frequencies, the PE condition was satisfied. In Figure 5, the parameter estimates converged to their true values, including the angular velocity. Standard adaptive controllers are not always robust in the presence of model uncertainties and external disturbances. Hence, if $-\rho \text{sgn}(e_2)$ in (13) was relieved, our proposed control would not perform that well. For example, a step signal with an amplitude of 100 was added at 20 s as an external disturbance. Figure 7 shows the tracking errors using the adaptive controller without the robust term. Figure 8 exhibits the improvement of tracking errors using our proposed controller with the robust term $-\rho \text{sgn}(e_2)$. Comparing Figure 7 with Figure 8, the robust term effectively suppressed the disturbances and the tracking error maintained a very small value.

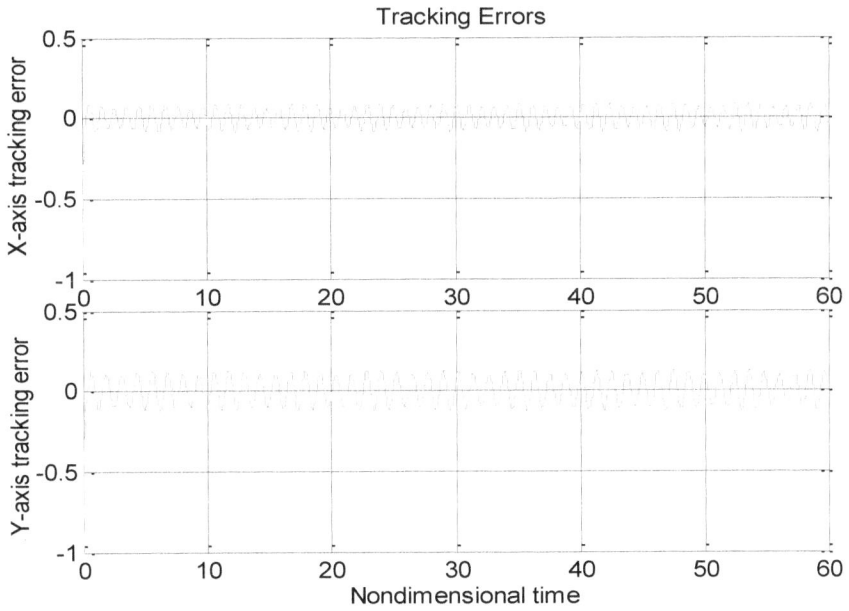

Figure 3. Tracking errors using a "dull" controller.

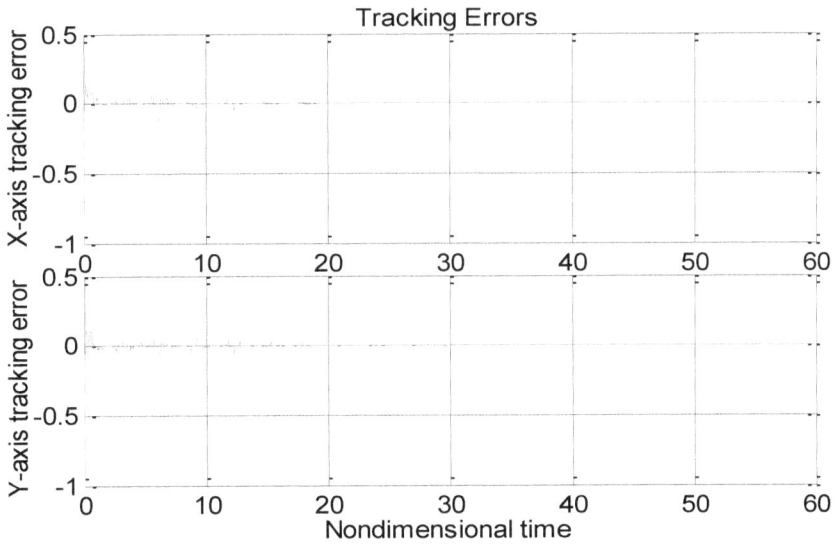

Figure 4. Tracking errors using the adaptive backstepping control (ABC) approach.

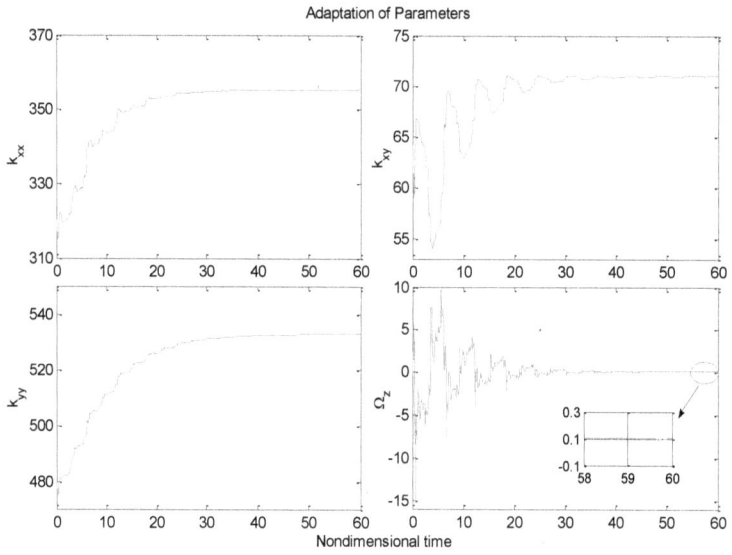

Figure 5. Adaptive parameter estimates using ABC.

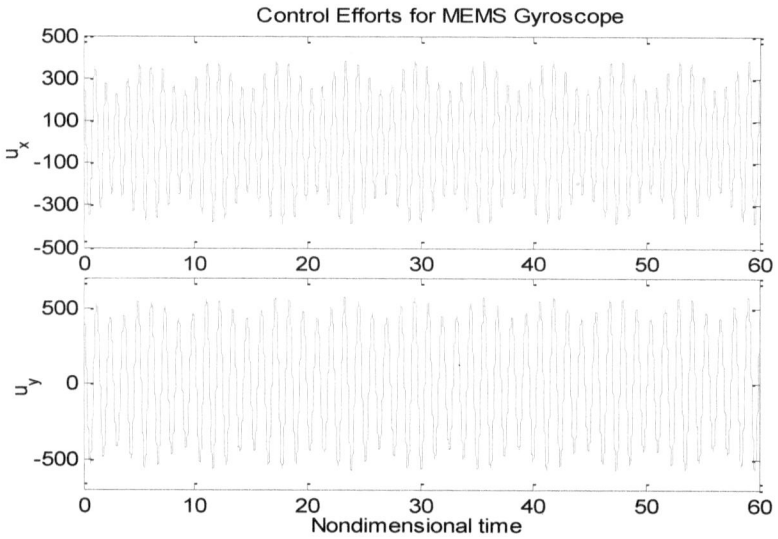

Figure 6. Control efforts for microgyroscope using ABC.

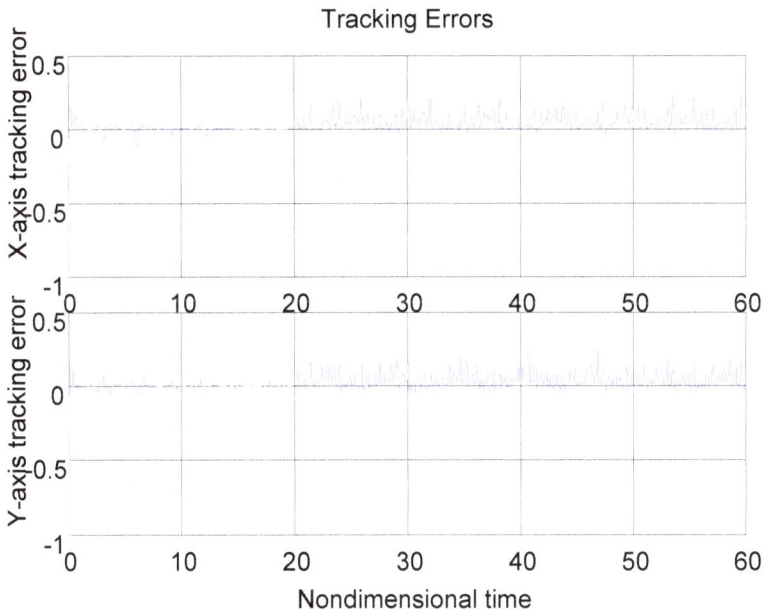

Figure 7. Tracking errors using ABC under step disturbances without robust term.

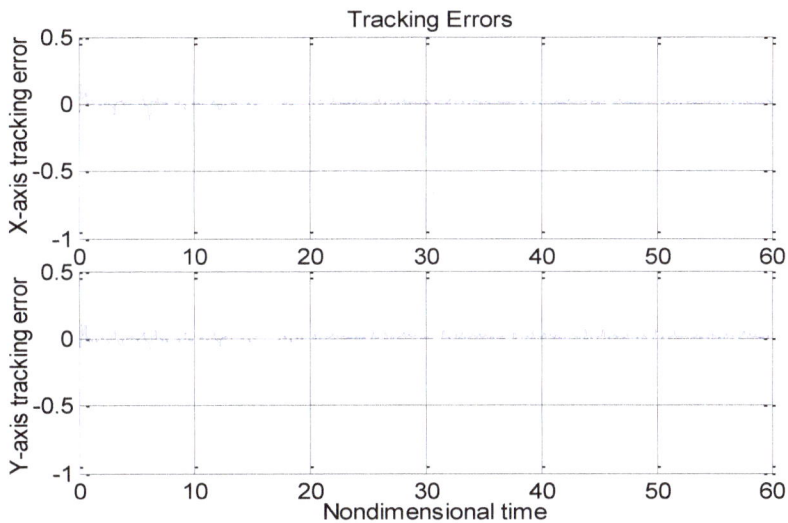

Figure 8. Tracking errors using ABC under step disturbances with robust term.

A well-known adaptive microgyroscope controller without the backstepping technique was presented in [2] by Park. The performance of our proposed ABC strategy was compared with the adaptive controller in [2]. Figures 9–11 show the dynamic response using the adaptive controller in [2] with the same nominal gyroscope parameters under the same model uncertainties and disturbances.

The tracking errors with the adaptive controller displayed quite a large overshot at the beginning, as did the control efforts. The settling time of tracking errors was also worse than our proposed adaptive backstepping controller. The advantage of our proposed controller over the adaptive controller in the performance of parameter estimation is clear. Put simply, the proposed adaptive backstepping controller could improve the dynamic and static performance of the microgyroscope.

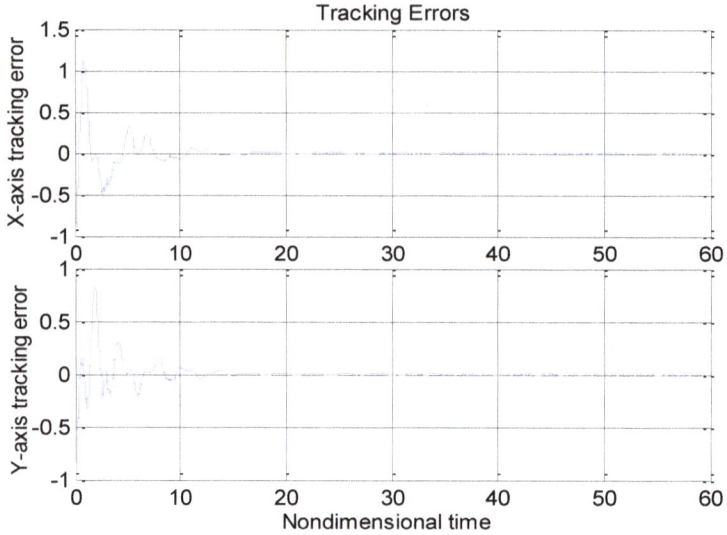

Figure 9. Tracking errors using the adaptive controller in [2].

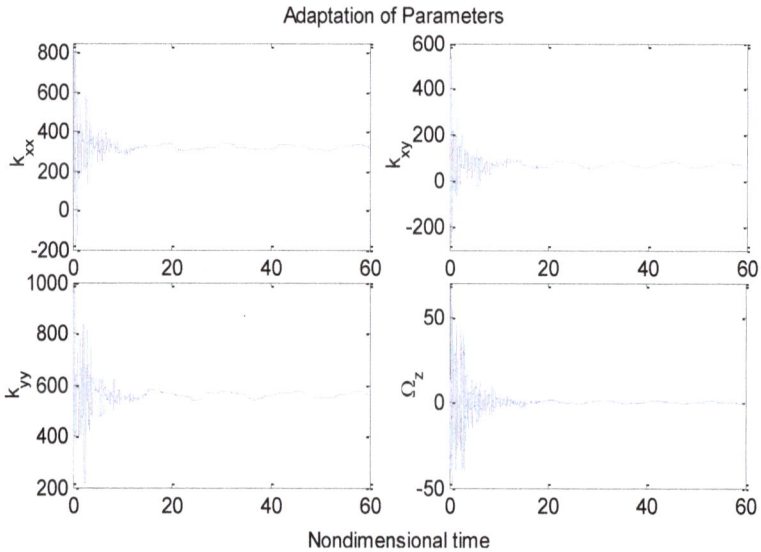

Figure 10. Adaptation of parameter estimates using the adaptive controller in [2].

Micromachines **2018**, *9*, 338

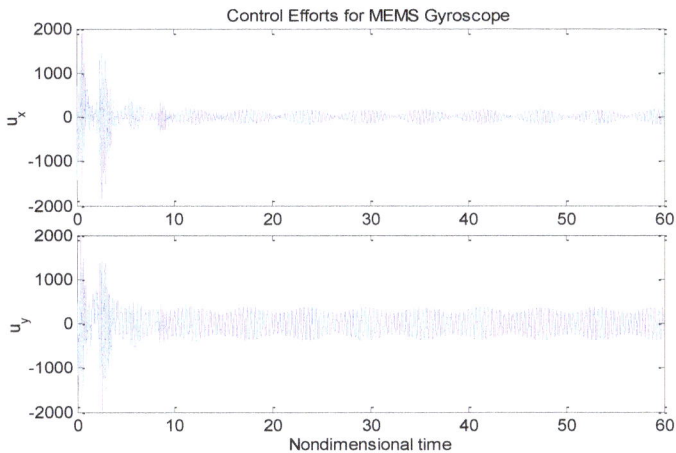

Figure 11. Control efforts for a microgyroscope using the adaptive controller in [2].

6. Conclusions

An adaptive control with backstepping technique for a z-axis microgyroscope was investigated and analyzed. The dynamics model of the microgyroscope was developed and transformed to aid in the backstepping control design. A backstepping approach and adaptive strategy were utilized to deal with the model uncertainties, disturbances, and unknown parameters of the microgyroscope. A controller was designed to recursively and progressively step back out of the subsystem, guaranteeing stability at each step until reaching the final external control step. Consistent parameter estimates, asymptotic stability, and tracking performance under the lumped disturbances were proved based on a Lyapunov analysis. Numerical simulation examples demonstrated the validity of the proposed ABC scheme, showing the improved performance and consistent parameter estimation.

In our study, we only emphasized the proposed adaptive backstepping control algorithm on the microgyroscope model. In the next step, the proposed adaptive backstepping controller should be implemented in a practical experimental system to verify its effectiveness.

Author Contributions: Conceptualization, J.F.; Methodology, Y.F. and J.F.; Software, Y.Y.; Validation, Y.Y.; Formal Analysis, Y.F.; Investigation, Y.F.; Resources, Y.F.; Data Curation, Y.Y.; Writing-Original Draft Preparation, Y.F.; Writing-Review & Editing, J.F.; Visualization, Y.F.; Supervision, J.F.; Project Administration, J.F.; Funding Acquisition, J.F.

Funding: This work was supported by Natural Science Foundation of Jiangsu Province under Grant No. BK20171198; The Fundamental Research Funds for the Central Universities under Grant No. 2017B20014. 2017B21214.

Acknowledgments: The authors thank the anonymous reviewers for their useful comments that improved the quality of the paper.

Conflicts of Interest: The authors declare no conflict of interest.

References

1. Apostolyuk, V. *Theory and Design of Micromechanical Vibratory Gyroscopes, MEMS/NEMS*; Springer: Berlin, Germany, 2006.
2. Park, S. Adaptive control of a vibratory angle measuring gyroscope. *Sensors* **2010**, *10*, 8478–8490. [CrossRef] [PubMed]
3. Fei, J.; Lu, C. Adaptive sliding mode control of dynamic systems using double loop recurrent neural network structure. *IEEE Trans. Neural Netw. Learn. Syst.* **2018**, *29*, 1275–1286. [CrossRef] [PubMed]
4. Xia, D.; Hu, Y.; Ni, P. A digitalized gyroscope system based on a modified adaptive control method. *Sensors* **2016**, *16*, 321. [CrossRef] [PubMed]

5. Fei, J.; Yan, W. Adaptive control of MEMS gyroscope using global fast terminal sliding mode control and fuzzy-neural-network. *Nonlinear Dyn.* **2014**, *78*, 103–116. [CrossRef]

6. Zhang, R.; Shao, T.; Zhao, W. Sliding mode control of MEMS gyroscopes using composite learning. *Neurocomputing* **2018**, *275*, 2555–2564. [CrossRef]

7. Rahmani, M. MEMS gyroscope control using a novel compound robust control. *ISA Trans.* **2018**, *72*, 37–43. [CrossRef] [PubMed]

8. Fei, J.; Lu, C. Adaptive fractional order sliding mode controller with neural estimator. *J. Frankl. Inst.* **2018**, *355*, 2369–2391. [CrossRef]

9. Fei, J.; Ding, H. Adaptive sliding mode control of dynamic system using RBF neural network. *Nonlinear Dyn.* **2012**, *70*, 1563–1573. [CrossRef]

10. Xu, B.; Zhang, P. Minimal-learning-parameter technique based adaptive neural sliding mode control of MEMS gyroscope. *Complexity* **2017**, *2017*, 1–8. [CrossRef]

11. Fei, J.; Zhou, J. Robust adaptive control of MEMS triaxial gyroscope using fuzzy compensator. *IEEE Trans. Syst. Man Cybern. Part B Cybern.* **2012**, *42*, 1599–1607.

12. Fei, J.; Xin, M. An adaptive fuzzy sliding mode controller for MEMS triaxial gyroscope with angular velocity estimation. *Nonlinear Dyn.* **2012**, *70*, 97–109. [CrossRef]

13. He, C.; Zhang, J.; Zhao, Q. An electrical-coupling-suppressing MEMS gyroscope with feed-forward coupling compensation and scalable fuzzy control. *Sci. China Inf. Sci.* **2017**, *60*, 042402. [CrossRef]

14. Hu, Z.; Gallacher, B. A mode-matched force-rebalance control for a MEMS vibratory gyroscope. *Sens. Actuators A-Phys.* **2018**, *273*, 1–11. [CrossRef]

15. Song, Z.; Li, H.; Sun, K. Adaptive dynamic surface control for MEMS triaxial gyroscope with nonlinear inputs. *Nonlinear Dyn.* **2014**, *78*, 173–182. [CrossRef]

16. Rigatos, G.; Zhu, G.; Yousef, H. Flatness-based adaptive fuzzy control of electrostatically actuated MEMS using output feedback. *Fuzzy Sets Syst.* **2016**, *290*, 138–157. [CrossRef]

17. Xu, K.; Jiang, F.; Zhang, W. Micromachined integrated self-adaptive nonlinear stops for mechanical shock protection of MEMS. *J. Micromech. Microeng.* **2018**, *28*, 064006. [CrossRef]

18. Krstic, M.; Kanellakopoulos, I.; Kokotovic, P. *Nonlinear and Adaptive Control Design*; John Willey & Sons, Inc.: Hoboken, NJ, USA, 1995.

19. Hu, Q.; Meng, Y. Adaptive backstepping control for air-breathing hypersonic vehicle with actuator dynamics. *Aerosp. Sci. Technol.* **2017**, *67*, 412–421. [CrossRef]

20. Zuniga, Y.; Langarica, D.; Leyva, J. Adaptive backstepping control for a fuel cell/boost converter system. *IEEE J. Emerg. Sel. Top. Power Electron.* **2018**, *6*, 686–695. [CrossRef]

21. Deutscher, J. Backstepping design of robust state feedback regulators for linear 2×2 hyperbolic systems. *IEEE Trans. Autom. Control* **2017**, *62*, 5240–5247. [CrossRef]

22. Fang, Y.; Fei, J.; Hu, T. Adaptive backstepping fuzzy sliding mode vibration control of flexible structure. *J. Low Freq. Noise Vib. Act. Control* **2018**, 1–18. [CrossRef]

23. Fei, J.; Wang, T. Adaptive fuzzy-neural-network based on RBFNN control for active power filter. *Int. J. Mach. Learn. Cybern.* **2018**, 1–12. [CrossRef]

24. Chu, Y.; Fei, J. Dynamic global PID sliding mode control using RBF neural compensator for three-phase active power filter. *Trans. Inst. Meas. Control* **2017**. [CrossRef]

25. Liu, Y.; Gong, M.; Tong, S. Adaptive fuzzy output feedback control for a class of nonlinear systems with full state constraints. *IEEE Trans. Fuzzy Syst.* 2018. [CrossRef]

26. Wu, N.; Feng, S. Mixed fuzzy/boundary control design for nonlinear coupled systems of ODE and boundary-disturbed uncertain beam. *IEEE Trans. Fuzzy Syst.* **2018**. [CrossRef]

27. Pan, Y.; Wang, H.; Li, X. Adaptive command-filtered backstepping control of robot arms with compliant actuators. *IEEE Trans. Control Syst. Technol.* **2018**, *26*, 1149–1156. [CrossRef]

28. Fang, Y.; Fei, J.; Yang, Y. Adaptive control of MEMS gyroscope using backstepping approach. In Proceedings of the 2014 14th International Conference on Control, Automation and Systems, Kintex, Korea, 22–25 October 2014; pp. 361–366.

micromachines

MDPI

Article

High Temperature AlGaN/GaN Membrane Based Pressure Sensors

Durga Gajula [1,*] ⬤**, Ifat Jahangir** [2] **and Goutam Koley** [1]

[1] Holcombe Department of Electrical and Computer Engineering, Clemson University,
   Anderson, SC 29625, USA; gkoley@clemson.edu
[2] Department of Electrical Engineering, University of South Carolina, Columbia, SC 29208, USA;
   ifat00@gmail.com
* Correspondence: gdraophy@gmail.com; Tel.: +1-864-656-2201

Received: 9 April 2018; Accepted: 25 April 2018; Published: 28 April 2018

✓ check for
updates

Abstract: A highly sensitive Gallium Nitride (GaN) diaphragm based micro-scale pressure sensor with an AlGaN/GaN heterostructure field effect transistor (HFET) deflection transducer has been designed and fabricated for high temperature applications. The performance of the pressure sensor was studied over a pressure range of 20 kPa, which resulted in an ultra-high sensitivity of ~0.76%/kPa, with a signal-to-noise ratio as high as 16 dB, when biased optimally in the subthreshold region. A high gauge factor of 260 was determined from strain distribution in the sensor membrane obtained from finite element simulations. A repeatable sensor performance was observed over multiple pressure cycles up to a temperature of 200 °C.

Keywords: MEMS; high temperature pressure sensors; AlGaN/GaN circular HFETs; GaN diaphragm

---

1. Introduction

In harsh environments, such as in the aerospace, automotive, nuclear power and petroleum industries, there is a great need for high temperature pressure sensors [1–3]. Silicon electrical properties degrade with the temperatures above 150 °C, due to the generation of thermal carriers and high leakage currents, which makes it less suitable for harsh environments. The corrosion resistance of silicon is also limited at high temperatures. This moved the researcher's interest to higher bang-gap materials like SiC, AlN, Gallium Nitride (GaN) and so forth [4–11]. Due to their higher band gap, these materials have excellent thermal stability at higher temperatures. Among these, AlGaN got a special interest due to its excellent piezo-electric properties [12]. It is also chemically inert, mechanically stable and radiation hardened, which makes it a promising device material for hostile environments. GaN layer is highly piezo electric and AlGaN/GaN heterostructure have a spontaneous polarization at the interface, which creates a 2DEG (two-dimensional electron gas) at the interface [13]. This 2DEG offers a great opportunity for using AlGaN/GaN in piezo-resistive and piezo-electric transducers, since both the 2DEG density and the mobility of the 2DEG can be modulated with the strain. The sensitivity for the applied strain for AlGaN/GaN transducers will have higher values than those of silicon, where only the carrier mobility gets modulated with the applied strain. These characteristics of AlGaN/GaN interface make them better suitable material for the sensing applications over silicon [7–15]. Silicon carbide based piezo resistive pressure sensors at temperatures up to 600 °C have been studied, however these devices have low output signals and low pressure sensitivity values [16–18]. Capacitive sensing is a dominant technique in pressure sensing, however the motion of the sensor is constrained to a small vertical and horizontal movements [19]. If the vertical displacement is large, the capacitance are not suitable for pressure sensing. So, the interest has been shifted to diaphragm based pressure sensors, where the sensors motion is dependent on the yield strength of the material than on the design constraints. The theoretical temperature limit of Gallium Nitride (GaN) can be as high as 600 °C and GaN heterostructure also shown higher mobility values and high critical breakdowns [20]. The III-V nitrides is also having a high potential for monolithic integration [21]. There have been various studies on the AlGaN/GaN based pressure sensors for room temperature sensing applications [22,23]. In this article, we have investigated high temperature pressure sensing behavior of AlGaN/GaN based devices and its electrical properties with the applied pressure have been studied with the temperature. The mechanical stress distribution across the circumference of the diaphragm with pressure will provide the change in piezoelectric charge in GaN HFET (hetero-structure field-effect transistor) and so thus the change in source-drain resistance. The pressure sensitivity values are significantly better than existing technologies, which underscores the prospect of these devices for high temperature pressure sensing applications.

2. Materials and Methods

The pressure transducers used in this study were fabricated on AlGaN/GaN epitaxial layers on (111) silicon wafer, purchased from NTT Advanced Technology Corporation, Japan. The wafer had a 2 nm i-GaN cap layer and 15 nm $Al_{0.25}Ga_{0.75}N$ on top of 1 μm i-GaN, with a 300 nm buffer layer separating the GaN layer from the 675–750 μm thick Si substrate. At the beginning of the fabrication process, the top 100 nm of AlGaN/GaN layer was etched using BCl_3/Cl_2 plasma chemistry to define the mesa region at the periphery of the diaphragm, followed by deposition of a Ti (20 nm)/Al (100 nm)/Ti (45 nm)/Au (55 nm) metal stack. A rapid thermal annealing process was performed at 825 °C for a minute, to form ohmic contacts for the source and drain regions for the HFET. After that, plasma enhanced chemical vapor deposition (PECVD) technique was used to deposit 100 nm thick SiO_2 to cover the open regions of the mesa, which served as the gate dielectric. This was followed by two consecutive stages of metallization, the first one had Ni (25 nm)/Au (200 nm) stack as the gate metal contacts and the second one had Ti (20)/Au (225 nm) stack to from the probe contacts. Finally Bosch process was used from the bottom face of the sample to perform through wafer etching of silicon to release the diaphragm. Figure 1a shows the schematic diagram of the pressures sensor (Appendix A: Figure A1 represents the diagrammatic representation of the process flow the fabrication of these pressure sensors). Figure 1b,c shows the scanning electron microscopic (SEM) images of a diaphragm from the topside and the backside of the sample.

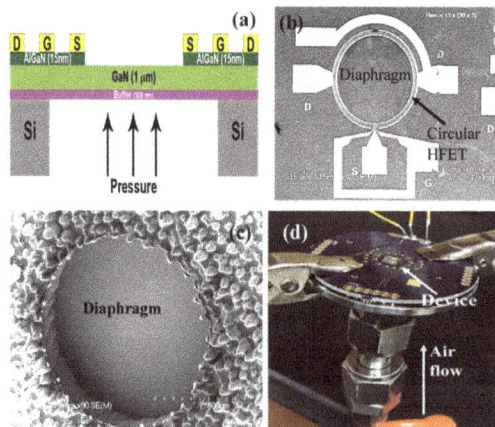

Figure 1. (**a**) Cross-sectional structure of AlGaN/GaN heterostructure field effect transistor (HFET) pressure sensor; (**b**) Diaphragm based AlGaN/GaN HFET pressure sensor, with a radius of 1000 μm diaphragm; (**c**) The back side of the diaphragm, where the pressure was applied; (**d**) the experimental set-up for pressure sensing.

3. Results and Discussion

The sheet resistance of the AlGaN layer and the contact resistance of the ohmic metal pads were found out to be 316 Ω/cm^2 and 19 Ω/cm^2 respectively, measured using transmission line measurement (TLM) technique. The transfer length, L_T, was found to be 18 μm and contact resistivity was calculated as $\rho_c = (L_T)^2 \times R_{sheet} = 0.57$ $\Omega \cdot$cm^2. However, due to the device contacts being of shorter length (10 μm) than the L_T, the estimated total contact resistance for both contacts is about 8.6 Ω, while the overall channel resistance is about 2.6 Ω, which means a significant amount of voltage (~77%) is dropped at the contacts, which needs to be accounted for in the calculations that follow.

For electrical and pressure transduction measurements, the HFET embedded diaphragms were glued to a printed circuit board (PCB) with a small pinhole in such a way that the pinhole is aligned with the diaphragm and the glue forms a vacuum seal. The contact pads were wire bonded and the PCB was mounted on a high pressure fixture shown in Figure 1d, where a gas line was installed beneath the diaphragm with high temperature O-rings for good vacuum sealing. A heater and a thermo-couple were also placed on top of the PCB, near the transducer chip, to carry out the experiments at higher temperatures. At first, we performed transistor measurements to calculate field effect mobility, given by

$$\mu_{FET} = \frac{g_m \times l_g}{q \times V_{ds,e} \times C_g} \tag{1}$$

where g_m is the drain-source transconductance, q is the electron charge, l_g is the gate length of the channel, C_g is the gate capacitance, $V_{ds,e}$ is the effective voltage drop at the channel (across the intrinsic transistor), estimated as a fraction of applied V_{ds} from the TLM measurements. Since the series combination of the oxide capacitance and the capacitance of the top AlGaN layers dominate the overall capacitance, we consider this constant capacitance as C_g, without performing a full gate capacitance-voltage (C-V) measurement.

Figure 2a shows the I_{ds}-V_{ds} characteristics the HFET at room temperature (RT). Here the non-linearity observed at the low bias range can be attributed to the non-ideality of the contacts that was evident from the relatively higher contact resistivity. The current eventually saturates or reduces at different V_{ds} depending on the V_{gs}, which is expected from a well-behaved HFET with high current driving capacity. Figure 2b shows the variation of I_{ds} at $V_{gs} = 0$ V as a function of temperature. As the temperature goes up, the peak Ids goes down and so does the corresponding V_{ds}, as the channel resistance increases as a result of increased scattering. However, at low V_{ds} (<1 V), the I_{ds}-V_{ds} curves become more linear at higher temperatures, which can be attributed to the improved thermionic emission at the non-ideal contacts. That is why we calculate the field effect mobility at a higher V_{ds} (~V), as it is less sensitive to the Schottky-like behavior at the lower V_{ds}. Figure 2c shows the I_{ds}-V_{gs} characteristics of the HFET measured at different temperatures and at $V_{ds} = 1$ V. The peak transconductance at room temperature was ~15 mA/V, which resulted in a raw value of $\mu_{FET} = $ ~300 cm^2/(V·s) and sheet carrier concentration $n_s = 7.19 \times 10^{11}$ cm^{-2} at $V_{gs} = -2$ V and 2.35×10^{11} cm^{-2} at $V_{gs} = -7$ V. This low mobility is likely attributable to significant voltage drop across the drain and source contact resistances due to their sub-optimal width (< transfer length) caused by spatial limitation. The true value of the mobility lies between these two extreme cases. The reduction of mobility at RT is generally associated with the enhanced defect scattering in HFET and significant carrier trapping at the 2DEG surface [24].

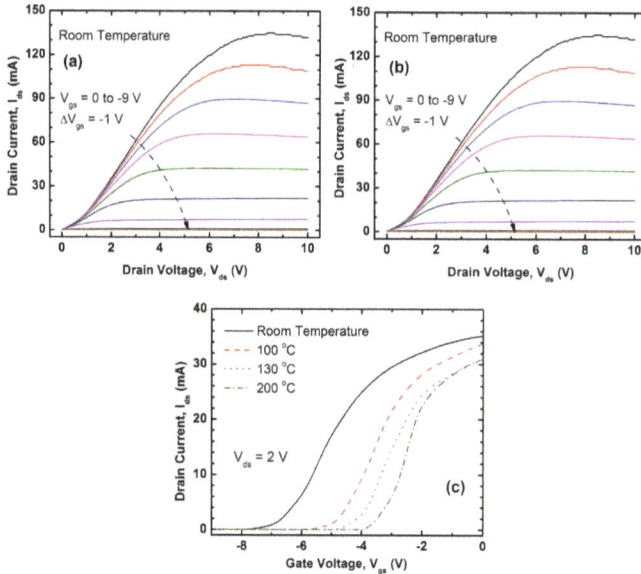

Figure 2. (a) I_{ds}-V_{ds} characteristics of the circular AlGaN/GaN HFET at room temperature; (b) variation of I_{ds} at $V_{gs} = 0$ V as a function of temperature and (c) I_{ds}-V_{gs} characteristics of the HFET measured at different temperatures and at $V_{ds} = 1$ V.

From Figure 2c, we observe that the threshold and turn off voltages of the HFET were reduced with the increase of temperature. The threshold voltages of HFET at RT, 100 °C, 130 °C and 200 °C are approximately -7.5 V, -5.7 V, -4.9 V and -4 V respectively. This is in good agreement with the study by Alim et al. [25] on the variation of the threshold with the temperature, where they also noticed that the positive shift in the Schottky barrier height along with trap-assisted phenomena shifted the threshold voltage towards positive values as the temperature was increased. At higher temperatures, the phonon scattering also plays a dominant role leading to the reduction in the mobility values.

Since AlGaN/GaN heterojunction has a spontaneous piezoelectric polarization at the interface, any external strain changes the density of the mobile carriers (2DEG) at the interface. The associated change in resistance with strain can be used as a direct measure of the strain that is being applied on the system [26,27]. Figure 3 shows the finite element (FE) simulations using COMSOL Multiphysics (version 4.3, COMSOL Inc., Stockholm, Sweden), which shows the (a) stress values across the diaphragm and the (b) displacement of the diaphragm at an applied pressure of 20 kPa above atmospheric pressure. From this computation, the maximum displacement of the diaphragm was estimated to be ~10 μm. The maximum stress in the diaphragm was at the circumference and because of we designed the HFETs to be at the periphery of the diaphragm to maximize the polarization-induced change in conductivity and hence the maximum sensitivity.

Figure 4a shows the variation of source drain resistance (R_{ds}) with 20 kPa of pressure difference being applied to the diaphragm in regular intervals, which resulted in the R_{ds} increasing. The pressure was applied and released quickly using a valve to reduce mechanical transients in the measurements. At each measurement point, the differential pressure was kept at 20 kPa for few seconds and then was reduced back to zero (atmospheric pressure) and repeated the experiments for a number of cycles. This was repeated for various temperatures and the results are compared in Figure 4a, where we kept the drain source voltage at 1.5 V but varied the gate voltage to achieve the highest sensitivity for each temperature. The signal to noise ratio is calculated for each dataset using the expression, $SNR_{dB} = 10 \log_{10}\left(\frac{R_{signal}}{R_{noise}}\right)$, where R_{signal} is the average change in resistance when the strain is applied and R_{noise} is the average variation in resistance when there is no strain. The calculated signal-to-noise ratio (SNR) is15–16 dB for these pressure sensors, for all the measured temperatures. The rise and fall times of the response are ~200 ms and ~600 ms respectively, which includes the mechanical transient arising from the time required for the pressure to reach the steady-state. Therefore, actual electrical transient is negligible, which is quite extraordinary for an AlGaN/GaN HFET without any surface passivation.

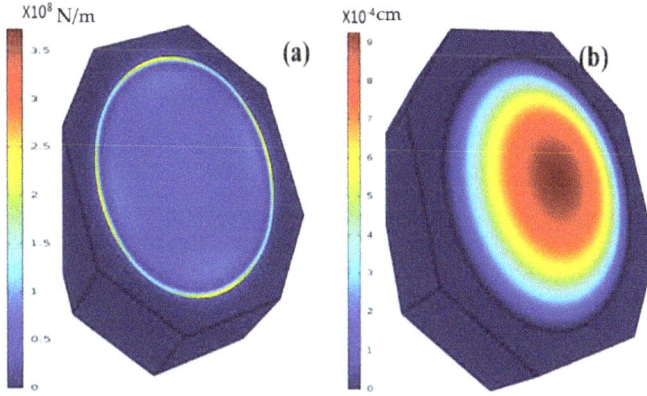

Figure 3. (**a**) Stress (N/m^2) simulations results on AlGaN/GaN diaphragm using finite element method using COMSOL (**b**) the displacement (cm) of the diaphragm with applied pressure (20 kPa).

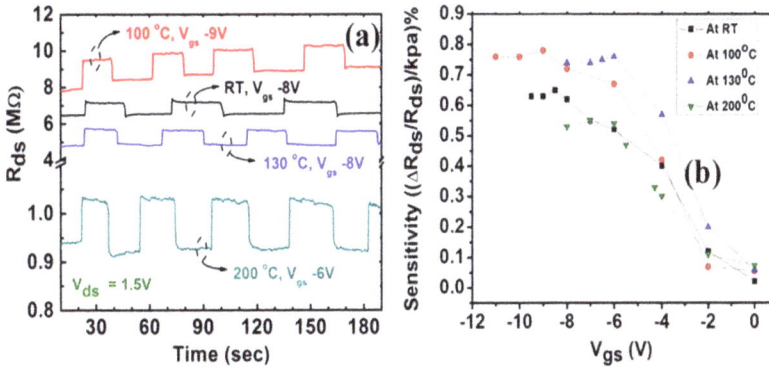

Figure 4. (**a**) The change in drain-source resistance, with applied pressure in regular intervals, indicated by the increased R_{ds}. The figure shows the measurement at different temperatures, V_{ds} is 1.5 V and pressure difference is 20 kPa; (**b**) The variation of pressure sensor sensitivity with gate voltage of HFET, at different temperature. V_{ds} is 1.5 V and the applied differential pressure is 20 kPa.

The sensitivity of the device is calculated from the equation, $S = \frac{\Delta R}{R_0}\frac{1}{P}\%$. Here S is defined as the percentage change in the resistance, with respect to the pressure difference (P). Gauge factor, GF of the device can be derived as

$$GF = \frac{\Delta R}{R_0}\frac{1}{\in} = \frac{1}{\in}\left(\frac{\Delta \mu_n}{\mu_n} + \frac{\Delta n_s}{n_s}\right) \tag{2}$$

where R_0 is the initial resistance and ΔR is the change in resistance with applied strain, $\in$. μ_n and n_s are the mobility and carrier concentrations respectively. From Equation (1), we see that the gauge factor depends on the changes in mobility and carrier concertation. In an HFET device, the gate voltage can be tuned deplete the 2DEG to bring n_s to a low level, also known as the subthreshold region, where a small change in n_s caused by the strain can significantly affect the $\frac{\Delta n_s}{n_s}$ ratio and increase GF [28]. As the temperature goes up, mobility decreases in general; but due to the imperfect contacts, the change in conductivity as a function of temperature does not follow the same trend as mobility in low V_{ds}, which causes a non-monotonous change in device response measured at $V_{ds} = 1.5$ V. In Figure 4a, we see that the response magnitude and sensitivity increase from RT to 100 °C and then gradually decrease through 200 °C. At higher temperature, μ_n decreases [29], as a result, overall sensitivity is expected to decrease. However, because of our contacts being shorter than the transfer length, we have variable contact resistance which improves with slightly higher temperature (Figure 2b) due to the increased efficiency in thermionic emission process. This increases the injection efficiency from the contacts into the channel, which allows the changes in Δn_s and ΔR to appear larger under applied pressure as well, due to the non-linearity in the I_{ds}-V_{ds} curves at the low field. However, as temperature keeps on increasing, the contact resistance is expected to reach an equilibrium at one point (L_T coming closer to the contact length) and the high temperature causes the sensitivity to drop.

Figure 4b, which shows the variation of the pressure sensitivity as a function of V_{gs} of the HFET at different temperatures, is in good agreement with the aforementioned explanation. For all four temperatures shown here, the sensitivity becomes nearly constant for $V_{gs} < -6$ V, which indicates that a V_{gs} just below -6 V is the optimal gate bias. The change in n_s is maximum when V_{gs} is between 0 V and about -6 V, which is why the large changes in sensitivity is only observed in this region. At zero gate bias, very low sensitivity is observed since the baseline carrier concentration n_s is very high (order of high 10^{12} cm^{-2}), while the change in carrier concentration Δn_s due to deflection related strain is very low. Due to this a gate control is required to reduce the 2DEG density which will automatically increase the sensitivity (proportional to $\Delta n_s/n_s$) [30]. This is in agreement with an earlier study by Zimmermann et al. which also showed that the pressure response increases at higher gate bias [31]. The sensitivity of our pressure sensors varied from 0.022%/kPa, at zero gate voltage, to 0.5–0.76%/kPa in the subthreshold region ($V_g \approx -6$ V, see discussion above) for different temperatures, with the maximum gauge factor (GF) being ~260. The corresponding sensitivity in terms of change in the drain voltage (assuming a constant current of $\approx 1 \times 10^{-7}$ amp) is ~7–18 mV/kPa, which is slightly higher than the value of 7.25–14.5 mV/kPa reported for commercial high sensitivity pressure sensors (IMI sensors) [32]. It is important to note that the sensitivity values obtained from our AlGaN/GaN pressure sensor are orders of magnitude higher than the sensitivity value of 0.02% change for 50 bar reported by Boulbar et al. on AlGaN/GaN heterojunction based pressure sensors fabricated on sapphire substrate [19]. Our results are also close to an order of magnitude better than the recently reported sensitivity value of 0.64%/psig (= 0.09%/kPa) measured on InAlN/GaN heterostructure based micro-pressure sensors [33].

4. Conclusions

In summary, we have demonstrated for the first time a diaphragm based AlGaN/GaN HFET embedded circular membrane pressure sensor for high temperature pressure sensing, with ultra-high sensitivity. Finite element simulation was utilized to determine the strain across the diaphragm and determine the gauge factor, which was found out to be ~260 in the sub-threshold region. A very high sensitivity of 0.76%/kPa was also measured, which is the highest reported so far for III-Nitride based pressure sensors. The pressure sensor performance was found to be quite repeatable and was maintained up to a temperature of 200 °C.

Author Contributions: Durga Gajula fabricated and characterized the sensor and also wrote the paper. Ifat Jahangir designed the sensor. Goutam Koley advised throughout this work. All authors have read and approved the final manuscript.

Acknowledgments: We thankfully acknowledge financial support for this work from the National Science Foundation through Grants # IIP-1512342, CBET-1606882 and IIP-1602006.

Conflicts of Interest: The authors declare no conflict of interest.

Appendix

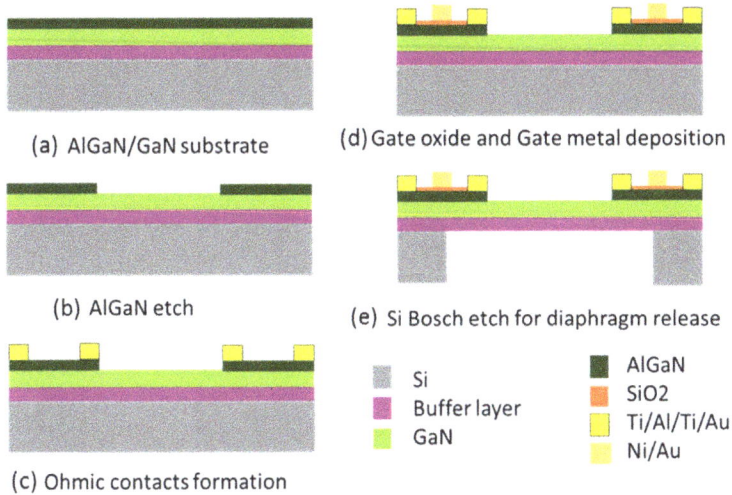

Figure A1. Diagrammatic representation of process flow for the fabrication of pressure sensors. (a)–(e) Different steps of the process flow shown in order, as discussed in the text.

References

1. Johnson, R.W.; Evans, J.L.; Jacobsen, P.; Thompson, J.R.; Christopher, M. The changing automotive environment: High-temperature electronics. *IEEE Trans. Electron. Packag. Manuf.* **2004**, *27*, 164–176. [CrossRef]

2. George, T.; Son, K.A.; Powers, P.A.; del Castillo, L.Y.; Okojie, R. Harsh environment microtechnologies for NASA and terrestrial applications. *IEEE Sens.* **2005**, 1253–1258. [CrossRef]

3. Werner, M.R.; Fahrner, W.R. Review on materials, microsensors, systems and devices for high-temperature and harsh-environment applications. *IEEE Trans. Ind. Electron.* **2001**, *48*, 249–257. [CrossRef]

4. Jiang, Y.; Li, J.; Zhou, Z.; Jiang, X.; Zhang, D. Fabrication of All-SiC Fiber-Optic Pressure Sensors for High-Temperature Applications. *Sensors* **2016**, *16*, 1660. [CrossRef] [PubMed]

5. Bongraina, A.; Rousseaub, L.; Valbinb, L.; Madaouib, N.; Lissorguesb, G.; Verjusa, F.; Chapona, P.A. A new technology of ultrathin AlN piezoelectric sensor for pulse wave measurement. *Procedia Eng.* **2015**, *120*, 459–463. [CrossRef]

6. Dzuba, J.; Vanko, G.; Držík, M.; Rýger, I.; Kutiš, V.; Zehetner, J.; Lalinský, T. AlGaN/GaN diaphragm-based pressure sensor with direct high performance piezoelectric transduction mechanism. *Appl. Phys. Lett.* **2015**, *107*, 122102. [CrossRef]

7. Jahangir, I.; Koley, G. Dual-channel microcantilever heaters for volatile organic compound detection and mixture analysis. *Sci. Rep.* **2016**, *6*, 28735. [CrossRef] [PubMed]

8. Jahangir, I.; Quddus, E.B.; Koley, G. Unique detection of organic vapors below their auto-ignition temperature using III–V Nitride based triangular microcantilever heater. *Sens. Actuators B Chem.* **2016**, *222*, 459–467. [CrossRef]

9. Jahangir, I.; Koley, G. Modeling the performance limits of novel microcantilever heaters for volatile organic compound detection. *J. Micromech. Microeng.* **2017**, *27*, 015024. [CrossRef]

10. Jahangir, I.; Quddus, E.B.; Koley, G. III-V Nitride based triangular microcantilever heater for selective detection of organic vapors at low temperatures. In Proceedings of the 72nd Device Research Conference, Santa Barbara, CA, USA, 22–25 June 2014; pp. 111–112.

11. Jahangir, I.; Koley, G. Dual Channel Microcantilever Heaters for Selective Detection and Quantification of a Generic Mixture of Volatile Organic Compounds. In Proceedings of the 2016 IEEE SENSORS, Orlando, FL, USA, 30 October–3 November 2016; pp. 1–3.

12. Cimalla, V.; Pezoldt, J.; Ambacher, O. Group III nitride and SiC based MEMS and NEMS: Materials properties, technology and applications. *J. Phys. D Appl. Phys.* **2007**, *40*, 6386–6434. [CrossRef]

13. Ambacher, O.; Foutz, B.; Smart, J.; Shealy, J.R.; Weimann, N.G.; Chu, K.; Murphy, M.; Sierakowski, A.J.; Schaff, W.J.; Eastman, L.F. Two dimensional electron gases induced by spontaneous and piezoelectric polarization in undoped and doped AlGaN/GaN heterostructures. *J. Appl. Phys.* **2000**, *87*, 334–344. [CrossRef]

14. Khan, D.; Bayram, F.; Gajula, D.; Talukdar, A.; Li, H.; Koley, G. Plasmonic amplification of photoacoustic waves detected using piezotransistive GaN microcantilevers. *Appl. Phys. Lett.* **2017**, *111*, 062102. [CrossRef]

15. Khan, D.; Bayram, F.; Li, H.; Gajula, D.R.; Koley, G. Plasmonic Enhancement of Photoacoustic Signal for Sensing Applications. In Proceedings of the 2017 75th Annual Device Research Conference (DRC), South Bend, IN, USA, 25–28 June 2017; pp. 1–2.

16. Young, D.J.; Du, J.; Zorman, C.A.; Ko, W.H. High-temperature single-crystal 3C-SiC capacitive pressure sensor. *IEEE Sens. J.* **2004**, *4*, 464–470. [CrossRef]

17. Ziermann, R.; von Berg, J.; Reichert, W.; Obermeier, E.; Eickhoff, M.; Krotz, G. A high temperature pressure sensor with /spl beta/-SiC piezoresistors on SOI substrates. In Proceedings of the IEEE International Conference on Solid State Sensors and Actuators, Chicago, IL, USA, 16–19 June 1997; pp. 1411–1414.

18. Okojie, R.S.; Ned, A.A.; Kurtz, A.D. Operation of α(6H)-SiC Pressure Sensor at 500 °C. *Sens. Actuators A Phys.* **1998**, *66*, 200–204. [CrossRef]

19. Kang, B.S.; Kim, J.; Jang, S.; Ren, F.; Johnson, J.W.; Therrien, R.J.; Rajagopal, P.; Roberts, J.C.; Piner, E.L.; Linthicum, K.L.; et al. Capacitance pressure sensor based on GaN high-electron-mobility transistor-on-Si membrane. *Appl. Phys. Lett.* **2005**, *86*, 253502. [CrossRef]

20. Mishra, U.K.; Shen, L.; Kazior, T.E.; Wu, Y.F. GaN-Based RF power devices and amplifiers. *Proc. IEEE* **2008**, *96*, 287–305. [CrossRef]

21. Rais-Zadeh, M.; Gokhale, V.J.; Ansari, A.; Faucher, M.; Theron, D.; Cordier, Y.; Buchaillot, L. Gallium nitride as an electromechanical material. *J. Microelectromech. Syst.* **2014**, *23*, 1252–1271. [CrossRef]

22. Le Boulbar, E.D.; Edwards, M.J.; Vittoz, S.; Vanko, G.; Brinkfeldt, K.; Rufer, L.; Johander, P.; Lalinsky, T.; Bowen, C.R.; Allsopp, D.W.E. Effect of bias conditions on pressure sensors based on AlGaN/GaN High Electron Mobility Transistor. *Sens. Actuators A Phys.* **2013**, *194*, 247–251. [CrossRef]

23. Lalinsky´, T.; Hudek, P.; Vanko, G.; Dzuba, J.; Srnánek, V.K.R.; Choleva, P.; Vallo, M.; Držík, M.; Matay, L.; Kostič, I. Micromachined membrane structures for pressure sensors based on AlGaN/GaN circular HEMT sensing device. *Microelectron. Eng.* **2012**, *98*, 578–581. [CrossRef]

24. Eastman, L.F.; Tilak, V.; Smart, J.; Green, B.M.; Chumbes, E.M.; Dimitrov, R.; Kim, H.; Ambacher, O.S.; Weimann, N.; Prunty, T.; et al. Undoped AlGaN/GaN HEMTs for microwave power amplification. *IEEE Trans. Electron Devices* **2001**, *48*, 479–485. [CrossRef]

25. Alim, M.; Rezazadeh, A.; Gaquiere, C. Temperature dependence of the threshold voltage of AlGaN/GaN/SiC high electron mobility transistors. *Semicond. Sci. Technol.* **2016**, *31*, 125016. [CrossRef]

26. Qazi, M.; DeRoller, N.; Talukdar, A.; Koley, G. III-V Nitride based piezoresistive microcantilever for sensing applications. *Appl. Phys. Lett.* **2011**, *99*, 193508. [CrossRef]

27. Talukdar, A.; Qazi, M.; Koley, G. Static and dynamic responses of GaN piezoresistive microcantilever with embedded AlGaN/GaN HFET for sensing applications. In Proceedings of the IEEE SENSORS 2013, Baltimore, MD, USA, 3–6 November 2013; pp. 1–4.

28. Dang, X.Z.; Asbeck, P.M.; Yu, E.T. Measurement of drift mobility in AlGaN/GaN heterostructure field-effect transistor. *Appl. Phys. Lett.* **1999**, *74*, 3890–3892. [CrossRef]

29. Ko, T.-S.; Lin, D.-Y.; Lin, C.-F.; Chang, C.-W.; Zhang, J.-C.; Tu, S.-J. High-temperature carrier density and mobility enhancements in AlGaN/GaN HEMT using AlN spacer layer. *J. Cryst. Growth* **2017**, *464*, 175–179. [CrossRef]

30. Talukdar, A.; Koley, G. Impact of Biasing Conditions on Displacement Transduction by III-Nitride Microcantilevers. *IEEE Electron Device Lett.* **2014**, *35*, 1299–1301. [CrossRef]
31. Zimmermann, T.; Neuburger, M.; Benkart, P.; Hernández-Guillén, F.J.; Pietzka, C.; Kunze, M.; Daumiller, I.; Dadgar, A.; Krost, A.; Kohn, E. Piezoelectric GaN Sensor Structures. *IEEE Electron Device Lett.* **2006**, *27*, 309–312. [CrossRef]
32. Sensors for Machinery Health Monitoring. Available online: http://www.imi-sensors.com (accessed on 8 April 2018).
33. Chapin, C.A.; Miller, R.A.; Dowling, K.M.; Chen, R.; Senesky, D.G. InAlN/GaN high electron mobility micro-pressure sensors for high-temperature environments. *Sens. Actuators A Phys.* **2017**, *263*, 216–223. [CrossRef]

micromachines

MDPI

Article

DC-25 GHz and Low-Loss MEMS Thermoelectric Power Sensors with Floating Thermal Slug and Reliable Back Cavity Based on GaAs MMIC Technology

Zhiqiang Zhang [1,*] and Yao Ma [2]

[1] Key Laboratory of MEMS of the Ministry of Education, Southeast University, Nanjing 210096, China
[2] College of Field Battle Engineering, People's Liberation Army University of Science and Technology, Nanjing 210007, China; eemaoma@gmail.com
* Correspondence: zqzhang@seu.edu.cn; Tel.: +86-25-8379-4642-8816

Received: 22 February 2018; Accepted: 27 March 2018; Published: 29 March 2018

check for
updates

Abstract: Wideband and low-loss microwave power measurements are becoming increasingly important for microwave communication and radar systems. To achieve such a power measurement, this paper presents the design and measurement of wideband DC-25 GHz and low-loss MEMS thermoelectric power sensors with a floating thermal slug and a reliable back cavity. In the sensors, the microwave power is converted to thermovoltages via heat. The collaborative design of the thermal slug and the back cavity, i.e., two thermal flow paths, is utilized to improve the efficiency of heat transfer and to ensure reliable applications. These sensors are required to operate up to 25 GHz. In order to achieve low microwave losses at the bandwidth, the floating thermal slug is designed instead of the grounded one. The effects of the floating slug on the reflection losses are analyzed by the simulation. The fabrication of these sensors is completed by GaAs monolithic microwave integrated circuits (MMIC) and micro-electro-mechanical systems (MEMS) technology. Measured reflection losses are less than −25.6 dB up to 12 GHz and −18.6 dB up to 25 GHz. The design of the floating thermal slug reduces the losses, which is equivalent to improving the sensitivity. At 10 and 25 GHz, experiments exhibit that the sensors result in sensitivities of about 51.13 and 35.28 µV/mW for the floating slug and 81.68 and 55.20 µV/mW for the floating slug and the cavity.

Keywords: thermoelectric power sensor; wideband; GaAs MMIC; MEMS; floating slug; back cavity; microwave measurement

1. Introduction

With the rapid development of multi-band microwave communication and radar systems, the wideband and low-loss power measurements become more and more important for microwave signals. In recent years, the commonly used power measurement methods include three types: diode-, thermistor-, and thermopile-based microwave power sensors [1,2]. The diode-based microwave power sensors are based on employing the square law region in the nonlinear I–V curve of diodes, where the input microwave power is proportional to the low-frequency power. Using the diode method, the peak RF power is measured. As for the diode-based sensors, they are active components and require an additional attenuator when measuring the high power of mW levels. The thermistor-based microwave power sensors convert the input microwave power into heat and result in the change in resistance of the thermistors based on Joule effect, where the thermistors are generally negative resistance temperance coefficient. Using the thermistor method, the average RF power is measured. But, as for the thermistor-based sensors, they need a feedback circuit of bridge balancer and are susceptible to the external environment. So, the two types of sensors are not suitable for broadband and low-loss applications. The thermopile-based microwave power sensors convert the microwave power into heat and finally into the thermovoltage, where the thermovoltage depends on the temperature at both ends of thermocouples, regardless of the temperature profile. As for the thermopile-based sensors, they are based on Seebeck effect and output DC voltages [3,4]. These thermoelectric power sensors become a preferred choice due to zero dc power consumption, wide operation frequency, high power handling, and high linearity. However, the disadvantage of the sensors is that their sensitivity is not high. This is mainly caused by heat losses of the substrate in the conversion process of microwave power-heat-electricity. The micro-electro-mechanical systems (MEMS) technology can reduce the heat losses through locally etching the substrate and forming the membrane structure. Thermopile-based MEMS microwave power sensors have been reported widely [5–8]. Following that, the effects of the thermopile's optimization [9], thermoelectric modeling [10–12], packaging [13,14], and temperature and humidity reliability [15,16] on the thermopile-based power sensors are studied. The heat losses of the substrate are reduced and the sensitivity of the sensors is increased as the substrate underneath the thermopile is thinned. It should be noted that the power sensors with a too thin membrane will bring challenges to the reliability and packaging. In order to choose the proper configuration and size of the membrane, the thermopile-based MEMS power sensors with dual thermal flow paths are proposed, where the grounded thermal slug and the cavity are included [17]. The thermal slug and the back cavity are regarded as the dual thermal flow paths. The cavity with a thicker membrane leads to good reliability, with acceptable sensitivity. Nevertheless, the grounded slug makes part of the two load resistors short circuit, resulting into smaller resistance and higher reflection losses at high frequencies. So, they only operate below 12 GHz due to the limitation of reflection losses.

In order to solve the above problem, the design of DC-25 GHz and low-loss MEMS thermoelectric power sensors with the floating thermal slug and the reliable back cavity is proposed in this paper. The back cavity has a robust membrane, and the robust membrane contributes to the packaging of the thermoelectric power sensors [13]. Here, the thermal slug is designed to be floating to obtain low microwave losses of up to 25 GHz. The microwave power sensors with the dual thermal flow paths are optimized by the simulation. The fabrication of the power sensors is based on GaAs monolithic microwave integrated circuits (MMIC) technology. Experiments demonstrate that these sensors can operate at the wide frequency range, with zero DC power consumption and low reflection losses. The design of the floating thermal slug reduces the microwave losses, which is equivalent to an improvement in sensitivity. Furthermore, the measured performances of the sensors with the floating slug are compared to the basic sensor and the reported sensors with the grounded slug. The main purpose of this work is to achieve the wideband and low-loss measurement for the sensors with dual thermal flow paths.

2. Structure and Design

Figure 1 shows the wideband and low-loss MEMS microwave power sensor with the dual thermal flow paths. It includes a coplanar waveguide (CPW) line, two load matching resistors, a thermopile, a floating thermal slug, a back cavity, and two pads. The CPW line is composed of a signal line and two ground lines, and is used to transmit microwave signals. Its characteristic impedance is 50 Ω for the following measurement, where the width of the signal line and the distance between the signal and ground lines are designed to be 100 and 58 µm, respectively. In order to achieve the good matching relationship, the two load resistors with the resistance of 100 Ω are in parallel connected to one end of the CPW line. The length and the width of each resistor are 58 and 14.5 µm, respectively. The thermopile is placed close to the load resistors, and it is composed of twelve thermocouples that are connected in series to obtain large output thermovoltages. In the GaAs MMIC process, each thermocouple is made of n$^+$ GaAs and AuGeNi/Au [17]. In the design, the distance between the resistors and the thermopile is 10 µm. When the microwave signal under test is input and transmitted to the CPW, the load matching resistors completely absorb the microwave power and generate heat based on the theory of Joule heat. The generated heat causes an increase in temperature around the resistors. So that, the temperature difference is formed at both ends of the thermopile. Finally, the resulting temperature difference is converted to the output thermovoltages by the thermopile, based on Seebeck effect. These power sensors do not require a dc bias during operation.

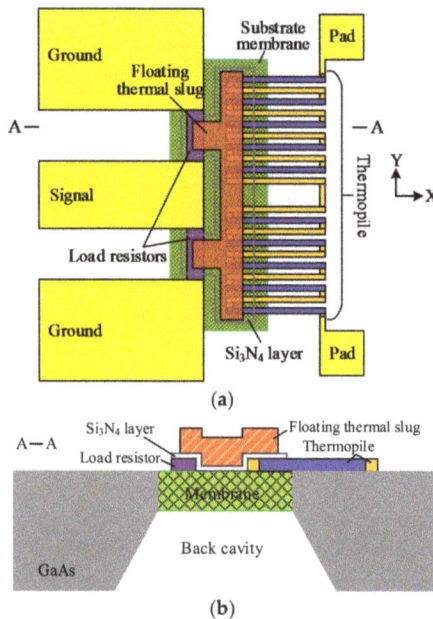

Figure 1. Schematic diagrams of the proposed micro-electro-mechanical systems (MEMS) thermoelectric microwave power sensor with dual thermal flow paths (floating thermal slug and back cavity) in the GaAs monolithic microwave integrated circuits (MMIC) process. (**a**) Top view; (**b**) Cross-sectional view.

Therefore, the power sensors utilize the conversion principle of microwave power-heat-electricity. For a certain microwave power, the thermovoltage is related to the temperature difference. That is, the sensitivity of the power sensor is determined by the efficiency of heat transfer from the resistors to the thermopile. In order to increase the temperature at the end of the thermopile in proximity to the resistors, two methods are adopted. The one, the floating thermal slug is placed on the load resistors and the thermopile, and electrically isolated from the resistors and thermopile through using a Si_3N_4 dielectric layer. The thermal slug is made of gold. The dielectric layer is thin (0.23 µm) and it has little effect on heat conduction. So, the thermal slug can transfer the heat from the resistors to the thermopile. The method is equivalent to arranging a heat conduction path above the substrate, which increases the efficiency of the heat transfer. In this paper, the thermal slug is designed to be in the floating state, and differed from the grounded slug in [17]. It means that the floating thermal slug is not connected to the CPW ground line. The floating thermal slug, the Si_3N_4 layer and the load resistor layer constitute a MIM capacitor, and the capacitor will cause parasitic capacitive reactance. But, the effect of the capacitance on the resistance of the load resistors is small, to less than 5% from the simulation. So the advantage of such design is that the floating slug almost does not affect the resistance of the load resistors, thereby achieving good impedance matching. Thus, the floating design of the thermal slug is ability to obtain low reflection losses at a wider frequency rang. More importantly, it shows that more power is dissipated to generate heat, which contributes to increasing the sensitivity of the sensors. So, the floating thermal slug can achieve the low reflection loss at the wideband and improve the sensitivity. The other, the back cavity is etched by the MEMS technique, and the substrate membrane underneath the resistors and the hot end of the thermopile is formed. The thin substrate leads to a reduction in the heat losses of the substrate. The method is equivalent to arranging another heat conduction path in the substrate, which increases the efficiency of heat transfer. The thinner the substrate is, the smaller the heat losses of the substrate are. However, if a too thin membrane is etched, then the power sensors will bring challenges to the packaging and reliability. In order to ensure the packaging and reliability, the back cavity with a robust stiffness of the membrane is fabricated, where the substrate membrane is about 20 μm in thickness.

Based on the Seebeck effect, the output thermovoltages that are generated by the twelve thermocouples and can be expressed as:

$$V_{out} = (\alpha_1 - \alpha_2) \sum_{i=1}^{i=N} \Delta T(i) \tag{1}$$

where α_1 and α_2 are Seebeck coefficients of n$^+$ GaAs and AuGeNi/Au, N is the number of the thermocouples and equal to 7, and $\Delta T(i)$ is the temperature difference between the hot and cold ends of the named (*i*) thermocouple. According to the theory of the heat transfer equation, $\Delta T(i)$ can be written as [11]:

$$\Delta T(x,y) = \sum_{n=1}^{+\infty} C_n (e^{\frac{n\pi}{W}x} - e^{\frac{n\pi}{W}(2L-x)}) \sin \frac{n\pi}{W} y \quad n = 1, 2 \cdots + \infty \tag{2}$$

where C_n is the coefficient that can be obtained, L and W are the length and width of the edge of the power sensors, respectively. Thus, the sensitivity of the thermoelectric power sensors is represented as:

$$S = \frac{V_{out}}{P_{in}} \tag{3}$$

where P_{in} is the input microwave power. Therefore, as seen in Equation (1), the output thermovoltages are proportional to the temperature difference between the hot and cold ends of the thermopile; as seen in Equation (2), the sensitivity of the power sensors is proportional to the thermovoltages.

In this paper, the MEMS thermoelectric power sensors with dual thermal flow paths are designed to operate up to 25 GHz. In order to analyze the effects of the floating thermal slug on the reflection losses, the power sensors are simulated and optimized by HFSS (High Frequency Structure Simulator). Figure 2a shows the simulated reflection losses of the GaAs MMIC-based power sensors with different overlapping lengths between the floating thermal slug and the load resistors. The simulated frequency range is from DC to 25 GHz. When the overlapping lengths between the floating thermal slug and the load resistors are 2, 10, and 14.5 μm, the corresponding percentages are 13.8%, 69.0%, and 100%, respectively. In general, the larger the overlapping length is, the higher the reflection loss at microwave frequencies is. This is because that the large overlapping length leads to the increase of the capacitance. For the overlapping lengths of 2, 10, and 14.5 μm, the simulated reflection losses are less than −24.8, −21.1, and −19.8 dB at DC-25 GHz, respectively. These results show that the proposed power sensors exhibit the low reflection losses, which verifies the design validity of the floating thermal slug. For the overlapping lengths of 2 and 10 μm, the optimized reflection losses are below −20 dB. Figure 2b shows simulated reflection losses versus the overlapping lengths between the floating thermal slug and the load resistors at the fixed frequencies of 5, 10, 20, and 25 GHz. At these frequencies, the reflection losses increase as the overlapping lengths increase. Figure 3 shows simulated electromagnetic field distribution of the power sensor for the overlapping length of 10 μm between the floating thermal slug and the resistors. As can be observed, a portion of the electromagnetic field is coupled to the thermopile through the floating thermal slug, but the amount is small. It means that there is a small effect on the sensing output of the thermopile.

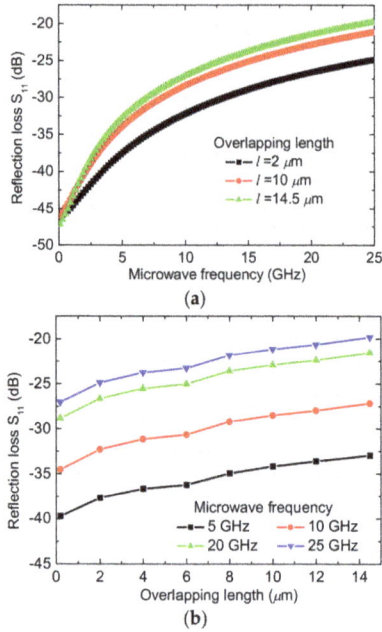

Figure 2. Simulated reflection losses of the GaAs MMIC-based power sensors. (**a**) S_{11} versus microwave frequency at the fixed overlapping length of 2, 10, and 14.5 μm and (**b**) S_{11} versus the overlapping lengths at the fixed frequencies of 5, 10, 20, and 25 GHz.

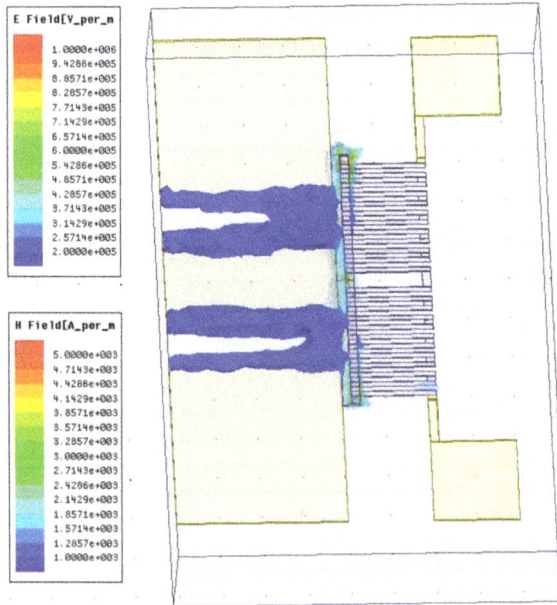

Figure 3. Simulated electromagnetic field distribution of the power sensor with the overlapping length of 10 μm between the floating thermal slug and the resistors.

In order to show the heat transfer aspect, these power sensors are simulated by using an ANSYS software. Figure 4 shows the simulated temperature distribution of the power sensor with the dual thermal flow paths under the power level of 100 mW. In order to show the design validity, the basic and improved MEMS thermoelectric power sensors are given together. Here, their common sizes are same. The basic sensor is no thermal slug and cavity. The improved sensors D1 and D2 only have the floating thermal slug, where the overlapping lengths between the floating slug and the resistors are 2 and 10 μm, respectively. They are called as the sensors with small and large floating thermal slugs. The improved sensor D3 includes the floating thermal slug and the back cavity. It is called as the sensor with the dual thermal flow paths. Figure 5 shows the temperature on the hot junctions of the several thermocouples of the sensors A1, D1, D2, and D3 when the power is 100 mW. By comparing the three sensors A1, D1, and D2, they show that the thermal slug can act as the heat conduction path above the substrate and increase the efficiency of heat transfer. By comparing the four sensors, the sensor D3 with the floating thermal slug and the back cavity shows the highest temperature. Such results verify the design validity of the dual thermal flow paths.

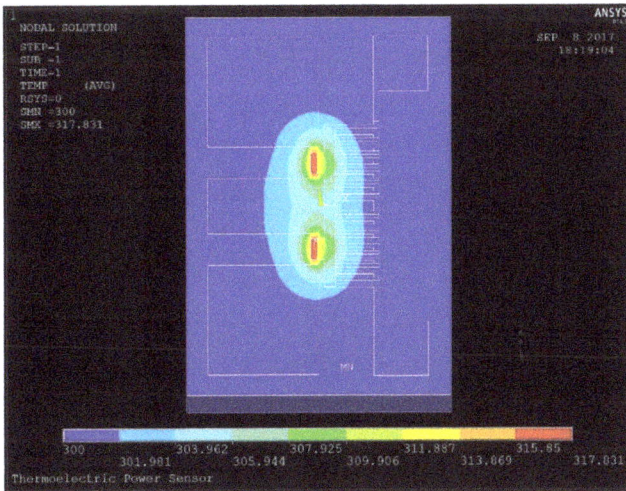

Figure 4. Simulated temperature distribution of the power sensor with the dual thermal flow paths under the power level of 100 mW.

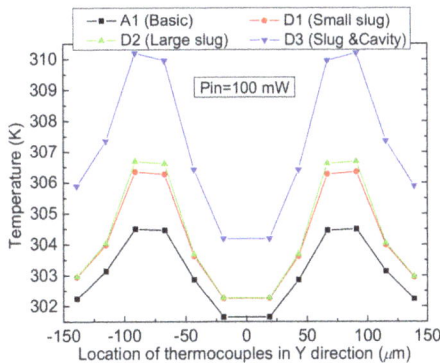

Figure 5. Temperature on hot junctions of the several thermocouples of the sensors A1, D1, D2, and D3 with respect to the location of the thermocouples in the Y direction when the power is 100 mW.

3. Measurement and Discussion

In order to facilitate the performance comparison under the same process and experimental conditions, the basic and improved MEMS thermoelectric microwave power sensors are given together. These thermoelectric power sensors are fabricated using the GaAs MMIC process [17,18].

(i). The sensors start on a 3-inch GaAs wafer, and n$^+$ GaAs is used to fabricate ohmic contact areas for the doping concentration of 1.0×10^{18} cm^{-3} and one leg of the thermopile for 1.0×10^{17} cm^{-3}.

(ii). AuGeNi/Au is sputtered and patterned to form the other leg of the thermopile by a liftoff process.

(iii). TaN (square resistance of 25 Ω/□) is sputtered and patterned as the load resistors.

(iv). Ti/Pt/Au/Ti (500/300/3500/500 Å) is evaporated and patterned to form the CPW and pads.

(v). Si$_3$N$_4$ (1000 Å) is deposited by PECVD (Plasma Enhanced Chemical Vapor Deposition) and etched as the dielectric layer.

(vi). Ti/Au/Ti is evaporated as a seed layer, and Au (2 μm) is electroplated to thicken the CPW and pads and to fabricate the floating slug.

(vii). GaAs is thinned to 100 μm in thickness, and the substrate membrane underneath the resistors and the hot end is implemented by a via-hole etching technique.

Figure 6 shows SEM (Scanning Electron Microscope) views of three wideband and low-loss thermopile-based microwave power sensors with the floating thermal slug and the back cavity in GaAs MMIC.

Microwave performances of the several power sensors are measured by the calibrated network analyzer and the probe station, and Figure 7 plots the results of the reflection losses. As for the basic sensor and the three improved power sensors, the measured reflection losses are less than −25.69 dB up to 12 GHz and −18.61 dB up to 25 GHz. It means that less than 0.28% (below 12 GHz) and 1.4% (below 25 GHz) microwave power is reflected back to the input port. So, they show the good impedance matching. For the basic sensor, the measured S$_{11}$ are about −31.03, −25.33, and −22.90 dB at 10, 20, and 25 GHz, respectively. For the sensor D1 with the overlapping length of 2 μm, the measured S$_{11}$ are about −29.05, −23.22, and −20.73 dB at 10, 20, and 25 GHz, respectively. For the sensor D2 with the overlapping length of 10 μm, the measured S$_{11}$ are about −26.79, −20.96, and −18.62 dB at 10, 20, and 25 GHz, respectively. For the sensor D3 with the floating slug and the back cavity, the measured S$_{11}$ are about −33.21, −27.41, and −24.73 dB at 10, 20, and 25 GHz, respectively. As can be observed in the sensors D1 and D2, the floating thermal slug with the overlapping length of 10 μm causes higher reflection losses than the slug with the overlapping length of 2 μm. However, when compared to the sensors with the grounded thermal slug in [17], they are smaller and acceptable for microwave applications. Such results verify the design validity of the floating thermal slug. In addition, the sensor D3 with the dual thermal flow paths exhibits the lowest reflection losses. The experiments demonstrate that the power sensors with the floating thermal slug can achieve low reflection losses in a wide frequency range. As for the thermopile-based power sensors, low reflection losses mean that less microwave power is reflected back and more power is dissipated to generate heat, which contributes to increasing the sensitivity of the sensors.

When the microwave power is input through the CPW line, the thermovoltage is output and measured on the pads. Figure 8 shows the measured thermovoltages as a function of the microwave power at different frequencies for the basic and improved MEMS power sensors. In Figure 8, the thermovoltage increases as the microwave power changes from 1 to 150 mW at the fixed frequency, where the linear relationships between them are obtained. It means that the effects of the electrometric filed coupled by the floating thermal slug on the thermopile and its output are small. In other words, the power sensors can achieve a stable sensing output. Furthermore, the sequence of the thermovoltages is the sensor A1 < D1 < D2 < D3, at a fixed power and frequency. For example, when the microwave power is 150 mW for the sensors A1, D1, D2, and D3, the thermovoltages are about 8.26, 9.39, 10.77, and 17.00 mV at 1 GHz, 7.49, 8.25, 8.98, and 15.34 mV at 5 GHz, 6.16, 7.05, 7.93, and 12.73 mV at 10 GHz, 5.39, 5.93, 6.51, and 11.09 mV at 15 GHz, and 4.62, 5.01, 5.43, and 9.41 mV at 20 GHz, respectively.

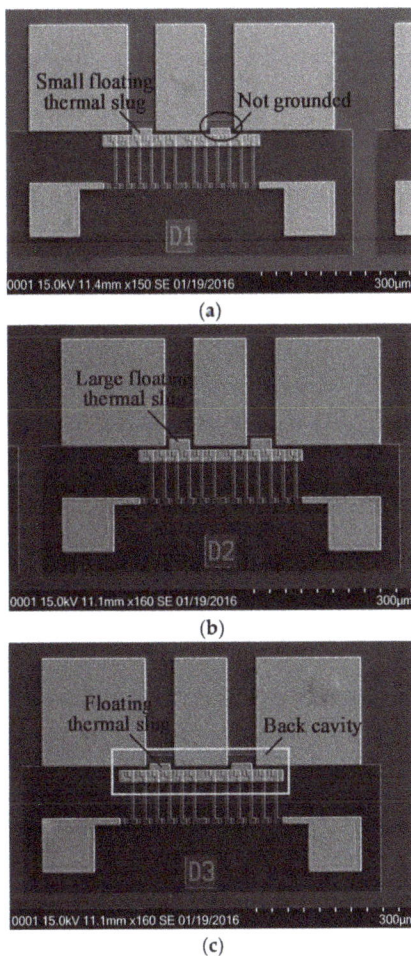

(a)

(b)

(c)

Figure 6. SEM views of three wideband and low-loss MEMS thermoelectric microwave power sensors in GaAs MMIC. (**a**) Sensor D1 with the overlapping length of 2 μm (small floating thermal slug); (**b**) Sensor D2 with the overlapping length of 10 μm (large floating thermal slug); and, (**c**) Sensor D3 with the floating thermal slug and the back cavity.

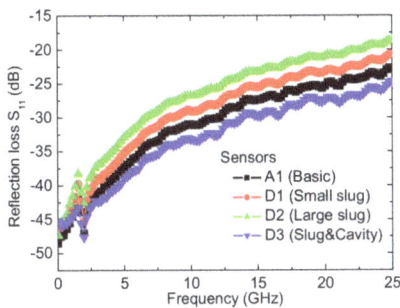

Figure 7. Measured reflection losses of the basic and improved thermopile-based microwave power sensors.

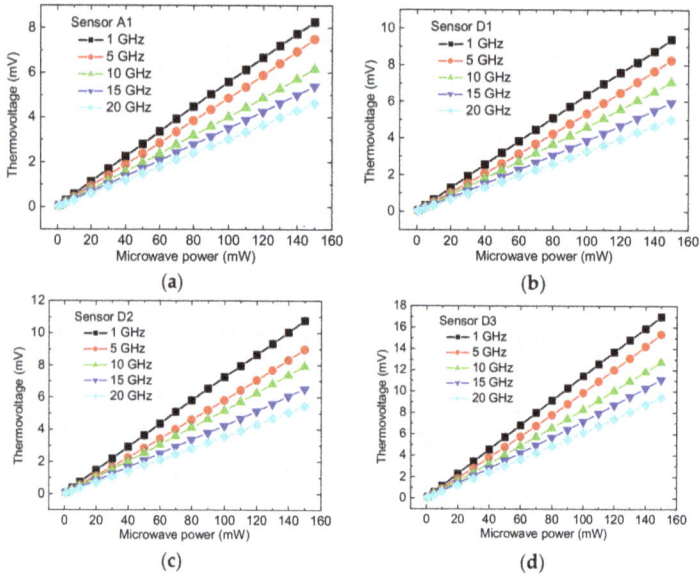

Figure 8. Measured thermovoltage as a function of the microwave power at 1, 5, 10, 15, and 20 GHz for the basic and improved MEMS power sensors. (**a**) Sensor A1 with the basic structure; (**b**) Sensor D1 with with the overlapping length of 2 μm; (**c**) Sensor D2 with the overlapping length of 10 μm; and, (**d**) Sensor D3 with the floating thermal slug and the back cavity.

Figure 9 shows the average sensitivity as a function of the microwave frequency for the basic and improved MEMS power sensors. At 1, 10, and 25 GHz, measured sensitivities are about 56.28, 39.68, and 27.36 μV/mW for the sensor A1 (basic), 63.33, 45.12, and 30.95 μV/mW for the sensor D1 (Overlapping length of 2 μm), 72.54, 51.13, and 35.28 μV/mW for the sensor D2 (Overlapping length of 10 μm), and 114.52, 81.68, and 55.20 μV/mW for the sensor D3 (floating slug and cavity). As can be observed in Figure 9, the sensors D1 and D2 have better sensitivities than the sensor A1. This shows the design validity of the floating thermal slug. Moreover, the sensitivities of the sensor D2 are higher than that of the sensor D1. It indicates that the large floating slug can better improve the heat transfer efficiency. At last, measurements show that the sensor D3 with the dual thermal flow paths (floating thermal slug and back cavity) produces the highest sensitivity. The results of these sensors are consistent with the design. When compared to the basic sensor A1, the proposed sensors D1, D2 and D3 in this paper generate improvements of 13.71%, 28.86%, and 105.85% at 10 GHz. The sensitivities of these relevant sensors with the floating slug are higher than those of the reported sensors with the grounded slug (e.g., 3.73% and 11.24% for the overlapping length of 2 and 10 μm (small and large grounded slugs), and 99.19% for the grounded slug and cavity at 10 GHz in [17]). This is because that the design of the floating slug in this paper leads to less reflection losses, which means that more power is dissipated to generate heat. This is equivalent to improving the sensitivity of the thermoelectric sensors.

Figure 10 shows the standard errors of the corresponding sensitivity in Figure 9. The error bars are small, which further shows the small effects of the electrometric filed on the output of the sensors. In other words, these sensors with the floating slug and the cavity can generate the stable output. In Figure 10, the errors in the sensitivity of the sensor D3 are a little higher than others. They result mainly from the air convection and the test environment. This is because all of the tests are performed at the room temperature and atmosphere condition, instead of a vacuum environment. For the thermopile-based sensors, a high temperature difference results in the large convective heat transfer, and the condition of the test environment affects the reference voltage of the multimeter (before the RF power is applied). Fortunately, as for the sensor D3, the errors relative to the sensitivity are small and acceptable. Table 1 shows the comparison of the MEMS thermoelectric power sensors in the GaAs process.

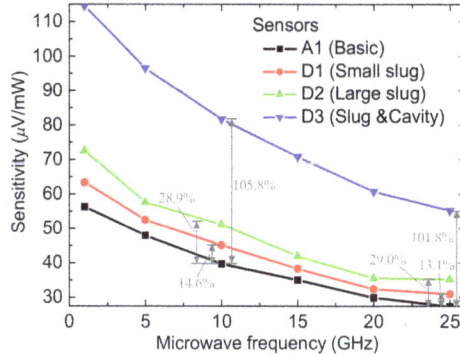

Figure 9. Average sensitivity as a function of the microwave frequency for the basic and improved MEMS power sensors.

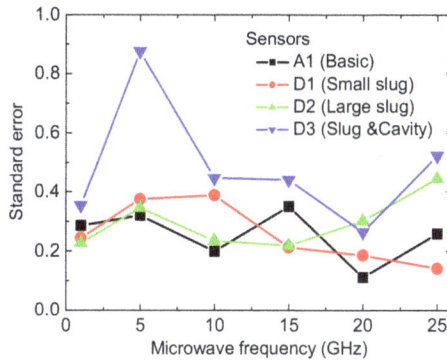

Figure 10. Standard error of the corresponding sensitivity in Figure 9.

Table 1. Comparison of the MEMS thermoelectric power sensors in the GaAs process.

Ref.	Operation Frequency	Reflection Loss (dB)	Sensitivity (μV/mW)	Thickness of Substrate Membrane/Reliability	Membrane Process
[5]	dc-26.5 GHz	−29.4	16,400	1.5 μm/not good	No standard
[11]	dc-10 GHz	−26@10 GHz	160@10 GHz	10 μm/general	Standard
[17]	0.01–12 GHz	−23.15@12 GHz	79.04@10 GHz	20 μm/good	Standard
This work	dc-25 GHz	−32.61@12 GHz	81.68@10 GHz	20 μm/good	Standard
		−24.73@25 GHz	55.20@25 GHz		

4. Conclusions

In this paper, the DC-25 GHz and low-loss MEMS thermoelectric power sensors with dual thermal flow paths are proposed for the GaAs MMIC and MEMS applications. In order to improve the sensitivity of the sensors, the collaboration usage of the floating thermal slug and the back cavity helps to increase the efficiency of heat transfer. The design of the floating slug achieves the low reflection of up to 25 GHz, accompanied by an equivalent improvement in sensitivity. The experiments demonstrate that these microwave power sensors produce low losses, wideband operation, and good sensitivity, with the enough stiffness membrane. The design method can be applied to similar thermopile-based devices.

Acknowledgments: This work was supported by the National Natural Science Foundation of China (NSFC: 61604039).

Author Contributions: Zhiqiang Zhang proposed the study and performed the experiments. Yao Ma contributed discussion and analysis. The final manuscript was approved by Zhiqiang Zhang and Yao Ma.

Conflicts of Interest: The authors declare no conflict of interest.

References

1. Brush, A.S. Measurement of microwave power. *IEEE Instrum. Meas. Mag.* **2007**, *10*, 20–25. [CrossRef]
2. Daullé, A.; Xavier, P.; Rauly, D. A power sensor for fast measurement of telecommunications signals using substitution method. *IEEE Trans. Instrum. Meas.* **2001**, *50*, 1190–1196. [CrossRef]
3. Scott, J.B.; Low, T.S.; Cochran, S.; Keppeler, B.; Staroba, J.; Yeats, B. New thermocouple-based microwave/millimeter-wave power sensor MMIC techniques in GaAs. *IEEE Trans. Microw. Theory Tech.* **2011**, *59*, 338–344. [CrossRef]
4. Zhang, Z.; Guo, Y.; Li, F.; Gong, Y.; Liao, X. A sandwich-type thermoelectric microwave power sensor for GaAs MMIC-compatible applications. *IEEE Electron Device Lett.* **2016**, *37*, 1639–1641. [CrossRef]
5. Dehé, A.; Krozer, V.; Chen, B.; Hartnagel, H.L. High-sensitivity microwave power sensor for GaAs-MMIC implementation. *Electron. Lett.* **1996**, *32*, 2149–2150. [CrossRef]
6. Dehé, A.; Krozer, V.; Fricke, K.; Klingbeil, H.; Beilenhoff, K.; Hartnagel, H.L. Integrated microwave power sensor. *Electron. Lett.* **1995**, *31*, 2187–2188. [CrossRef]
7. Milanovic, V.; Gaitan, M.; Zaghloul, M.E. Micromachined thermocouple microwave detector by commercial CMOS fabrication. *IEEE Trans. Microw. Theory Tech.* **1998**, *46*, 550–553. [CrossRef]
8. Milanovic, V.; Gaitan, M.; Bowen, E.D.; Tea, N.H.; Zaghloul, M.E. Thermoelectric power sensor for microwave applications by commercial CMOS fabrication. *IEEE Electron Device Lett.* **1997**, *18*, 450–452. [CrossRef]
9. Wang, D.-B.; Liao, X.-P.; Liu, T. Optimization of indirectly-heated type microwave power sensors based on GaAs micromachining. *IEEE Sens. J.* **2012**, *12*, 1349–1355. [CrossRef]
10. Yi, Z.; Liao, X. A 3D model of the thermoelectric microwave power sensor by MEMS technology. *Sensors* **2016**, *16*, 921. [CrossRef] [PubMed]
11. Yi, Z.; Liao, X.; Wu, H. Modeling of the terminating-type power sensors fabricated by GaAs MMIC process. *J. Micromech. Microeng.* **2013**, *23*, 085003. [CrossRef]
12. Yi, Z.; Yao, H.; Liao, X. Theoretical and experimental investigation of cascade microwave power sensor. *IEEE Trans. Electron Devices* **2017**, *64*, 1728–1734. [CrossRef]
13. Zhang, Z.; Liao, X. Packaging-test-fixture for in-line coupling RF MEMS power sensors. *J. Microelectromech. Syst.* **2011**, *20*, 1231–1233. [CrossRef]
14. Wang, D.-B.; Liao, X.-P.; Liu, T. A thermoelectric power sensor and its package based on MEMS technology. *J. Microelectromech. Syst.* **2012**, *21*, 121–131. [CrossRef]
15. Wang, D.-B.; Liao, X. Research on temperature characteristic of thermoelectric microwave power sensors based on GaAs MMIC technology. *Electron. Lett.* **2013**, *49*, 1462–1464. [CrossRef]
16. Wang, D.-B.; Gao, B.; Zhang, Y.; Zhao, J.; Zhang, C.; Guo, Y. The research of indirectly-heated type microwave power sensors based on GaAs MMIC technology. *Microsyst. Technol.* **2016**, *22*, 2233–2239. [CrossRef]
17. Zhang, Z.; Liao, X. n$^+$ GaAs/AuGeNi-Au thermocouple-type RF MEMS power sensors based on dual thermal flow paths in GaAs MMIC. *Sensors* **2017**, *17*, 1426. [CrossRef] [PubMed]
18. Zhang, Z.; Liao, X. GaAs MMIC fabrication for the RF MEMS power sensor with both detection and non-detection states. *Sens. Actuat. A Phys.* **2012**, *188*, 29–34. [CrossRef]

MDPI

St. Alban-Anlage 66

4052 Basel

Switzerland

Tel. +41 61 683 77 34

Fax +41 61 302 89 18

www.mdpi.com

Micromachines Editorial Office

E-mail: micromachines@mdpi.com

www.mdpi.com/journal/micromachines